Microplastics

Also by John Pichtel

Fundamentals of Site Remediation, Third Edition
Terrorism and WMDs: Awareness and Response, Second Edition
Waste Management Principles: Municipal, Hazardous, Special Wastes, Second Edition

Microplastics

Behavior, Fate, and Remediation

John Pichtel and Mathew Simpson

Lanham • Boulder • New York • London

Published by Bernan Press
An imprint of The Rowman & Littlefield Publishing Group, Inc.
4501 Forbes Boulevard, Suite 200, Lanham, Maryland 20706
www.rowman.com

86-90 Paul Street, London EC2A 4NE

British Library Cataloguing in Publication Information Available

Library of Congress Cataloging-in-Publication Data

Names: Pichtel, John, 1957- author. | Simpson, Mathew E., author.
Title: Microplastics : behavior, fate, and remediation / John Pichtel and Mathew Simpson.
Description: Lanham : Bernan Press, [2023] | Includes bibliographical references and index.
Identifiers: LCCN 2023003071 (print) | LCCN 2023003072 (ebook) | ISBN 9781636710808 (paperback) | ISBN 9781636710815 (epub)
Subjects: LCSH: Microplastics.
Classification: LCC TD427.P62 P53 2023 (print) | LCC TD427.P62 (ebook) | DDC 363.738—dc23/eng/20230202
LC record available at https://lccn.loc.gov/2023003071
LC ebook record available at https://lccn.loc.gov/2023003072

∞™ The paper used in this publication meets the minimum requirements of American National Standard for Information Sciences—Permanence of Paper for Printed Library Materials, ANSI/NISO Z39.48-1992.

Contents

Preface

The generation of plastic waste worldwide has more than doubled over the past two decades.

Globally, a significant percentage of waste plastic is mismanaged, with a substantial proportion released to terrestrial and freshwater ecosystems, and ultimately the oceans. Focused research has revealed that natural processes such as sunlight, abrasion, microbial action, and other mechanisms can convert large plastic debris to extremely fine fragments, fibers, and films—microplastics (MPs).

The concept of *microplastics* emerged on the radar screen of environmental scientists only in the 1990s; since then, water and land pollution by MPs has emerged as a significant environmental and public health concern. It is now recognized that minute plastic particles are pervasive in the environment from the equator to the poles, from the Himalayas to the deep oceans. MPs are known to sorb an array of organic and inorganic contaminants, and be transported through food chains and occur in our food supply. Research reports have only recently identified the presence of MPs in the blood of adults and in stool samples of infants.

In order to effectively control pollution of the biosphere by MPs, there is an urgent need to formulate efficient technologies for their removal and, ideally, their destruction. In the last five years an enormous compendium of articles has been published on regional and global distribution of MPs, behavior and fate, threats to organisms, and methods of detection and quantification. In 2020 alone, an estimated 12,000 papers were published concerning MPs. Substantially less data, however, exists regarding technologies for effective recovery of MPs from water and sediment. The same can be stated for methods to transform MPs into environmentally benign compounds.

Unit operations at conventional drinking water and wastewater treatment facilities have been documented to capture a substantial percentage of MPs from surface water; in contrast, technologies for MP recovery from sediment are primarily at the pilot scale. Additional research is needed to develop and improve upon techniques for MP recovery from both liquid and solid media.

Knowledge is limited regarding methods for the destruction of MPs; however, an array of chemical, physical, and biological methods are currently under investigation. The degree of success varies substantially, and mechanisms responsible for MP destruction are only partly understood. A number of technologies show promise: at the laboratory scale, several have resulted in conversion of MP particles to water-soluble organic compounds and carbon dioxide. Certain innovative methods have transformed organic by-products into useful products such as fuels. These technologies offer promise for long-term water security and ecological stability, and deserve further attention by scientists.

This book provides a comprehensive overview of the MPs issue with emphasis on removal from water and sediment. In addition, the authors present prospective technologies for the destruction of MPs. Part I of this book provides background relevant to understanding the physicochemical properties and environmental behavior of MPs. Chapters address

- polymer chemical composition including the use of additives;
- description of MP types, both primary and secondary;
- MPs distribution in the biosphere;
- sorption of various contaminants by MPs; and
- detection and quantification of these particles.

It is not the intent of this book to review exhaustively the global distribution of MPs, nor their abilities in adsorbing and transporting environmental contaminants—many outstanding publications have already addressed these topics in rich detail. Therefore, only a limited review of the distribution of MPs and their ability to sorb environmental contaminants is presented.

Part II focuses on removal of MPs from water including wastewater, and removal from solid media (sediment). Technologies include those currently in use for treatment of municipal and industrial wastewater, and for recovery of minerals and certain solid waste components. Several innovative approaches are included.

Part III addresses technologies for the destruction of recovered MPs. Most methods are available only at the laboratory or pilot scale; however, some have shown great potential for transforming seemingly recalcitrant plastic particles to simple and innocuous products.

For the benefit of the reader, all terms shown in italics are defined in the glossary at the end of this book. A list of acronyms is included at the end of the book as well.

John Pichtel
Mathew Simpson

Acknowledgments

The authors wish to thank Holly McGuire and Janice Braunstein at Rowman & Littlefield who have been indispensable in preparing this work for publication. Many thanks to Arun Rajakumar at Deanta Publishing Services for his meticulous work in editing the manuscript. We are grateful to Parisa Ebrahimbabaie for technical support and to Brandon Smith of the Ball State University Digital Corps for production of many outstanding-quality drawings.

The authors express sincere thanks to our family members for the support they provided during the preparation of this work: Theresa, Leah, and Yozef Pichtel (JP); and Katherine, Gwenevere, and Sebastian Simpson (MS). Deep gratitude is expressed to Werner Erhard, a modern-day genius who assisted JP in opening countless avenues for producing quality results.

Part I

CHARACTERIZATION OF MICROPLASTICS

Microplastics (MPs) are commonly defined as synthetic particles measuring less than 5 mm. The category of particles termed *primary microplastics* is intentionally manufactured for industrial, commercial, and consumer applications. In contrast, *secondary microplastics* are inadvertently created by weathering and fragmentation of larger plastic articles. When released into the environment, both categories of particles have aroused significant concern among scientists, governments, and citizens worldwide.

MPs have been identified, often in large quantities, in environmental media including freshwater, seawater, sediment, wastewater effluent, compost piles, soil, and air. Their presence has been detected in the organs and tissue of hundreds of marine, freshwater, and terrestrial species; they are furthermore suspected to be migrating through food chains and impacting adversely on organisms. Recent reports suggest that MP pollution can pose a threat to human food safety.

It is essential that the unique and diverse properties of MPs be carefully studied and clarified so that their behavior in aquatic and terrestrial environments is better understood. Such knowledge can ultimately be applied for effective removal of MPs from the environment and, ideally, for their destruction under controlled conditions. Addressing these knowledge gaps will support efforts to mitigate the impacts of MP pollution in the biosphere.

Part I of this book provides the reader with the broader context of the MPs issue. Plastics waste generation and management are presented in chapter 1. Background information about MPs pollution is also presented, including potential hazards to ecosystems. Different synthetic polymers possess a wide range of properties which influence their behavior in the environment. In chapter 2, the chemical composition and selected modes of synthesis of polymers is provided. Chapter 3 provides a detailed description of both primary and secondary MPs, with emphasis on those influences that promote the conversion of macroplastics to MPs.

As stated in the Preface, it is not the intent of this book to review exhaustively the global distribution of MPs, nor their abilities in adsorbing and transporting environmental contaminants. Many excellent publications have already addressed these topics in great detail; therefore, only a limited review of the distribution of MPs (chapter 4) and their ability to sorb environmental contaminants (chapter 5) are presented herein. In order to draw comprehensive, consistent, and accurate conclusions about MPs behavior and fate in different environmental compartments, proper sample collection, preparation, and identification is required. Chapter 6 presents many of the necessary tools to accomplish these goals.

Chapter 1

Introduction

BACKGROUND

The term *plastic* was coined in the seventeenth century to describe a substance that is malleable and could be shaped or re-formed; it originates from the classical Greek work *plastikos*, defined as "able to be molded into different shapes" (Kaushik 2019). Synthetic plastics have existed since the early twentieth century. The modern-day plastics industry had its origins with the development of the first fully synthetic polymer known as Bakelite in 1907, attributed to Belgian-American chemist Leo Hendrick Baekeland (Chalmin 2019).

Through the course of two world wars, chemical technology has made spectacular advances in the development of new polymer formulations. The rapid progress in this new enterprise set the stage for the mass production of plastics for domestic, commercial, and industrial applications in the 1950s and 1960s.

Plastics comprise a large and diverse assemblage of unique and versatile materials that are now embraced by modern society as indispensable for countless applications. In 1950 the annual global production of plastics was a meager 1.5 million tons (Chalmin 2019). Global plastics manufacture has since increased dramatically to meet ever-growing demands. A few minor downturns occurred in recent decades, for example, during the 1973 oil crisis, the 2007 financial crisis, and the Covid-19 pandemic which began in 2019; regardless, global production of plastics continues to flourish. Current plastic production is approximately 368 million tons worldwide (PlasticsEurope 2020), and it is estimated that 8.3 billion tons of virgin plastics have been manufactured since 1950. According to long-term forecasts based on trends of consumer use and demographics, manufacture and consumption of plastic

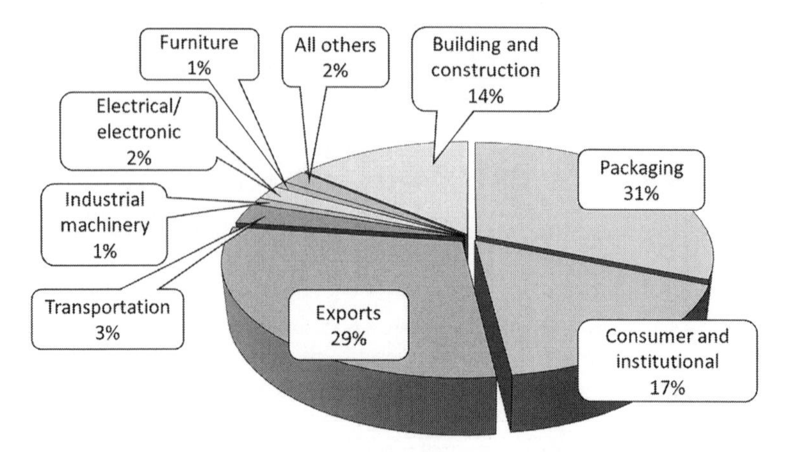

Figure 1.1 Market Sectors for Plastics Worldwide
Source: Reproduced with kind permission from the American Chemistry Council.

will expand further (Auta et al. 2017)—production is predicted to increase at an annual rate of 4% (PlasticsEurope 2018).

The main market sectors for plastics are packaging (31%), consumer and institutional products (17%), and building and construction (14%). Transportation, electrical/electronic equipment, and industrial machinery each comprise less than 10% of the global market share (Hu et al. 2021; Geyer 2017). About 13% of plastics is for other uses (figure 1.1).

Among various resin types, polyethylene (PE), polypropylene (PP), polystyrene (PS), and polyvinyl chloride (PVC) are the most consumed products. Other important resin types include polyethylene terephthalate (PET), polyurethane (PUR), polyamides (PA), and others (table 1.1) (Hu et al. 2021; PlasticsEurope 2020).

THE PLASTICS WASTE PROBLEM

The desirable properties of synthetic plastics such as being lightweight, durable, and resistant to breakdown make them the favored material for countless applications by domestic, commercial, industrial, military, and other consumers. Unfortunately, however, the myriad advantages and usefulness of plastics have given rise to a throw-away mentality in modern society. The majority of consumer plastic products are designed to be disposable after a single use or have only a short usage life; such products are typically disposed within 12 months following manufacture (Koelmans 2014). Other plastics are designed for extended use for products such as water pipes, electric cables, and household commodities (Hassanpour and Unnisa 2017).

Table 1.1 Common Applications of Plastics by Resin Type

Polymer Type	*Abbreviation*	*Applications*
Acrylonitrile butadiene styrene	ABS	Cases for computer monitors, printers and keyboards, drainage pipe
Low-density polyethylene	LDPE	Supermarket bags, food packaging film, bottles, drinking straws, fishing nets, outdoor furniture, house siding, floor tiles
High-density polyethylene	HDPE	Detergent bottles, milk jugs, toys, plastic crates, irrigation and drainage pipes, wire insulation
Polycarbonate/ acrylonitrile butadiene styrene	PC/ABS	Automobile interior and exterior parts, mobile phone bodies
Polypropylene	PP	Yogurt containers, bottle caps, drinking straws, medicine bottles, car batteries, car bumpers, disposable syringes, appliances, carpet backings, rope, fishing nets
Polystyrene	PS	Food containers, tableware, disposable plates and cups
Expanded polystyrene	EPS	Packaging foam, disposable cups, building materials (insulation)
Polyvinyl chloride	PVC	Plumbing, electrical conduits, shower curtains, raincoats, bottles, window frames, gutters, garden hoses, flooring
Polyurethane	PUR	Tires, gaskets, bumpers, refrigerator insulation, sponges, furniture cushioning, life jackets, surface coatings, insulation, paints, packing
Polyethylene terephthalate	PET	Soft-drink bottles, processed meat packages, microwavable packaging, pillow and sleeping bag filling, textile fibers, tubes, pipes, insulation molding
Polyamides (Nylon 6,6)	PA	Fibers, toothbrush bristles, fishing line, food packaging, inks, clothing, parachute fabrics, rainwear, cellophane
Polycarbonate	PC	Compact discs, eyeglass lenses, security windows, street lighting, safety visors, baby bottles
Polytetrafluoroethylene	PTFE	Industrial applications such as specialized chemical paints, electronics and bearings. Nonstick kitchen saucepans and frying pans
Polyvinylidene chloride (Saran®)	PVDC	Food packaging

The unique physical and chemical characteristics that polymers possess are responsible for plastic wastes persisting for long periods in both aquatic and terrestrial ecosystems (Habib et al. 2020). The vast majority of synthetic polymers are only slowly degradable, whether by biological or physical/ chemical processes. Due to personal lifestyles, the resilience of polymers, and inadequate or improper waste management, land and waterways worldwide have received staggering volumes of plastic waste. Of the billions of tons of plastics manufactured since the 1950s, almost 80% have been either disposed in landfills or pollute the natural environment (Geyer et al. 2017) (figure 1.2). Lebreton and Andrady (2019) calculated that between 60 and 99 million metric tons of mismanaged plastic waste were produced globally in a single year, and estimated that this figure could triple to 155–265 Mt y^{-1} by 2060. The combination of demand for single-use plastic products and improper management of plastics worldwide has resulted in approximately 7.5 billion metric tons of cumulative plastic waste generated up to 2021 (Recycle Coach 2022).

Floating on the surfaces of the world's oceans is an estimated 5 trillion plastic items, weighing 227,000 metric tons (Eriksen et al. 2014); current activities may be adding as much as 11 million metric tons of total plastics annually (Jambeck et al. 2015). About one-third of plastic waste is mismanaged, which accelerates losses to oceans (Ballerini et al. 2018). An estimated 80% of marine plastic debris originates from terrestrial sources (Jambeck et al. 2015; Mani et al. 2015; Wagner et al. 2014). The major activities responsible include littering, illegal dumping, cargo shipping, harbor and fishery operations, and discharge from wastewater treatment plants (Claessens et al. 2011; Habib et al. 2020; Xu et al. 2021). Freshwater bodies such as rivers and other drainages are also contaminated with substantial loads of

Figure 1.2 Pollution by Plastics: (a) Beach in northern UK and (b) Drainage Ditch in Asia
Source: Figure 1.2a: Photo by Andy Waddington, Camus Daraich / CC BY-SA 2.0 (Wikipedia) https://commons.wikimedia.org/wiki/File:Sea_washed_plastic_debris,_Camus_Daraich_-_geograph.org.uk_-_1188625.jpg. Figure 1.2b: https://commons.wikimedia.org/wiki/File:India_-_Sights_%26_Culture_-_garbage-filled_canal_(2832914746).jpg

improperly managed plastics, much of which eventually enters the oceans (Driedger et al. 2015). The largest inputs of plastic waste to the oceans originate from the coastlines of Asia (primarily China and India) and the United States (Lebreton and Andrady 2019; Jambeck et al. 2015). With rising production and consumption of plastics, and with limited regulatory oversight and control in many countries, plastic litter is anticipated to continue to accumulate in marine, freshwater, and terrestrial ecosystems.

The majority of published research which address the environmental threats posed by plastics focus on oceans (Thompson et al. 2009), where waste plastics tend to accumulate (Barnes et al. 2009; Ryan et al. 2009; Ryan 2015). The earliest reports of floating plastic in the ocean appeared in the scientific literature in the early 1970s (Carpenter and Smith 1972; Carpenter et al. 1972). In the Sargasso Sea, Carpenter and Smith (1972) estimated an average concentration of 3,500 plastic pieces km^{-2} over a distance of 1300 km.

One of the most notorious revelations involving marine pollution by plastics was the discovery of massive "garbage patches," or *gyres*, in several of the world's oceans. The best known is the Great Pacific Garbage Patch located in the North Pacific. In recent years researchers have discovered four more giant patches of concentrated marine debris in the South Pacific Ocean, the North Atlantic, South Atlantic, and the Indian Ocean. Recent research indicates that these patches are growing (Lebreton et al. 2018). The Great Pacific Garbage Patch is believed to have increased tenfold each decade since 1945 (Maser 2014). Plastic debris can circulate in these gyres for years, posing significant risks to marine biota.

By the 1960s, scientists were reporting that northern fur seals (*Callorhinus ursinus*) were becoming entangled in netting and other articles in the Bering Sea (Fowler 1987). At about the same time, researchers discovered that seabirds were ingesting plastic litter (Ryan 2015; Kenyon and Kridler 1969). Plastic was identified in the digestive tracts of prions (*Pachyptila* spp.) in New Zealand (Harper and Fowler 1987), and plastic particles were found in stomachs of Leach's storm petrels (*Oceanodroma leucorhoa*) from Newfoundland, Canada (Rothstein et al. 1973). Atlantic puffins (*Fratercula arctica*) were collected from 1969 to 1971 and elastic threads were detected in their stomachs (Berland 1971; Parslow and Jefferies 1972), and plastic particles (mostly PE) were detected in great shearwaters (*Puffinus gravis*) in the south Atlantic Ocean (Randall et al. 1983). Many other types of plastics from bottle caps, plastic sheets, and toys have been found in bird gizzards (Harper and Fowler 1987). Following these early investigations, scientists came to recognize the extensive presence of microplastics (MPs) in marine habitats and their adverse effect on marine life.

MANAGEMENT OF PLASTIC WASTE

Of all municipal waste generated worldwide, approximately 12% is plastic (World Bank 2022). In the United States, plastics generation has grown from 8.2% of waste in 1990 to 12.2% in 2018. Total plastic waste generation was 35.7 million tons in 2018 (US EPA 2021).

The issue of plastics pollution is best addressed proactively, that is, during the consumer decision-making process. In other words, a decline in plastics demand should ultimately be followed by reduced production. Once plastic products are purchased and used, however, a comprehensive plastics separation and recycling program and infrastructure is necessary for effective management. Unfortunately, with few exceptions, the rate of plastics recycling is low in both developed and developing nations worldwide—only a small proportion of plastic waste (9%) is recycled (Brooks et al. 2018). The recycling rate of plastic waste in the United States is 8.7% (US EPA 2021) and 53% is landfilled (Luo et al. 2021). In Europe plastics recycling ranges between 26% and 52% (PlasticsEurope 2020) and 31% is landfilled (Luo et al. 2021; Paço et al., 2017).

A number of practical obstacles to plastics recycling exist. First, recycling plastics is more expensive than using raw petroleum feedstock. There is about a 20% increase in manufacturing costs associated with relying upon recycled plastics as compared to costs for utilizing virgin feedstock (Crawford and Quinn 2017). In many less-developed nations, suitable infrastructure is not available for effective waste collection, transport, and management, including recycling and disposal. Furthermore, in many countries there may be little incentive to recycle if the perception exists that landfill space is inexhaustible (and, ideally, far from home). This is compounded by lack of public education about waste generation and recycling.

Another reason for low plastic recycling rates is that different resin types tend to be incompatible in new products; in other words, if recovered HDPE is mixed with PVC, the resulting new product will lack many desired physical properties. Separation of resin types for recycling by the consumer is desirable but not common. The presence of different additives in various plastic products adds to the difficulty of producing a uniform new product.

Lastly, a typical recyclable plastic product can be recycled only about three times. The repeated melting and remolding of the plastic results in loss of mechanical properties: flexibility decreases and the plastic becomes brittle and discolors. With the loss of mechanical properties, the plastic can no longer be applied to its original use and is discarded (Crawford and Quinn 2017).

Only a modest proportion of plastic waste is incinerated. This is partly a consequence of public opposition due to odor generation, and the presence of potentially toxic compounds in gaseous emissions and ash. When certain plastic materials burn they release hazardous organic compounds such as polychlorinated dibenzodioxins ("*dioxins*") and polychlorinated dibenzofurans ("*furans*"); both are associated with serious human health impacts.

If current plastic production and waste management trends continue, an estimated 13 billion tons of plastic waste will occur in landfills or in the natural environment by 2050 (Geyer et al. 2017).

While the problem of plastic wastes on land and in the oceans has received attention from governments and citizens over several decades, only recently have smaller plastic fragments, termed *microplastics*, emerged as a pollutant of concern (Sharma and Chatterjee 2017).

MICROPLASTICS

Pollution caused by MPs is considered more widespread and hazardous compared with larger litter because of its vastly greater quantities and fine particle sizes (Hahladakis et al. 2018). It is not certain when the term *microplastics* was first proposed to describe the smallest plastic fragments occurring in ecosystems and in organisms. This term was used by Ryan and Moloney (1990) during research activities at South African beaches. Thompson et al. (2004) used it when describing the extent of contamination of plastic fragments in seawater and coasts of the North Atlantic Ocean. MP size was not defined in these early reports, however.

MPs are composed of many polymer types which correspond to common end-use markets for plastics. The most common MP resin types are PE, PP, and PVC. These are followed by PET, PS, and PA (Andrady et al. 2011; Guo et al. 2019; Alimba and Faggio 2019; Hidalgo-Ruz et al. 2012; Hu et al. 2021). The quantities of plastic waste produced from different polymers is shown in figure 1.3.

MPs are documented polluting oceans, rivers, lakes, and agricultural land (Duis and Coors 2016; Eerkes-Medrano et al. 2015; Van Cauwenberghe et al. 2015), and even occur in the atmosphere in measurable quantities (Chen et al. 2020). MPs in near-shore areas originate mostly from the mainland via wastewater effluent, runoff, rivers, and even air flow. Ships and offshore platforms are sources of MPs to the deep oceans. MPs have been detected at all latitudes and even in polar regions (Mishra et al. 2021). As many as 5.25 trillion plastic particles are estimated to contaminate the global sea surface (Eriksen et al. 2014). Nearly 10% of the total mass of the Great Pacific Garbage Patch is calculated to be composed of MPs. According to Lebreton et al. (2018), over 90% of the small plastic pieces floating on the surface of the gyre (1.1–3.6 trillion pieces) are MPs. A large volume of marine MPs also occurs

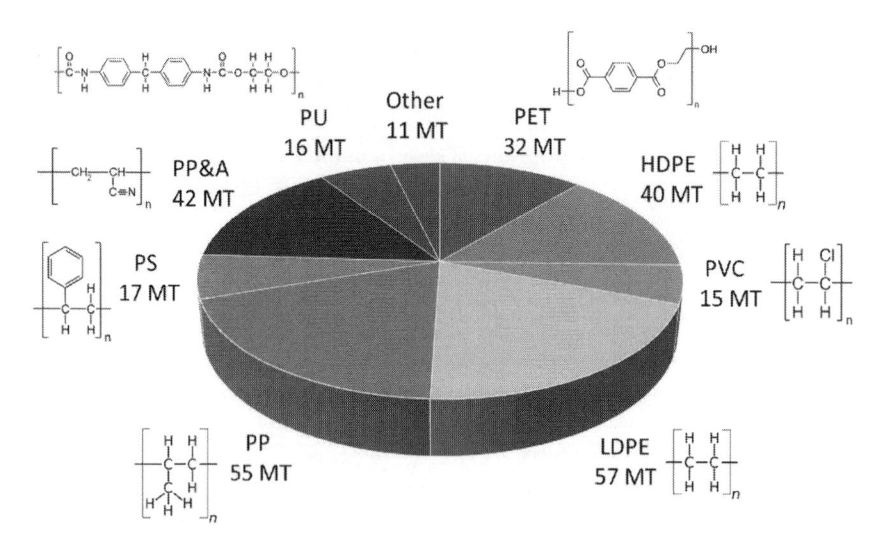

Figure 1.3 Proportion of Different Polymer Types in Plastic Wastes
Source: Data from Geyer, Jambeck, Law Sci. Adv. 2017; 3: e1700782. Distributed under CC BY-NC 4.0 License.

in sediment (Koelmans et al. 2017; Woodall et al. 2014). The extent of MPs contamination of freshwater is considered to be as severe as that in oceans (Wagner et al. 2014). MPs are widely detected in inland lakes and estuaries (Driedger et al. 2015). (The distribution of MPs in marine and freshwater environments is presented in chapter 4.)

Size Ranges of MPs

The number of scientific papers that address MP pollution has increased markedly in recent years. Disagreement exists, however, as to the exact size range of MPs, and several definitions have been proffered to categorize them. Particle size is a dominant characteristic as regards determining environmental behavior and fate of MPs (Song et al. 2019; He et al. 2018; Tong et al. 2020; Wang et al. 2021). Finer-sized particles possess a greater surface area compared with larger particles and are thus expected to be more reactive with mineral and organic materials, including living tissue, occurring in the local environment. Bioaccumulation and toxic effects to organisms may also be size-dependent (Kim et al. 2020; Lei et al. 2018; Sendra et al. 2019; Wang et al. 2021). There is general agreement that plastic items larger than 5 mm (originating from the definition provided by Arthur et al. [2009]) be termed *macroplastics* (figure 1.4).

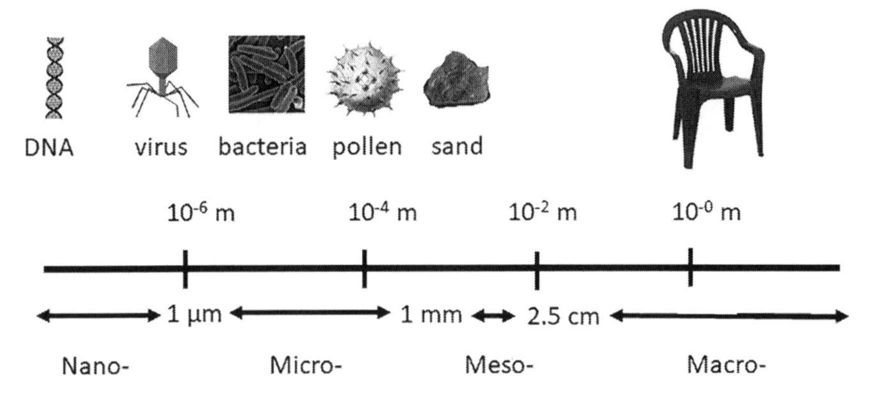

Figure 1.4 Plastic Particles as Defined by Size Range

In studies using size ranges between 1 μm and 5 mm, those particles measuring less than 1 mm have been determined to represent a significant proportion (i.e., from 35 to 90%) of all MPs (McDermid and McMullen 2004; Browne et al. 2010; Eriksen et al. 2013; Song et al. 2014; Zhao et al. 2014). The Joint Group of Experts on the Scientific Aspects of Marine Protection (Kershaw and Rochman 2015) suggests that MPs be defined to occur within the 1 nm to 5 mm size range. This lower limit, however, includes what some studies have defined as *nanoplastics* (Koelmans 2014). A more generally accepted lower cutoff for MPs is 1 μm (Van Melkebeke 2019). A further category was proposed by MSFD Technical Subgroup on Marine Litter (2013) which distinguishes between small MPs (from 1 μm to 1 mm) and large MPs (from 1 mm to 5 mm). The rationale for the cutoff is that 1 mm marks the upper size boundary of particles that are available for uptake by planktonic species. Given that planktonic species comprise the base of the marine food web, concerns exist regarding serious repercussions to ecosystems from small MPs (Moore 2008; Van Cauwenberghe 2015).

The term *nanoplastic* has been introduced to describe particles ≤ 1 μm in size (Bergmann et al. 2015); some authors, however, define nanoplastics as particles measuring up to 100 nm (Ter Halle et al. 2017). These particles comprise a relatively new component of the science of plastics pollution. Nanoplastics are the least understood type of marine litter—and they may potentially be more hazardous than MPs (Hahladakis et al. 2018; Koelmans 2014).

Morphology of MPs

MPs occur in numerous forms, some of which influence their migration in the biosphere and translocation within organisms. Particle morphology also affects the capabilities of engineered systems to capture and remove them.

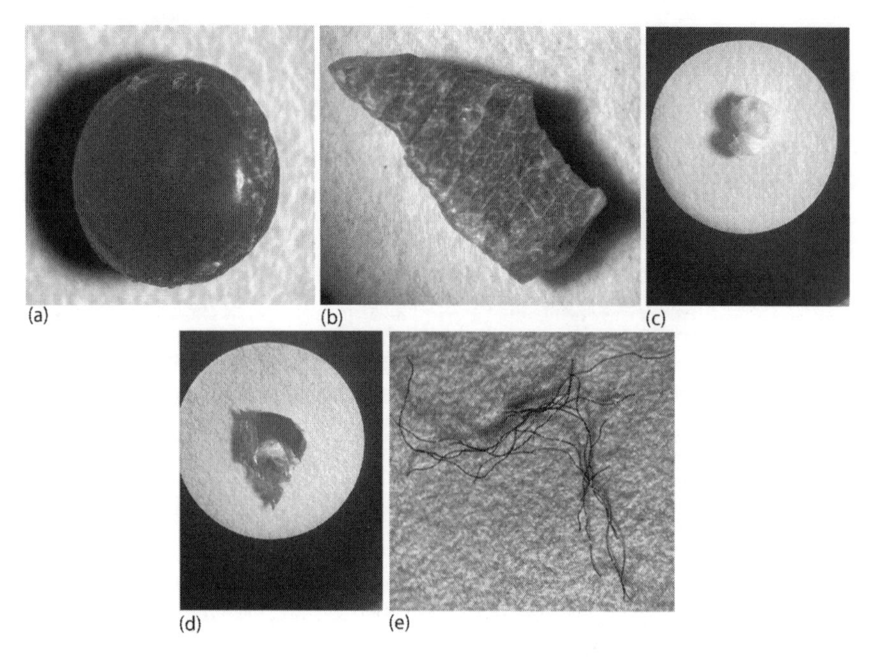

Figure 1.5 The Varying Morphology of MP Particles: (a) Pellets, (b) Fragments, (c) Foams, (d) Films, and (e) Fibers

MPs are generally categorized into the following morphological groups: pellets (i.e., nurdles or mermaid tears), fragments, granules, flakes, films, fibers, and foams (figure. 1.5) (Hidalgo-Ruz et al. 2012). Each category exhibits unique physical and/or chemical behavior in nature (Hu et al. 2021). Details on the types of MPs and modes of generation appear in chapter 3.

Hazards of MPs

Over the past decade researchers have described several key aspects of the MPs problem, including sources (Bradney et al. 2019), migration (Guo et al. 2020), regional and global distribution (Fu et al. 2020), bioaccumulation (Xu et al. 2020), toxicity (Chen et al. 2020), and ecological risks (Ma et al. 2020). One of the overriding ecological concerns associated with MPs is ingestion by organisms, resulting in reduced foraging ability, loss of nutrition, and blockage of the digestive tract (Bakir et al. 2012; Graham and Thompson 2009).

The Food Web

Recent reports confirm that MPs are bioavailable to a wide range of biota. As MPs contaminate the environment, their presence has been demonstrated throughout the food chain. At lower trophic levels in the marine environment, MPs have been detected in zooplankton, Chaetognatha, ichthyoplankton, copepods, and salps (Cverenkárová et al. 2021). At higher trophic levels, MP contamination has been reported in invertebrates (polychaetes, crustaceans, echinoderms, bivalves) and vertebrates (fish, seabirds, and mammals) (Bakir et al. 2014; Thompson et al. 2004; Ward and Shumway 2004; Graham and Thompson 2009; Cau et al. 2019; Murray and Cowie 2011; Lusher et al. 2013). These organisms ingest MPs either through direct consumption or via trophic transfer (Hollman et al. 2013). As a consequence of the persistent nature of MPs, there is concern that transfer through the food web will result in biological accumulation. This could also include transfer and accumulation of contaminants, whether as additives to the original plastic formulation, or toxins sorbed to MP surfaces.

Hazards and Humans

Entry of MPs into the food web ultimately poses a hazard to public health (Habib et al. 2020). In recent years concerns have been raised regarding the potentially toxic impacts of MPs on humans. Unfortunately, research in this area remains limited (Revel et al. 2017). The human body is exposed to MPs via ingestion, inhalation, and dermal exposure (Yee et al. 2021). MPs have been identified in commercially available foods and beverages and public water supplies (Carbery et al. 2018), as well as in the atmosphere. Ingestion is considered the most significant means of human exposure (Yee et al. 2021; Prata et al. 2020).

MPs have been reported in food items such as mussels and commercial fish. Van Cauwenberghe and Janssen (2014) determined that the MP content in the soft tissue of commercially grown blue mussel (*Mytilus edulis*) and Japanese oyster (*Crassostrea gigas*) was 0.36 particles$\cdot g^{-1}$ and 0.47 particles$\cdot g^{-1}$ ww (wet weight), respectively. MPs have been detected in table salt: Iniguez et al. (2017) measured a MP content of 50–280 particles$\cdot kg^{-1}$ salt. The most frequently found polymer was PET, followed by PP and PE. Yang et al. (2015) obtained several brands of sea salts from supermarkets in China and found MPs content of 550–681 particles$\cdot kg^{-1}$ salt. Fragments and fibers were prevalent MP types. MPs have also been detected in sugar: Liebezeit and Liebezeit (2013) found that MPs in the form of fibers and fragments were detected at levels of 217 and 32 particles$\cdot kg^{-1}$, respectively. In a study by Mason et al. (2018), eleven globally sourced brands of bottled water, purchased in nine different countries, were tested for MP contamination. Of 259 total bottles processed, 93% showed some MP contamination.

In recent work, MPs have been detected in human feces. In one study (Schwabl et al. 2019), stool samples from all subjects tested positive for MPs.

A median value of 20 MPs (50 to 500 μm in size) per 10 g of human stool was determined. Nine plastic types were detected, with PP and PET being the most abundant. PET and PC MPs were detected in all infant stool specimens examined by Zhang et al. (2021) (PET: 5700–82,000 $ng \cdot g^{-1}$, median 36,000 $ng \cdot g^{-1}$; PC: 49–2100 $ng \cdot g^{-1}$, median 78 $ng \cdot g^{-1}$). MPs were also found in most (PET) or all (PC) adult stool samples; however, concentrations were an order of magnitude lower in adults than in infants for PET MPs. In a recent study (Leslie et al. 2022), nano-sized plastics were discovered in human blood. PET, PE, and polymers of styrene were the most widely encountered particles in the blood of 22 healthy volunteers. The mean of concentrations for each donor was 1.6 μg total plastic particles•ml $blood^{-1}$. This data reveals that at least a portion of the plastic particles that humans come in contact with are bioavailable, and that the rate of elimination is slower than the rate of absorption into the blood.

Inhaled airborne MPs originate from urban dust which contains synthetic textiles and rubber tire particles (Prata 2018; Yee et al. 2021). A study by Amato-Lourenço (2021) determined the presence of MPs in human lung tissue. MP particles and fibers were observed in 13 of 20 tissue samples. All particles measured < 5.5 μm in size, and fibers ranged from 8.12 to 16.8 μm. The most frequently detected polymers were PE and PP.

Dermal contact of MPs is possible via use of certain consumer products and textiles, and by exposure to dust (Revel et al. 2018). Dermal contact occurs when consumers use facial or body scrubs which contain MPs (Hernandez et al. 2017). Contact is also possible when washing with water contaminated with MPs. Absorption through the skin is unlikely to occur, however, as transfer of particles across skin requires penetration of striatum corneum (the outermost layer of the epidermis) which is limited to particles below 100 nm. However, nanoplastics (measuring less than 1 μm) could possibly penetrate human skin (Sykes et al. 2014; Revel et al. 2018).

According to calculations based on 26 studies (Cox et al. 2019), the US population is estimated to consume from 39,000 to 52,000 MP particles per year, depending on age and gender. These estimates were obtained by evaluating a fraction of American's caloric intake and included the content of MPs in seafood, sugar, salt, honey, alcohol, and bottled and tap water. Values of total intake increase to 74,000 and 121,000 particles per year when inhalation is included. As of this writing, data regarding the physiological and toxicological effects from ingestion and inhalation of MPs by organisms (including humans) is scarce.

MPs Sorb and Transport Contaminants

MPs are documented to sorb a wide array of water pollutants and can transport them great distances. Sorbed contaminants include heavy metals such as Cd, Cu, Zn, and Pb (Hodson et al. 2017; Ashton et al. 2010; Brennecke et al. 2016; Holmes et al. 2012, 2014); toxic organic compounds including polychlorinated biphenyls (PCBs), PAHs, PBDEs, and other persistent bioaccumulative and toxic substances such as pesticides and halogenated flame retardants (Guo et al. 2019; Wang et al. 2018; Fries and Zarfl 2012; Mato et al. 2001; Ogata et al. 2009; Hirai et al. 2011; Heskett et al. 2012); antibiotics (Li et al. 2018); and pathogenic microorganisms (Rodrigues et al. 2019; Kirstein et al. 2016; Eiler et al. 2006; Masó et al. 2003). The potential for sorption of pollutants by MPs is presented in chapter 5.

CLOSING STATEMENTS

To manage the environmental and public health issues associated with MPs, it is essential to understand their chemical and physical properties, their sources to different environmental compartments, and their behavior in the biosphere.

To limit MP pollution of terrestrial and aquatic ecosystems, a proactive approach to waste management is critical. According to waste generation models (Lebreton and Andrady 2019), current demand for plastics is greatest in nations in the temperate climate regions, that is, the United States and Europe. However, projections suggest that consumer demand for plastics in South Asia, Southeast Asia, and Africa will increase exponentially in future decades, with concomitant rise in plastic waste generation. Municipalities in these nations must be prepared for the inevitable upsurge in plastic waste and the necessity for its appropriate management.

A Chinese proverb states, "If we do not change our direction, we are likely to end up where we are headed." To effectively mitigate the release of plastic waste into the environment, several issues must be addressed, with urgency, by citizens as well as municipalities in both industrialized and less developed nations. These include: (1) reduce demand for single-use plastics; (2) avoid purchasing products that contain plastic microbeads; (3) improving the waste management infrastructure; and (4) increasing the capacity for municipal waste recycling. Implicit in these issues is the absolute necessity of educating the public regarding the importance of sound waste management.

REFERENCES

AFT Fluorotec. (2022). Protective PTFE anti-corrosion functional coatings; [accessed 2022 January 5]. https://www.fluorotec.com/what-we-do/coating-solutions/anti-corrosion-coatings.

Alimba, C.G., and Faggio, C. (2019). Microplastics in the marine environment: Current trends in environmental pollution and mechanisms of toxicological profile. *Environmental Toxicology and Pharmacology* 68:61–74.

Amato-Lourenço, L.F., Carvalho-Oliveira, R., Ribeiro Júnior, G., Santos Galvão, L., Augusto Ando, R., and Mauad, T. (2021). Presence of airborne microplastics in human lung tissue. *Journal of Hazardous Materials* 416:126124.

Andrady, A.L. (2011). Microplastics in the marine environment. *Marine Pollution Bulletin* 62(8):1596–1605.

Arthur, C., Baker, J., and Bamford, H. (2009). Proceedings of the International Research Workshop on the Occurrence, Effects and Fate of Microplastic Marine Debris. NOAA Technical Memorandum NOS-OR&R-30, January:530.

Ashton, K., Holmes, L., and Turner, A. (2010). Association of metals with plastic production pellets in the marine environment. *Marine Pollution Bulletin* 60(11):2050–2055.

Auta, H.S., Emenike, C., and Fauziah, S. (2017). Distribution and importance of microplastics in the marine environment: A review of the sources, fate, effects, and potential solutions. *Environment International* 102:165–176.

Bakir, A., Rowland, S.J., and Thompson, R.C. (2012). Competitive sorption of persistent organic pollutants onto microplastics in the marine environment. *Marine Pollution Bulletin* 64(12):2782–2789.

Bakir, A., Rowland, S.J., and Thompson, R.C. (2014). Transport of persistent organic pollutants by microplastics in estuarine conditions. *Estuarine, Coastal and Shelf Science* 140:14–21.

Ballerini, T., Andrady, A.L., Cole, M., and Galgani, F. (2018, March). Plastic pollution in the ocean: What we know and what we don't know about. DOI: 10.13140/RG.2.2.36720.92160.

Barnes, D.K., Galgani, F., Thompson, R.C., and Barlaz, M. (2009). Accumulation and fragmentation of plastic debris in global environments. *Philosophical Transactions of the Royal Society B: Biological Sciences* 364(1526):1985–1998.

Bergmann, M., Gutow, L., and Klages, M. (2015). *Marine Anthropogenic Litter*. Berlin (Germany): Springer.

Berland, B. (1971). Piggha og lundefugl med gummistrik. *Fauna* 24:35–37.

Bradney, L., Wijesekara, H., Palansooriya, K.N., Obadamudalige, N., Bolan, N.S., Ok, Y.S., Rinklebe, J., Kim, K.-H., and Kirkham, M.B. (2019). Particulate plastics as a vector for toxic trace-element uptake by aquatic and terrestrial organisms and human health risk. *Environment International* 131:104937.

Brennecke, D., Duarte, B., Paiva, F., Caçador, I., and Canning-Clode, J. (2016). Microplastics as vector for heavy metal contamination from the marine environment. *Estuarine, Coastal and Shelf Science* 178:189–195.

Brooks, A.L., Wang, S., and Jambeck, J.R. (2018). The Chinese import ban and its impact on global plastic waste trade. *Science Advances* 4(6):eaat0131. DOI: 10.1126/sciadv.aat0131.

Browne, M.A., Galloway, T.S., and Thompson, R.C. (2010). Spatial patterns of plastic debris along estuarine shorelines. *Environmental Science & Technology* 44(9):3404–3409.

Carbery, M., O'Connor, W., and Palanisami, T. (2018). Trophic transfer of microplastics and mixed contaminants in the marine food web and implications for human health. *Environment International* 115:400–409.

Carpenter, E.J., Anderson, S.J., Harvey, G.R., Miklas, H.P., and Peck, B.B. (1972). Polystyrene spherules in coastal waters. *Science* 178(4062):749–750.

Carpenter, E.J., and Smith, K.L. (1972). Plastics on the Sargasso Sea surface. *Science* 175(4027):1240–1241.

Cau, A., Avio, C.G., Dessì, C., Follesa, M.C., Moccia, D., Regoli, F., and Pusceddu, A. (2019). Microplastics in the crustaceans *Nephrops norvegicus* and *Aristeus antennatus*: Flagship species for deep-sea environments? *Environmental Pollution* 255:113107.

Chalmin, P. (2019). The history of plastics: from the Capitol to the Tarpeian Rock: Field actions science reports. *The Journal of Field Actions* 19:6–11.

Chen, G., Feng, Q., and Wang, J. (2020). Mini-review of microplastics in the atmosphere and their risks to humans. *Science of the Total Environment* 703:135504.

Claessens, M., De Meester, S., Van Landuyt, L., De Clerck, K., and Janssen, C.R. (2011). Occurrence and distribution of microplastics in marine sediments along the Belgian coast. *Marine Pollution Bulletin* 62(10):2199–2204.

Cox, K.D., Covernton, G.A., Davies, H.L., Dower, J.F., Juanes, F., and Dudas, S.E. (2019). Human consumption of microplastics. *Environmental Science & Technology* 53:7068–7074.

Crawford, C.B., and Quinn, B. (2016). *Microplastic Pollutants.* Amsterdam (Netherlands): Elsevier.

Cverenkárová, K., Valachovičová, M., Mackuľak, T., Žemlička, L., and Bírošová, L. (2021). Microplastics in the food chain. *Life* 11:1349.

Driedger, A.G., Dürr, H.H., Mitchell, K., and Van Cappellen, P. (2015). Plastic debris in the Laurentian Great Lakes: A review. *Journal of Great Lakes Research* 41(1):9–19.

Duis, K., and Coors, A. (2016). Microplastics in the aquatic and terrestrial environment: Sources (with a specific focus on personal care products), fate and effects. *Environmental Sciences Europe* 28(1):1–25.

Eerkes-Medrano, D., Thompson, R.C., and Aldridge, D.C. (2015). Microplastics in freshwater systems: A review of the emerging threats, identification of knowledge gaps and prioritisation of research needs. *Water Research* 75:63–82.

Eiler, A., Johansson, M., and Bertilsson, S. (2006). Environmental influences on *Vibrio* populations in northern temperate and boreal coastal waters (Baltic and Skagerrak Seas). *Applied and Environmental Microbiology* 72(9):6004–6011.

Eriksen, M., Lebreton, L.C., Carson, H.S., Thiel, M., Moore, C.J., Borerro, J.C., Galgani, F., Ryan, P.G., and Reisser, J. (2014). Plastic pollution in the world's oceans: more than 5 trillion plastic pieces weighing over 250,000 tons afloat at sea. *PLoS ONE* 9(12):e111913.

Eriksen, M., Mason, S., Wilson, S., Box, C., Zellers, A., Edwards, W., Farley, H., and Amato, S. (2013). Microplastic pollution in the surface waters of the Laurentian Great Lakes. *Marine Pollution Bulletin* 77(1–2):177–182.

Fowler, C.W. (1987). Marine debris and northern fur seals: A case study. *Marine Pollution Bulletin* 18:326–335.
Fries, E., and Zarfl, C. (2012). Sorption of polycyclic aromatic hydrocarbons (PAHs) to low and high density polyethylene (PE). *Environmental Science and Pollution Research* 19(4):1296–1304.
Fu, D., Chen, C.M., Qi, H., Fan, Z., Wang, Z., Peng, L., and Li, B. (2020). Occurrences and distribution of microplastic pollution and the control measures in China. *Marine Pollution Bulletin* 153:110963.
Geyer, R., Jambeck, J.R., and Law, K.L. (2017). Production, use, and fate of all plastics ever made. *Science Advances* 3(7):e1700782.
Graham, E.R., and Thompson, J.T. (2009). Deposit- and suspension-feeding sea cucumbers (Echinodermata) ingest plastic fragments. *Journal of Experimental Marine Biology and Ecology* 368(1):22–29.
Guo, J.-J., Huang, X.-P., Xiang, L., Wang, Y.-Z., Li, Y.-W., Li, H., Cai, Q.-Y., Mo, C.-H., and Wong, M.-H. (2020). Source, migration and toxicology of microplastics in soil. *Environment International* 137:105263.
Guo, X., and Wang, J. (2019). The chemical behaviors of microplastics in marine environment: A review. *Marine Pollution Bulletin* 142:1–14.
Habib, R., Thiemann, T., and Kendi, R. (2020). Microplastics and wastewater treatment plants: A review. *Journal of Water Resources Protection* 12:1–35.
Hahladakis, J.N., Velis, C.A., Weber, R., Iacovidou, E., and Purnell, P. (2018). An overview of chemical additives present in plastics: Migration, release, fate and environmental impact during their use, disposal and recycling. *Journal of Hazardous Materials* 344:179–199.
Harper, P., and Fowler, J. (1987). Plastic pellets in New Zealand storm-killed prions (*Pachyptila* spp.), 1958–1998. *Notornis* 34:65–70.
Hassanpour, M., and Unnisa, S. (2017). Plastics; applications, materials, processing and techniques. *Plastic Surgery and Modern Techniques* 2017(2):1–5.
He, D., Luo, Y., Lu, S., Liu, M., Song, Y., and Lei, L. (2018). Microplastics in soils: Analytical methods, pollution characteristics and ecological risks. *TrAC Trends in Analytical Chemistry* 109:163–172.
Hernandez, L.M., Yousefi, N., and Tufenkji, N. (2017). Are there nanoplastics in your personal care products? *Environmental Science & Technology Letters* 4(7):280–285. DOI: 10.1021/acs.estlett.7b00187.
Heskett, M., Takada, H., Yamashita, R., Yuyama, M., Ito, M., Geok, Y.B., Ogata, Y. Kwan, C., Heckhausen, A., Taylor, H., Powell, T., Morishige, C., Young, D., Patterson, H., Robertson, B., Bailey, E., and Mermozk, J. (2012). Measurement of persistent organic pollutants (POPs) in plastic resin pellets from remote islands: Toward establishment of background concentrations for International Pellet Watch. *Marine Pollution Bulletin* 64(2):445–448.
Hidalgo-Ruz, V., Gutow, L., Thompson, R.C., and Thiel, M. (2012). Microplastics in the marine environment: a review of the methods used for identification and quantification. *Environmental Science & Technology* 46(6):3060–3075.
Hirai, H., Takada, H., Ogata, Y., Yamashita, R., Mizukawa, K., Saha, M., Kwan, C., Moore, C., Gray, H., Laursen, D., Zettler, D.R., Farrington, J.W., Reddy,

C.M., Peacock, E.E., and Ward, M.W. (2011). Organic micropollutants in marine plastics debris from the open ocean and remote and urban beaches. *Marine Pollution Bulletin* 62(8):1683–1692.

Hodson, M.E., Duffus-Hodson, C.A., Clark, A., Prendergast-Miller, M.T., and Thorpe, K.L. (2017). Plastic bag derived–microplastics as a vector for metal exposure in terrestrial invertebrates. *Environmental Science & Technology* 51(8):4714–4721.

Hollman, P.C.H., Bouwmeester, H., and Peters, R.J.B. (2013). *Microplastics in Aquatic Food Chain: Sources, Measurement, Occurrence and Potential Health Risks*. Wageningen (Netherlands): RIKILT Wageningen UR.

Holmes, L.A., Turner, A., and Thompson, R.C. (2012). Adsorption of trace metals to plastic resin pellets in the marine environment. *Environmental Pollution* 160:42–48.

Holmes, L.A., Turner, A., and Thompson, R.C. (2014). Interactions between trace metals and plastic production pellets under estuarine conditions. *Marine Chemistry* 167:25–32.

Hu, K., Tian, W., Yang, Y., Nie, G., Zhou, P., Wang, Y., Duan, X., and Wang, S. (2021). Microplastics remediation in aqueous systems: Strategies and technologies. *Water Research* 198:117144.

Iñiguez, M.E., Conesa, J.A., and Fullana, A. (2017). Microplastics in Spanish table salt. *Scientific Reports* 7:8620. DOI: 10.1038/s41598-017-09128-x.

Jambeck, J.R., Geyer, R., Wilcox, C., Siegler, T.R., Perryman, M., Andrady, A., Narayan, R., and Law, K.L. (2015). Plastic waste inputs from land into the ocean. *Science* 347(6223):768–771.

Kaushal, J., Khatri, M., and Arya, S.K. (2021). Recent insight into enzymatic degradation of plastics prevalent in the environment: A mini-review. *Cleaner Engineering and Technology* 2:100083.

Kaushik, U. (2019). Chapter 7: Plastic industry waste: Sources, management and recycling. In Jayakumar, R. (Ed.), *Research Trends in Multidisciplinary Research, Vol. 5*. New Delhi (India): AkiNik Publications.

Kenyon, K.W., and Kridler, E. (1969). Laysan albatrosses swallow indigestible matter. *The Auk* 86(2):339–343.

Kershaw, P., and Rochman, C. (2015). Sources, fate and effects of microplastics in the marine environment: part 2 of a global assessment. Reports and Studi es-IMO/FAO/Unesco-IOC/WMO/IAEA/UN/UNEP Joint Group of Experts on the Scientific Aspects of Marine Environmental Protection (GESAMP) Eng No. 93.

Kim, S.W., Kim, D., Jeong, S.-W., and An, Y.-J. (2020). Size-dependent effects of polystyrene plastic particles on the nematode *Caenorhabditis elegans* as related to soil physicochemical properties. *Environmental Pollution* 258:113740.

Kirstein, I.V., Kirmizi, S., Wichels, A., Garin-Fernandez, A., Erler, R., Löder, M., and Gerdts, G. (2016). Dangerous hitchhikers? Evidence for potentially pathogenic *Vibrio* spp. on microplastic particles. *Marine Environmental Research* 120:1–8.

Koelmans, A.A. (2014). The Challenge: Plastics in the marine environment. ET&C perspectives. *Environmental Toxicology and Chemistry* 33:5–10.

Koelmans, A.A., Kooi, M., Law, K.L., and Van Sebille, E. (2017). All is not lost: Deriving a topdown mass budget of plastic at sea. *Environmental Research Letters* 12(11):114028.

Lebreton, L., and Andrady, A. (2019). Future scenarios of global plastic waste generation and disposal. *Palgrave Communications* 5:6. DOI: 10.1057/s41599-018-0212-7.

Lebreton, L., Slat, B., Ferrari, F., Sainte-Rose, B., Aitken, J., Marthouse, R., Hajbane, S., Cunsolo, S., Schwarz, A., Levivier, A., Noble, K., Debeljak, P., Maral, H., Schoeneich-Argent, R., Brambini, R., and Reisser, J. (2018). Evidence that the great pacific garbage patch is rapidly accumulating plastic. *Scientific Reports* 8(1):1–15.

Lei, L., Liu, M., Song, Y., Lu, S., Hu, J., Cao, C., Xie, B., Shi, H., and He, D. (2018). Polystyrene (nano) microplastics cause size-dependent neurotoxicity, oxidative damage and other adverse effects in *Caenorhabditis elegans*. *Environmental Science: Nano* 5(8):2009–2020.

Leslie, H.A., van Velzen, M.J.M., Brandsma, S.H., Vethaak, D., Garcia-Vallejo, J.J., and Lamoree, M.H. (2022). Discovery and quantification of plastic particle pollution in human blood. *Environment International*; [accessed 2022 April 6]. DOI: 10.1016/j.envint.2022.107199.

Li, J., Zhang, K., and Zhang, H. (2018). Adsorption of antibiotics on microplastics. *Environmental Pollution* 237:460–467.

Liebezeit, G., and Liebezeit, E. (2013). Non-pollen particulates in honey and sugar. *Food Additives and Contaminants Part A* 30:2136–2140.

Luo, H., Zeng, Y., Zhao, Y., Xiang, Y., Li, Y., and Pan, X. (2021). Effects of advanced oxidation processes on leachates and properties of microplastics. *Journal of Hazardous Materials* 413:125342.

Lusher, A.L., Mchugh, M., and Thompson, R.C. (2013). Occurrence of microplastics in the gastrointestinal tract of pelagic and demersal fish from the English channel. *Marine Pollution Bulletin* 67(1–2):94–99.

Ma, H., Pu, S., Liu, S., Bai, Y., Mandal, S., and Xing, B. (2020). Microplastics in aquatic environments: Toxicity to trigger ecological consequences. *Environmental Pollution* 261:114089.

Mani, T., Hauk, A., Walter, U., and Burkhardt-Holm, P. (2015). Microplastics profile along the Rhine River. *Scientific Reports* 5(1):1–7.

Maser, C. (2014). *Interactions of Land, Ocean and Humans: A Global Perspective*. Boca Raton (FL): CRC Press.

Masó, M., Garcés, E., Pagès, F., and Camp, J. (2003). Drifting plastic debris as a potential vector for dispersing harmful algal bloom (HAB) species. *Scientia Marina* 67(1):107–111.

Mason, S.A., Welch, V.G., and Neratko, J. (2018). Synthetic polymer contamination in bottled water. *Frontiers in Chemistry*. DOI: 10.3389/fchem.2018.00407.

Mato, Y., Isobe, T., Takada, H., Kanehiro, H., Ohtake, C., and Kaminuma, T. (2001). Plastic resin pellets as a transport medium for toxic chemicals in the marine environment. *Environmental Science & Technology* 35(2):318–324.

McDermid, K.J., and McMullen, T.L. (2004). Quantitative analysis of small-plastic debris on beaches in the Hawaiian archipelago. *Marine Pollution Bulletin* 48(7–8):790–794.

Mishra, A.K., Singh, J., and Mishra, P.P. (2021). Microplastics in polar regions: An early warning to the world's pristine ecosystem. *Science of the Total Environment* 784:147149.

Moore, C.J. (2008). Synthetic polymers in the marine environment: A rapidly increasing, long-term threat. *Environmental Research* 108(2):131–139.

Moore, C.J., Lattin, G., and Zellers, A. (2011). Quantity and type of plastic debris flowing from two urban rivers to coastal waters and beaches of Southern California. *Revista de Gestão Costeira Integrada – Journal of Integrated Coastal Zone Management* 11(1):65–73.

MSFD Technical Subgroup on Marine Litter. (2013). *Guidance on Monitoring of Marine Litter in European Seas*. Number 1. European Union: Institute Edition.

Murray, F., and Cowie, P.R. (2011). Plastic contamination in the decapod crustacean *Nephrops norvegicus* (Linnaeus, 1758). *Marine Pollution Bulletin* 62(6):1207–1217.

Ogata, Y., Takada, H., Mizukawa, K., Hirai, H., Iwasa, S., Endo, S., Nakashima, A., and Murakami, M. (2009). International pellet watch: Global monitoring of persistent organic pollutants (POPs) in coastal waters. 1. Initial phase data on PCBs, DDTs, and HCHs. *Marine Pollution Bulletin* 58(10):1437–1446.

Paço, A., Duarte, K., da Costa, J.P., Santos, P.S., Pereira, R., Pereira, M., Freitas, A.C., Durate, A.C., and Rocha-Santos, T.A. (2017). Biodegradation of polyethylene microplastics by the marine fungus *Zalerion maritimum*. *Science of the Total Environment* 586:10–15.

Parslow, J., and Jefferies, D. (1972). Elastic thread pollution of puffins. *Marine Pollution Bulletin* 3(3):43–45.

PlasticsEurope. (2018). Plastics—The facts 2018. https://www.plasticseurope.org.

PlasticsEurope. (2020). Plastics—The facts 2020. *PlasticsEurope* 1:1–64.

Prata, J.C. (2018). Airborne microplastics: Consequences to human health? *Environmental Pollution* 234:115–126.

Prata, J.C., da Costa, J.P., Lopes, I., Duarte, A.C., and Rocha-Santos, T. (2020). Environmental exposure to microplastics: An overview on possible human health effects. *Science of the Total Environment* 702:134455.

Randall, B., Randall, R., and Rossouw, G. (1983). Plastic particle pollution in great shearwaters (*Puffinus gravis*) from Gough Island. Vol. 4. *Antarctic Legacy Archive*. http://hdl.handle.net/123456789/7557.

Recycle Coach. (2022). 7+ revealing plastic waste statistics (2021); [accessed 2022 February 9]. https://recyclecoach.com/resources/7-revealing-plastic-waste-statistics-2021.

Revel, M., Châtel, A., and Mouneyrac, C. (2018). Micro(nano)plastics: A threat to human health? *Current Opinion in Environmental Science and Health* 1:17–23.

Rodrigues, A., Oliver, D.M., McCarron, A., and Quilliam, R.S. (2019). Colonisation of plastic pellets (nurdles) by *E. coli* at public bathing beaches. *Marine Pollution Bulletin* 139:376–380.

Rothstein, S.I. (1973). Plastic particle pollution of the surface of the Atlantic Ocean: Evidence from a seabird. *The Condor* 75(3):344–345.

Ryan, P., and Moloney, C. (1990). Plastic and other artefacts on South African beaches: Temporal trends in abundance and composition. *South African Journal of Science* 86(7):450–452.

Ryan, P.G. (2015). A brief history of marine litter research. In: Bergmann, M., Gutow, L., and Klages, M., editors. *Marine Anthropogenic Litter.* Berlin (Germany): Springer, pp. 1–25.

Ryan, P.G., Moore, C.J., Van Franeker, J.A., and Moloney, C.L. (2009). Monitoring the abundance of plastic debris in the marine environment. *Philosophical Transactions of the Royal Society B: Biological Sciences* 364(1526):1999–2012.

Schwabl, P., Köppel, S., Königshofer, P., Bucsics, T., Trauner, M., Reiberger, T., and Liebmann, B. (2019). Detection of various microplastics in human stool: A prospective case series. *Annals of Internal Medicine* 171(7):453–457.

Sendra, M., Staffieri, E., Yeste, M.P., Moreno-Garrido, I., Gatica, J.M., Corsi, I., and Blasco, J. (2019). Are the primary characteristics of polystyrene nanoplastics responsible for toxicity and ad/absorption in the marine diatom *Phaeodactylum tricornutum*? *Environmental Pollution* 249:610–619.

Shah, A.A., Hasan, F., Hameed, A., and Ahmed, S. (2008). Biological degradation of plastics: A comprehensive review. *Biotechnology Advances* 26:246–265.

Sharma, S., and Chatterjee, S. (2017). Microplastic pollution, a threat to marine ecosystem and human health: A short review. *Environmental Science and Pollution Research* 24(27):21530–21547.

Song, Y.K., Hong, S.H., Jang, M., Kang, J.-H., Kwon, O.Y., Han, G.M., and Shim, W.J. (2014). Large accumulation of micro-sized synthetic polymer particles in the sea surface microlayer. *Environmental Science & Technology* 48(16):9014–9021.

Song, Z., Yang, X., Chen, F., Zhao, F., Zhao, Y., Ruan, L., Wang, Y., and Yang, Y. (2019). Fate and transport of nanoplastics in complex natural aquifer media: Effect of particle size and surface functionalization. *Science of the Total Environment* 669:120–128.

Sykes, E.A., Dai, Q., Tsoi, K.M., Hwang, D.M., and Chan, W.C.W. (2014). Nanoparticle exposure in animals can be visualized in the skin and analyzed via skin biopsy. *Nature Communications* 5:3796. DOI: 10.1038/ncomms4796.

Ter Halle, A., Jeanneau, L., Martignac, M., Jardé, E., Pedrono, B., Brach, L., and Gigault, J. (2017). Nanoplastic in the North Atlantic subtropical gyre. *Environmental Science & Technology* 51(23):13689–13697.

Thompson, R.C., Moore, C.J., Vom Saal, F.S., and Swan, S.H. (2009). Plastics, the environment and human health: Current consensus and future trends. *Philosophical Transactions of the Royal Society B: Biological Sciences* 364(1526):2153–2166.

Thompson, R.C., Olsen, Y., Mitchell, R.P., Davis, A., Rowland, S.J., John, A.W., McGonigle, D., and Russell, A.E. (2004). Lost at sea: Where is all the plastic? *Science* 304(5672):838.

Tong, M., He, L., Rong, H., Li, M., and Kim, H. (2020). Transport behaviors of plastic particles in saturated quartz sand without and with biochar/Fe_3O_4-biochar amendment. *Water Research* 169:115284.

US EPA (US Environmental Protection Agency). (2021). Plastics: Material-specific data [accessed 2021 August 5]. https://www.epa.gov/facts-and-figures-about-materials-waste-and-recycling/plastics-material-specific-data.

Van Cauwenberghe, L., Claessens, M., Vandegehuchte, M.B., and Janssen, C.R. (2015). Microplastics are taken up by mussels (*Mytilus edulis*) and lugworms (*Arenicola marina*) living in natural habitats. *Environmental Pollution* 199:10–17.

Van Cauwenberghe, L., and Janssen, C.R. (2014). Microplastics in bivalves cultured for human consumption. *Environmental Pollution* 193:65–70.

Van Melkebeke, M. (2019). Exploration and Optimization of Separation Techniques for the Removal of Microplastics from Marine Sediments During Dredging Operations [Master's Thesis]. Ghent (Belgium): Ghent University.

Wagner, M., Scherer, C., Alvarez-Muñoz, D., Brennholt, N., Bourrain, X., Buchinger, S., Fries, E., Grosbois, C., Klasmeier, J., Marti, T., and Rodriguez-Mozaz, S. (2014). Microplastics in freshwater ecosystems: What we know and what we need to know. *Environmental Sciences Europe* 26(1):1–9.

Wang, L., Wu, W.-M., Bolan, N.S., Tsang, D.C.W., Lie, Y., Qin, M., and Hou, D. (2021). Environmental fate, toxicity and risk management strategies of nanoplastics in the environment: Current status and future perspectives. *Journal of Hazardous Materials* 401:123415.

Wang, W., and Wang, J. (2018). Comparative evaluation of sorption kinetics and isotherms of pyrene onto microplastics. *Chemosphere* 193:567–573.

Ward, J.E., and Shumway, S.E. (2004). Separating the grain from the chaff: Particle selection in suspension- and deposit-feeding bivalves. *Journal of Experimental Marine Biology and Ecology* 300(1–2):83–130.

Woodall, L.C., Sanchez-Vidal, A., Canals, M., Paterson, G.L., Coppock, R., Sleight, V., Calafat, A., Rogers, A.D., Narayanaswamy, B.E., and Thompson, R.C. (2014). The deep sea is a major sink for microplastic debris. *Royal Society Open Science* 1(4):140317.

World Bank. (2022). What a waste 2.0. A global snapshot of solid waste management to 2050; [accessed 2022 January 5]. https://datatopics.worldbank.org/what-a-waste/tackling_increasing_plastic_waste.html.

Xu, Z., Bai, X., and Ye, Z. (2021). Removal and generation of microplastics in wastewater treatment plants: A review. *Journal of Cleaner Production* 291:125982.

Xu, S., Ma, J., Ji, R., Pan, K., and Miao, A.-J. (2020). Microplastics in aquatic environments: Occurrence, accumulation, and biological effects. *Science of the Total Environment* 703:134699.

Yang, D., Shi, H., Li, L., Li, J., Jabeen, K., and Kolandhasamy, P. (2015). Microplastic pollution in table salts from China. *Environmental Science & Technology* 49:13622–13627.

Yee , M.S.-L., Hii, L.-W., Looi, C.K., Lim, W.-M., Wong, S.-F., Kok, Y.-Y., Tan, B.-K., Wong, C.-Y., and Leong, C.O. (2021). Impact of microplastics and nanoplastics on human health. *Nanomaterials* 11:496. DOI: 10.3390/nano11020496.

Zhang, J., Wang, L., Trasande, L., and Kannan, K. (2021). Occurrence of polyethylene terephthalate and polycarbonate microplastics in infant and adult feces. *Environmental Science and Technology Letters* 8:989–994.

Zhao, S., Zhu, L., Wang, T., and Li, D. (2014). Suspended microplastics in the surface water of the Yangtze Estuary System, China: First observations on occurrence, distribution. *Marine Pollution Bulletin* 86(1–2):562–568.

Chapter 2

Polymer Chemistry and Properties

INTRODUCTION

The remarkable physical and chemical properties of plastics have made them indispensable for countless industrial, commercial, and domestic applications. Regardless of their diverse characteristics, however, all synthetic plastics share a common origin, in that they are constructed of long-chain molecules (i.e., *macromolecules* or *polymers*). The term *polymer* is derived from the classical Greek words *poly* (many) and *meros* (parts). A synthetic polymer is constructed via the linkage of many repeating individual units, or *monomers* of the identical structure in sequence, thus forming an extremely long chain. This backbone structure is a key characteristic among polymer types, ultimately determining numerous physical and chemical properties of the finished plastic product (table 2.1).

Plastics are manufactured from natural gas and feedstocks derived from natural gas processing and crude oil refining. The process begins with extraction of these fossil fuels, which are pumped from underground to the surface and transported via pipeline to refineries. In the case of natural gas, the liquid brought to the surface becomes a gas under ambient surface temperature and pressure. Natural gas contains ethane, propane, butane, isobutene, pentane, and also a small quantity of heavier hydrocarbons such as hexane, heptane, and octane. Natural gas can be cracked and transformed for the synthesis of plastics.

Only a modest fraction of crude petroleum is used in the manufacture of plastics (figure 2.1). Crude petroleum is composed of hundreds of different hydrocarbon groups of varying molecular weights, thus imparting unique properties to each group. These groups must be separated in a distillation tower at a petroleum refinery. Petroleum is loaded into the base of a

Table 2.1 Physical Properties of Monomers as Affected by Chain Length

Name	*Molecular Formula*	*Melting Point (°C)*	*Boiling Point (°C)*	*State at 25°C*
Methane	CH_4	−183	−164	Gas
Ethane	C_2H_6	−183	−89	Gas
Propane	C_3H_8	−190	−42	Gas
Butane	C_4H_{10}	−138	−0.5	Gas
Pentane	C_5H_{12}	−130	36	Liquid
Hexane	C_6H_{14}	−95	69	Liquid
Heptane	C_7H_{16}	−91	98	Liquid
Octane	C_8H_{18}	−57	125	Liquid
Nonane	C_9H_{20}	−51	151	Liquid
Decane	$C_{10}H_{22}$	−30	174	Liquid
Undecane	$C_{11}H_{24}$	−25	196	Liquid
Dodecane	$C_{12}H_{26}$	−10	216	Liquid
Eicosane	$C_{20}H_{42}$	37	343	Solid
Triacontane	$C_{30}H_{62}$	66	450	Solid
Polyethylene	$(C_2H_4)_n$	115–135	—	Solid

distillation tower and heated. During the heating process, the different hydrocarbon groups rise to specific levels in the tower based on their Molecular weight (which is directly related to their boiling point) and subsequently separated. The lightest compounds such as methane (CH_4), ethane (C_2H_5), and propane (C_3H_7) rise to the top of the tower. The heaviest components such as bitumen do not boil and remain at the base. Hydrocarbons recovered from the middle of the tower (middle distillates) include gasoline, jet fuel, diesel fuel, and many other mixtures. Another compound in this group is naphtha, which will become the primary feedstock for making plastic. Using methods such as thermal cracking, certain components of naphtha such as ethylene and propylene are recovered and eventually used to construct the backbone of plastics.

Following the initial transformations of natural gas and crude petroleum, the individual monomer ingredients are combined chemically into new arrangements to produce the long repeating chains (figure 2.2).

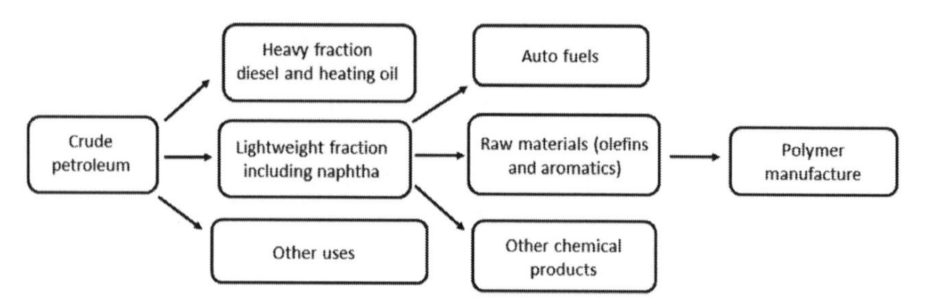

Figure 2.1 Proportion of Petroleum Conversion to Fuels, Petrochemicals, and Plastics

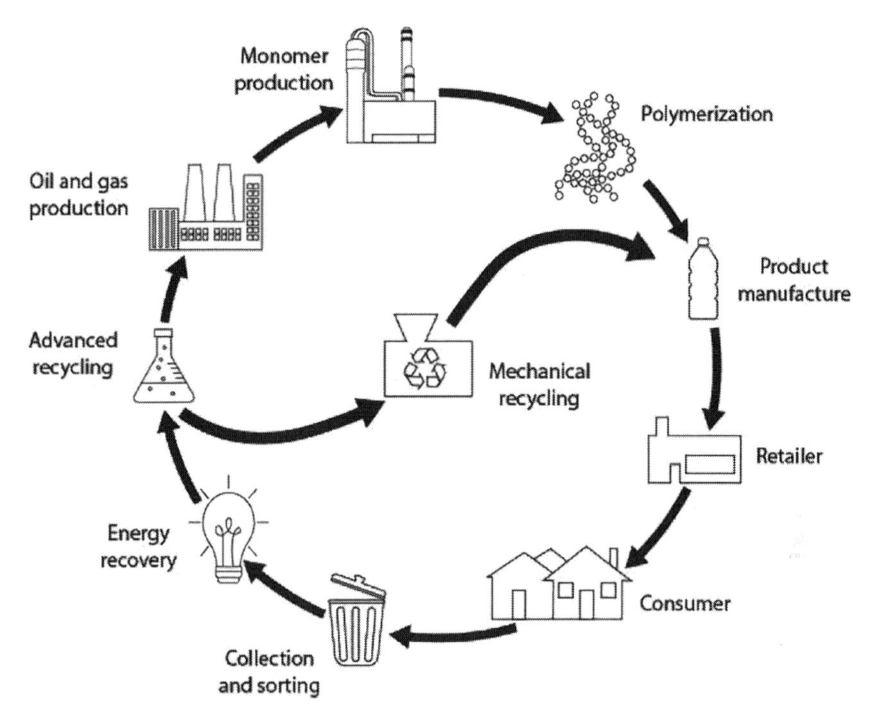

Figure 2.2 The Circular Plastics Economy Showing the Transformation of Petroleum and Natural Gas Hydrocarbons to Plastic Products
Source: Reproduced with kind permission of Nova Chemicals.

The chemical process of joining monomers in a chain to form a polymer is termed *polymerization*. Figure 2.3 presents the conversion of the ethylene monomer to polyethylene via this process. Once these long chains are formed, they are molded and shaped to form solid products.

Polymers are, however, much more complex than described thus far. The specific arrangements of the millions of atoms that comprise a synthetic polymer impart a direct impact on the unique properties of the finished plastic product (table 2.2). The way in which the atoms are arranged directs properties such as rigidity, durability, and glass transition temperature. The types of atoms and the bonds between atoms affect the resistance of the plastic to ultraviolet (UV)

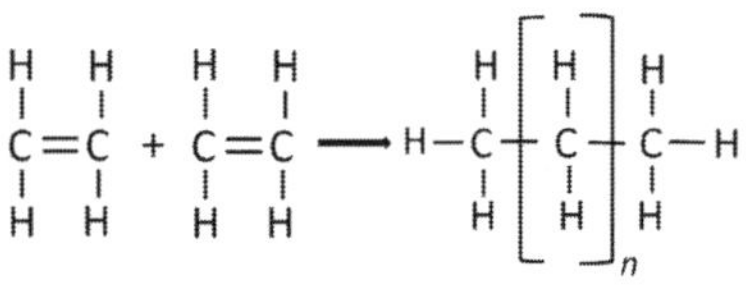

$$\mathrm{H_2C{=}CH_2 + H_2C{=}CH_2 \longrightarrow H{-}CH_2{-}\left[CH_2\right]_n{-}CH_2{-}H}$$

Figure 2.3 Linking of Ethylene Monomers to Form a Polymer Chain (Polyethylene)

Table 2.2 Relevant Properties of Common Polymers

Polymer Type	*Abbreviation*	*Chemical Formula*	*Density* (g/cm^3)	T_g (°C)	T_m (°C)	*Polymerization Mechanism*	*Recycling Code**
Polyethylene terephthalate	PET	$(C_{10}H_8O_4)_n$	1.30–1.40	69	256	Condensation	1
High-density polyethylene	HDPE	$(C_2H_4)_n$	0.93–0.97	−70	135	Addition	2
Polyvinyl chloride	PVC	$(C_2H_3Cl)_n$	1.20–1.70	85	190	Addition	3
Low-density polyethylene	LDPE	$(C_2H_4)_n$	0.91–0.93	−100	120	Addition	4
Polypropylene	PP	$(C_3H_6)_n$	0.89–0.92	−30	165	Addition	5
Polystyrene	PS	$(C_8H_8)_n$	1.04–1.10	63–112	264	Addition	6
Polyvinyl alcohol	PVA	$(C_2H_4O)_n$	1.19–1.35	85	—	Addition	7
Polyamide	PA	CO-NH	1.02–1.15	50	255	Condensation	7
Polycarbonate	PC	$C_{16}H_{18}O_5$	1.20–1.22	155	235	Condensation	7
Polyurethane	PUR	$C_{27}H_{36}N_2O_{10}$	0.87–1.42	−60	146	Addition	7

T_g: Glass transition temperature
T_m: Crystalline melting temperature
* Society of Plastics Industry Recycling Code

light and heat. The crystallinity of polymer chains (i.e., the structural order of atoms and molecules) controls properties such as density, hardness, and transparency. Furthermore, different polymers may be combined, and a wide range of additives may be incorporated in order to produce desirable characteristics. Taken together, these different properties control the persistence or degradability of microplastics (MPs) in the aquatic and soil environment.

To best understand the ways in which plastics behave in water, wastewater, and sediment, it is important to examine the chemical and physical characteristics of plastic materials in more detail. This chapter covers key aspects of polymer/plastic properties that affect MPs behavior in the biosphere, potential for removal from water and sediment, and possible mechanisms for their ultimate decomposition.

CLASSIFICATION OF POLYMERS

Many thousands of polymers have been synthesized, and more continue to enter the market. Several methods can be used to classify polymers, some of which are described below.

Thermoplastics and Thermosets

All polymers can be assigned into one of two major categories based on their thermal processing behavior or type of polymerization mechanism, that is, thermoplastics and thermosets. *Thermoplastics* are composed of linear or lightly branched chains that are capable of sliding past one another when exposed to increased temperature and pressure. Thermoplastics can be softened by heating in order to prepare into a desired form, and are hardened when cooled. These characteristics are reversible; that is, thermoplastics can be reheated, reshaped, and cooled repeatedly (PlasticsEurope 2020). Such polymers have the advantage of being more readily recycled, as waste thermoplastics can be refabricated by application of heat and pressure into new products. PE, PP, PS, PVC, and PET are important thermoplastics.

Thermosets, in contrast, are polymers which possess extensive crosslinked networks (figure 2.4). Polymer chains are chemically linked by covalent bonds during polymerization or by chemical or thermal treatment during fabrication. After they are heated and formed, thermoplastics cannot be re-melted and re-formed. Thermosets are, therefore, suitable as adhesives and coatings. Examples of thermosets include unsaturated polyesters used in manufacture of Fiberglass and other glass-reinforced composites, urea-formaldehyde, polyurethane, epoxy resins, acrylic, vinyl esters, and

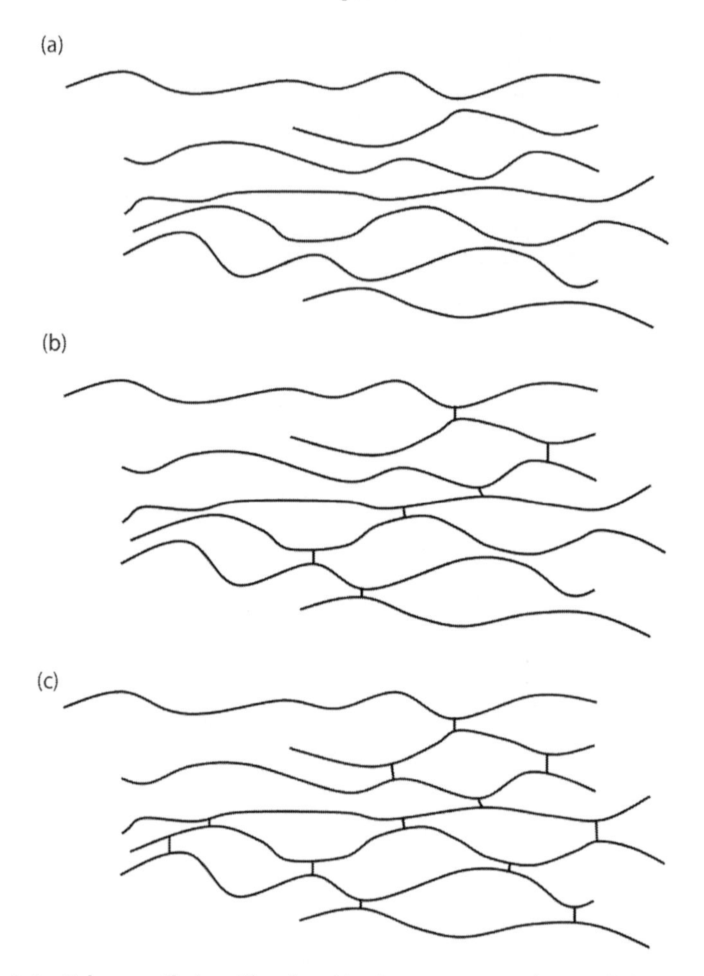

Figure 2.4 Polymer Chains Showing Varying Degrees of Crosslinking: (a) No Crosslinking, as Occurs in a Thermoplastic; (b) Lightly Crosslinked; and (c) Fully Crosslinked (Thermoset)

phenolic resins. As thermosets do not melt when heated, it is difficult to recycle them.

Ultimately, there are gradations between these two vast categories of plastics: polymers exist with no crosslinks (polyethylene) to lightly crosslinked, to fully crosslinked (thermosets).

Examples of common thermoplastics and thermosets appear in table 2.3.

Elastomers

The term *elastomer* means "elastic polymer." An elastomer is a polymer material that can be stretched and deformed under tension, but quickly returns to its

Table 2.3 Examples of Common Thermoplastics and Thermosets

Thermoplastics	*Thermosets*
Polyethylene (PE)	Polyurethane (PUR)
Polypropylene (PP)	Polyesters
Polyvinyl chloride (PVC)	Silicone
Polyethylene terephthalate (PET)	Epoxy resins
Polystyrene (PS)	Vinyl esters
Expanded polystyrene (EPS)	Melamine resins
Acrylonitrile butadiene styrene (ABS)	Phenol formaldehyde resins
Styrene-acrylonitrile copolymer (SAN)	Urea-formaldehyde resins
Polyamides (PA)	Phenolic resins
Polycarbonate (PC)	Acrylic resins
Polymethyl methacrylate (PMMA)	
Thermoplastic elastomers (TPE)	
Polyarylsulfone (PSU)	
Fluoropolymers	
Polyether ether ketone (PEEK)	
Polyoxymethylene (POM)	
Polybutylene terephthalate (PBT)	
Ethylene vinyl alcohol copolymer (EVOH)	

original shape and size when the tension is removed (Harper and Petrie 2003). Elastomers are usually thermosets (requiring vulcanization) but may also be thermoplastic. Examples of elastomers are styrene-butadiene copolymers, polyisoprene, silicone, polyurethane, neoprene (chloroprene), nitrile rubber, ethylene propylene rubber, as well as natural rubber. Elastomers are commonly used as material for tires, rubber bands, belts, prosthetics, and other items.

Polymerization Method

Polymers can be classified based on the method of polymerization, which is divided into two groups, that is, addition polymerization and condensation polymerization. In the addition process (also known as *chain-growth polymerization*), individual monomers, each containing a double bond, are added to one another in sequence to form the chain. The reactions may be initiated by free radicals which attack the carbon-carbon double bond (figure 2.5). An example initiator is an organic peroxide. Upon heating, the weak O–O peroxide bond breaks to form two radicals which attack the C=C bonds of the monomer, thereby creating a free-radical intermediate. This intermediate then reacts with the C=C of another monomer, linking the two. The chain reaction continues until a long chain is formed and the reaction terminates, often by reaction of two radicals with one another. In addition, reactions of the entire monomer are retained, that is, there is no production of by-products. A typical example of addition polymerization is shown in figure 2.5. Important addition polymers are PE, PP, PVC, PS, and Teflon® (table 2.4).

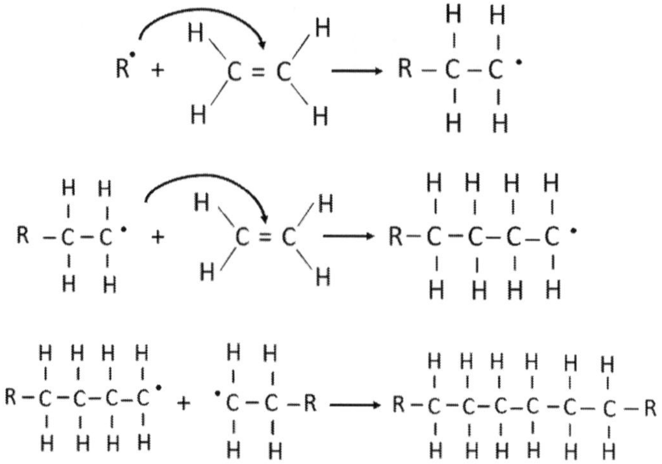

$R^{\bullet} + H_2C{=}CH_2 \longrightarrow R{-}CH_2{-}CH_2^{\bullet}$ Initiation

$R{-}CH_2{-}CH_2^{\bullet} + H_2C{=}CH_2 \longrightarrow R{-}CH_2{-}CH_2{-}CH_2{-}CH_2^{\bullet}$ Propagation

$R{-}CH_2{-}CH_2{-}CH_2{-}CH_2^{\bullet} + {}^{\bullet}CH_2{-}CH_2{-}R \longrightarrow R{-}CH_2{-}CH_2{-}CH_2{-}CH_2{-}CH_2{-}CH_2{-}R$ Termination

Figure 2.5 Addition Polymerization Process for Polyethylene

In condensation polymerization (aka *step growth*), two different functional groups react to form a new chemical group. During the reaction, small molecules such as water, methanol, or HCl may be evolved. This is in contrast to addition growth, where every atom of the constituent monomers is retained. An example condensation reaction, showing the formation of a polyester, is shown in figure 2.6. Plastics formulated by condensation polymerization include PET, nylon, and Bakelite (table 2.5).

Chain Type

Polymers can be classified based upon the chemical structure of their backbones. Polymers that have solely carbon atoms along the backbone are termed homochain polymers. In contrast, in heterochain polymers, a number of other elements may be linked in the backbone including oxygen, nitrogen, sulfur, and silicon. Typically, two types of atoms are arranged in an alternating pattern. In certain cases, three or four different types of atoms occur in a single backbone.

Some important heterochain polymers are shown in table 2.6.

Copolymer Blends

The polyethylene chain shown in figure 2.3 is termed a homopolymer, as all monomers in the chains are identical. If the monomer is vinyl chloride, then the homopolymer would be named polyvinyl chloride. Polymers are

Table 2.4 Examples of Addition Polymers

Polymer	*Repeating Unit*
Polyethylene	$-[CH_2-CH_2]-$
Polypropylene	$-[CH_2-CH(CH_3)]-$
Polyvinyl chloride	$-[CH_2-CH(Cl)]-$
Polystyrene	$-[CH_2-CH(C_6H_5)]-$
Polyvinyl alcohol	$-[CH_2-CH(OH)]-$
Polyvinyl acetate	$-[CH_2-CH_2(O-C(=O)-CH_3)]-$
Polyacrylonitrile	$-[CH_2-CH(C\equiv N)]-$
Poly (methyl methylacrylate)	$-[CH_2-C(CH_3)(C=O-O-CH_3)]-$
Poly (vinylidene chloride)	$-[CH_2-C(Cl)(Cl)]-$
Polytetrafluro-ethylene	$-[CF_2-CF_2]-$
cis-polyisoprene	$-[CH_2-CH=C(Cl)-CH_2]-$

often formulated, however, to impart new and beneficial properties by linking two or three different repeating units during or after the polymerization process. This practice avoids the expense and effort required for developing an entirely new polymer type (UNEP 2015). Polymers having two or more different monomer types in their chains are termed *copolymers*. Some of the most important commercial copolymers are derived from vinyl monomers

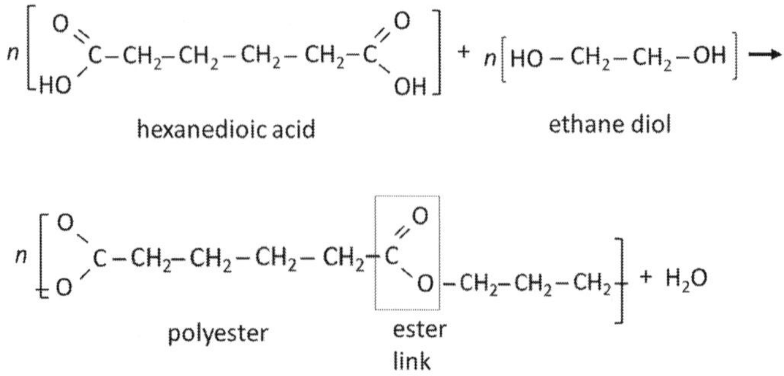

$$n\left[HO(O=)C-CH_2-CH_2-CH_2-CH_2-C(=O)OH\right] + n\left[HO-CH_2-CH_2-OH\right] \rightarrow$$

hexanedioic acid ethane diol

$$n\left[-O-(O=)C-CH_2-CH_2-CH_2-CH_2-C(=O)-O-CH_2-CH_2-CH_2-\right] + H_2O$$

polyester ester link

Figure 2.6 Condensation Reaction Showing the Formation of a Polyester

such as styrene, ethylene, acrylonitrile, and vinyl chloride (Fried 1995). Examples include acrylonitrile butadiene styrene, styrene/butadiene copolymer, nitrile rubber, styrene-acrylonitrile, styrene-isoprene-styrene, and ethylene-vinyl acetate. Copolymers have many applications, for example in cling film, food storage containers, and natural gas distribution pipes (Peacock and Calhoun 2012).

Copolymers are synthesized either via addition or condensation polymerization. The different constituent units of a copolymer can be arranged in several different ways, each imparting unique properties to the finished product. Even relatively minor changes in monomer content will result in substantial

Table 2.5 Examples of Thermoplastics Formed by the Condensation Method

Polymer	*Repeating Unit*
Polyethylene terephthalate	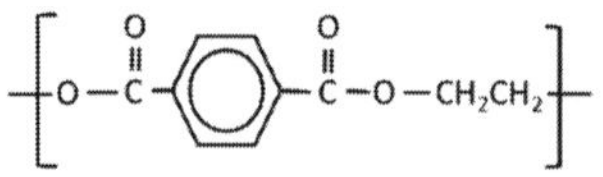 $-[O-C(=O)-C_6H_4-C(=O)-O-CH_2CH_2]-$
Polyurethane	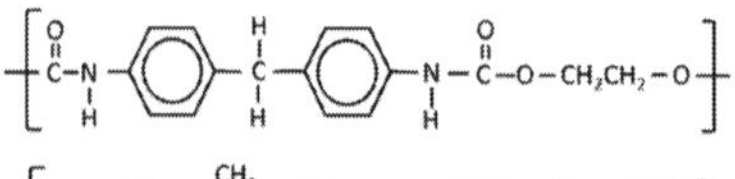 $-[C(=O)-NH-C_6H_4-CH_2-C_6H_4-NH-C(=O)-O-CH_2CH_2-O]-$
Polysulfone	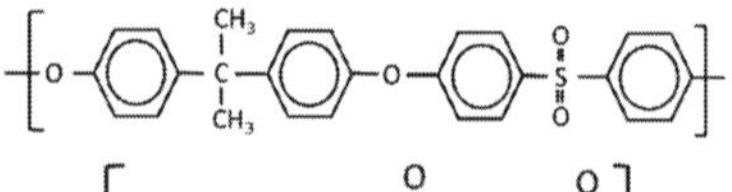 $-[O-C_6H_4-C(CH_3)_2-C_6H_4-O-C_6H_4-SO_2-C_6H_4]-$
Poly(hexamethylene sebacamide) (Nylon 6,10)	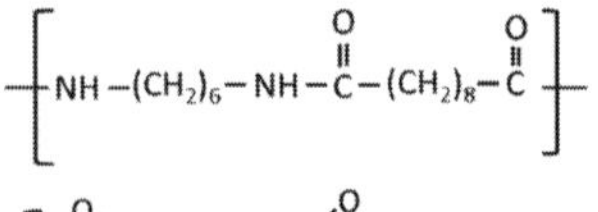 $-[NH-(CH_2)_6-NH-C(=O)-(CH_2)_8-C(=O)]-$
Poly(ethylene pyrometallitimide)	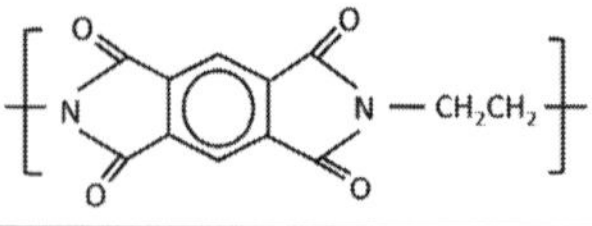 $-[N(CO)_2C_6H_2(CO)_2N-CH_2CH_2]-$

Table 2.6 Main Chain Structures of Some Important Heterochain Polymers

Component Atoms	Chemical Group	Structure
Carbon-oxygen	Polyether	–C–O–
	Polyesters of carboxylic acids	–C(=O)–O–
	Polyanhydrides of carboxylic acids	–C(=O)–O–C(=O)–
	Polycarbonates	–O–C(=O)–O–
Carbon-sulfur	Polythioethers	–S–C–
	Polysulfone	–S(=O)(=O)–C–
Carbon-nitrogen	Polyamines	–C–N–
	Polyimines	–C=N–
	Polyamides	–C(=O)–N–
	Polyureas	–N–C(=O)–N–

changes in physical or chemical properties including rigidity, solvent resistance, and heat distortion temperature (i.e., resistance to deformation under a given load at elevated temperature.).

Copolymers are classified based upon the ordering of their monomers in the final chain. It is possible that monomer placement is totally random or alternates perfectly; however, this is not very common. The four main types of copolymers are random, alternating, block, and graft (figure 2.7). In random copolymers, there is no consistency to the sequence of the different monomers along the chain. In alternating copolymers, the two different monomers occur in a regular alternating sequence. In block copolymers, one monomer will occur in a long chain, followed by a large portion of another monomer. In a graft copolymer, blocks of a different monomer are attached to the backbone chain as branches. Attachment occurs along functional sites on the main chain.

Crystallinity

As we stated earlier in this chapter, a polymer is defined as a very long chain of monomers linked together in sequence. However, these chains do

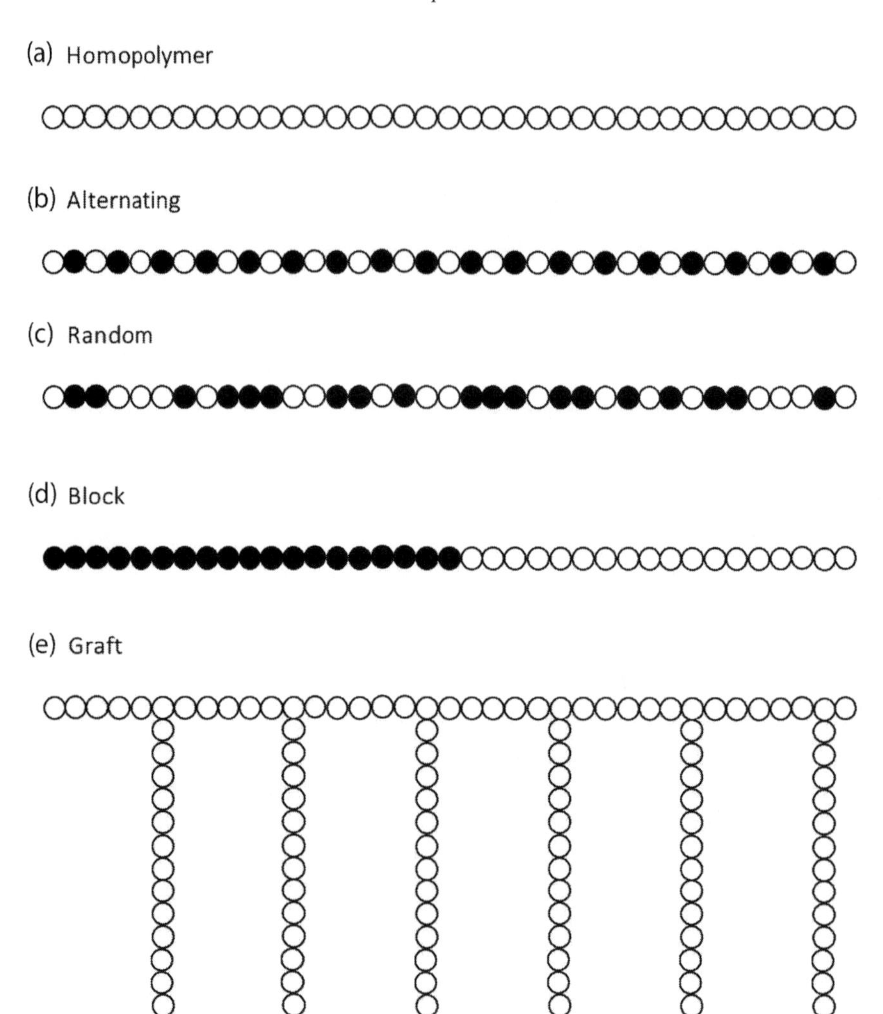

Figure 2.7 The Main Types of Copolymers

not all necessarily occur in an ordered fashion within a plastic. Polymers exist as a combination of chains having a definite geometric form (i.e., are *crystalline*), along with those that are less ordered (*amorphous*) (figure 2.8). The relative distribution of crystalline and amorphous regions strongly influences the chemical and physical properties of the plastic product—it provides for good strength (conveyed primarily by crystalline regions) and flexibility (derived from the amorphous portions) (Carraher 2017).

If the substituents on each molecular chain are small and/or identical, the chains can be tightly packed into an ordered structure, essentially locked

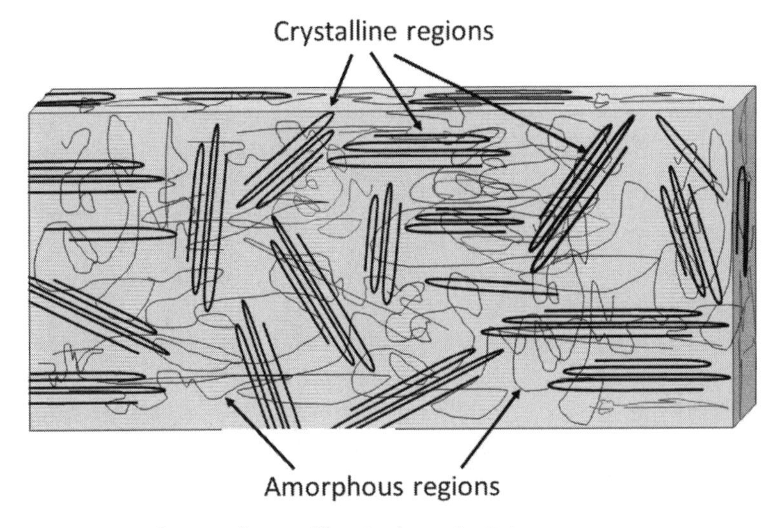

Figure 2.8 Amorphous and Crystalline Regions of a Polymer

against each other, and crystallinity can be highly developed. Molecular chains can fold and be bundled into *lamellae*, which are ribbon-like configurations about 10 μm thick, 1 μm long, and 0.1 μm wide (Harper and Petrie 2003). In some resins, the degree of crystallinity serves as a measure of hardness, rigidity, environmental stress-crack resistance, and heat resistance (Harper and Petrie 2003). Examples of crystalline polymers include PE, nylon, polyvinylidene formamide (PVF), and acetal.

In an amorphous polymer, the chains occur in a random and unordered structure. An example of an amorphous polymer is PS.

Crystalline polymers are typically opaque, as the folded chains packed in the lamellae reflect light. Amorphous polymers, having chains entangled in a random fashion, tend to let light pass through. With increased crystallinity, a polymer becomes progressively less transparent (mcpolymers.com 2020).

Random blocks will inevitably be present in the chains; complete crystallization, therefore, does not occur. Most polymers consist of a combination of crystalline and amorphous regions (figure 2.8). If branching occurs on any regions of the chain, there will be a loss of regularity and an inhibition of crystallization. Polymers in this family are termed semi-crystalline.

ADDITIVES

Most synthetic polymers are formulated with the goal of optimal functioning and long-term resilience and durability; in other words, they are designed so that they do not readily break down. This property is, in many cases, one of the primary functions provided by plastic products during its intended use (UNEP 2014). Polymer function and durability may be significantly improved by incorporation of additives.

Many plastic products are not composed solely of polymer chains. A wide range of chemicals is incorporated to modify certain physical and/or chemical properties. This includes enhancing the desirable characteristics of the product or reducing unwanted properties (Murphy 2001). Additives may also impart entirely new properties to a material. Additives improve the performance of polymers during processing and fabrication (e.g., injection molding, extrusion, blow molding, vacuum molding).

In some cases, the raw polymer chains alone possess the necessary properties to carry out the required application in the plastic product. In other cases, only a modest amount of additive is required. For example, one-time use packaging materials requires only a small quantity of additives, such as to reduce oxidation.

Additives can be categorized based on their functional and structural components into four classes (Gunaalan et al. 2020; Hansen et al. 2013):

- **Functional**. These substances are incorporated to modify the physicochemical properties of polymers. Examples include plasticizers, flame retardants, antioxidants, stabilizers, curing agents, lubricants, slip agents, antistatic agents, foaming agents, and biocides.
- **Colorants**. Pigments and dyes are used to provide the desired color.
- **Fillers**. These compounds improve the properties of polymer coatings and include mica, talc, kaolin, clay, calcium carbonate, barium sulfate, and others.
- **Reinforcements**. These increase mechanical resistance of the plastic product (e.g., glass fibers, carbon fibers).

Most additives occur in low percentages by weight of the polymer. Antiozonants, antistatic agents, biocides, and odorants often comprise up to 1–2% of polymer weight, and colorants from 1–4%. Other additives occur at a much higher weight: thermal stabilizers may comprise up to 8%; flame retardants from 10 to 20%; plasticizers from 10 to 70%; and fillers up to 50% of polymer weight (Andrady and Rajapakse 2016; Gunaalan et al. 2020).

A wide range of additives in plastic products are documented as hazardous to human health and the environment. Additive ingredients are often not chemically bound to the polymer matrix. As a result, they or their degradation products may leach substantially during the life span of a plastic product via use, attrition, and chemical and biological weathering (Lithner et al. 2011, 2012; Rochman 2015). Hazardous additives such as phthalates, brominated flame retardants, bisphenol-A (BPA), formaldehyde, acetaldehyde, 4-nonylphenol, and many volatile organic compounds have been detected in leachates from plastic products (Lithner et al. 2012). Zimmerman et al. (2021) performed toxicological and chemical analyses on 24 plastic items. A migration experiment in water demonstrated that "migrates" (i.e., leached compounds) contained a low of 17 to a high of 8,681 chemical features. As much as 88% of the plastic chemicals associated with one product were found to be migrating. Some of the most commonly used additives in plastic materials are shown in table 2.7 and are described below.

Plasticizers

According to ASTM D-883 (2020), a plasticizer is defined as a material incorporated into a plastic to increase its workability and flexibility or dispensability. Plasticizers are inert organic substances with low vapor pressures which should be relatively nonmobile, nontoxic, inexpensive, and compatible with the material to be plasticized. Many react physically with the polymer via swelling or dissolving to form a homogeneous network. During the mixing steps in polymer production, plasticizers are important for reducing shear. These additives also serve to reduce melt flow (Cano et al. 2002; Bhunia et al. 2013): that is, the ease of flow of the melt of a thermoplastic polymer. They also improve impact resistance in the final plastic film. Another beneficial function of plasticizers is providing the final product with flaccid and tacky or clinging properties (Bhunia et al. 2013; Robertson 2006; Sablani and Rahman 2007; Hahladakis et al. 2018). Approximately 300 plasticizers are currently manufactured, and at least 100 are of commercial importance (Cadogan and Howick 2012).

Polymers can be *flexibilized* by either internal or external plasticization. Internal plasticization requires chemical modification of the polymer or monomer. A polymer is externally plasticized by mixing a plasticizing agent, that is, without chemically changing the polymer. With external plasticization, the manufacturer is able to tailor various formulations for a polymer mixture, thus creating plastics having differing levels of flexibility (Cadogan and Howick 2000).

Flame Retardants

Plastic will thermally degrade once exposed to excess heat for a sufficient time, and most thermoplastics burn when heated to a sufficiently high temperature. The function of flame retardants is to inhibit or hinder the ignition and propagation of fire, suppress smoke formation, and prevent a polymer from dripping. By delaying the ignition and burning of materials, these additives

Table 2.7 Common Classes of Polymer Additives with Examples and Their Functions

Additive Type	*Typical Content, %w/w*	*Examples*	*Function*	*Comments*
Plasticizers	10–70	For PVC: phthalic esters such as DEHP; For PET: DPP, DEHA, DOA, DEP, diisobutyl phthalate, and DBP. For PVDC-based cling-films: acetyltributyl citrate. Other common plasticizers: DBP, DEHP, DHA, DCHP, BBP, HAD, HOA.	Soften polymer to increase its workability and flexibility.	Approx. 80% used in PVC.
Flame retardants	10–20	Four categories: (1) halogenated organic compounds (polybrominated diphenyl ethers, PBDEs; ecabromodiphenylethane; Tetrabromobisphenol A, TBBPA). (2) phosphorus-based compounds [Tris(2-chloroethyl) phosphate, TCEP); Tris(2-chlorisopropyl) phosphate, TCPP]. (3) metallic oxides (aluminum trihydrate, magnesium hydroxide). (4) inert fillers (calcium carbonate, talc, zinc borate).	Prevent or delay ignition and/or flame propagation.	
UV stabilizers	0.1–2	Benzophenones, benzotriazoles, nickel dibutyl dithiocarbamate, hindered amine light stabilizers (HALS).	Limit degradation of plastic by absorbing and dissipating the energy from UV radiation.	
Thermal stabilizers	0.1–8	Salts of Pb, Ba, Cd. Organotin compounds. Metallic salts of fatty acids (calcium and magnesium stearate). Dialkyl maleates or laureates, and dialkyl mercaptides.	Limit degradation during processing.	Used primarily in PVC. Based on Pb, Sn, Ba, Cd and Zn compounds. Pb is the most efficient and it is used in lower amounts.

Colorants	1–4	(1) pigments: organic (Co(II) diacetate); inorganic: Cd, Cr, Co, Pb compounds. (2) organic dyes	To impart desired color to plastic product.	Pigments and dyes comprise the major categories of colorants.
Curing agents	0.1–2	Amines, amides, isocyanates, silanes, anhydrides, aziridines, marcaptos	Toughening or hardening of a polymer material.	
Slip agents	0.1–3	Fatty acid amides (primary erucamide and oleamide), fatty acid esters, metallic stearates (e.g., zinc stearate), waxes.	Provide lubrication to film surface; offer antistatic properties, reduce melt viscosity, provide better mold release and anti-sticking properties.	Amounts depend upon chemical structure of slip agent and polymer type.
Antiblocking agents	0.1–5	Natural silicas and minerals (clay, talc, and quartz predominate; limestone) synthetic silicas (zeolites and silica gels).	Minimizes adhesion between surfaces by reducing the smoothness of the plastics surface	Adhesion is a common manufacturing problem when PA, LDPE, LLDPE, PP, and PET films are wound/ unwound from rolls.
Antistatic agents	0.1–1	Quaternary ammonium salts, aliphatic amines, phosphate esters, ethylene glycols.		Most types are Hydrophilic, can migrate to water.
Biocides	0.001–1	Chlorinated nitrogen sulfur heterocycles and compounds based on mercury, arsenic, copper, antimony, and tin (e.g. tributyltin).		Inhibit bacterial establishment and growth.
Fillers (as powders, fibers, or nanotubes)	Up to about 50	Calcium carbonate, talc, clay, zinc oxide, alumina, rutile, asbestos, barium sulfate, glass microspheres, siliceous earth, wood powder.	Reinforcement, reduction of cost	
Reinforcements	15–30	Glass fibers, carbon fibers, aramide fibers.	Provide physical strength to final product.	

Source: Adapted from Crawford and Quinn (2017), Gunaalan et al. (2020), Hahladakis et al. (2018), Sablani and Rahman (2007).BBP—Benzyl butyl phthalate; DBP—Dibutyl phthalate; DCHP—Dicyclohexyl phthalate; DEHP—Bis (2-ethylhexyl)phthalate; DEP—Diethyl phthalates; DHA—Diheptyl adipate; DOA—Di-octyladipate; DPP—Dipentyl phthalate; HAD—Heptyl adipate; HALS—Hindered amine light stabilizers; HOA—Heptyl octyl adipate.

ultimately allow occupants time to escape from a structure. A second benefit is limiting of property damage.

A wide range of flame-retardant formulations are added to polyolefins, polycarbonates, polyamides, polyester, and other polymers. Common commercially available flame retardants are brominated and chlorinated organic compounds, phosphorus-based compounds, and metallic oxides (table 2.7). The halogenated organic compounds, particularly bromine-based types, are the most popular flame retardants. Brominated flame retardants are preferred where tensile and elongation properties are important, as they allow resins to retain their mechanical and physical properties (Ampacet.com 2021). Chlorinated flame retardants are less thermally stable and more corrosive to equipment than are brominated types. A synergist is included in the formulation to optimize the action of brominated and chlorinated flame retardants. Antimony trioxide, zinc borate, and zinc molybdate are commonly used. The synergist itself has no flame-retardant action; however, it improves the action of the halogenated flame retardant (Ampacet.com 2021).

Phosphorus-based flame retardants include both organic and inorganic formulations. Examples include phosphates, organic phosphate esters, halogenated phosphorus compounds, and inorganic phosphorus-containing salts (Ampacet.com 2021). These compounds modify how the polymer degrades when exposed to heat. They also promote char formation which limits further decomposition of the polymer.

Combustion is a chemical chain reaction that is propagated by free radicals. Halides and phosphorus radicals bind to free radicals produced during combustion and can therefore terminate the reaction. Halide-containing compounds suppress flame propagation via scavenging and stabilizing free radicals as shown in Equations 2.1–2.4 (Carraher 2003).

$$\sim\!\sim RCH_2^{\bullet} + O_2 \rightarrow \sim\!\sim RCHO + {}^{\bullet}OH \tag{2.1}$$

$$^{\bullet}OH + RCH_2H \rightarrow \sim\!\sim RCH_2^{\bullet} + H_2O \tag{2.2}$$

$$HX + {}^{\bullet}OH \rightarrow H_2O + X^{\bullet} \tag{2.3}$$

$$X^{\bullet} + \sim\!\sim RCH_2^{\bullet} \rightarrow RCH_2X \tag{2.4}$$

Another category of flame retardants is metal hydroxides. Popular formulations include aluminum trihydrate (ATH, also known as gibbsite, $Al(OH)_3$) and magnesium hydroxide ($Mg(OH)_2$). At elevated temperatures these compounds form the corresponding oxides via an endothermic reaction that absorbs heat from the fire (Ampacet.com 2021). ATH releases water when heated and thus reduces the temperature of combustion.

Other flame retardants include inert fillers such as calcium carbonate, talc, and zinc borate ($Zn[BO_2]_2 \cdot 2H_2O$). These inhibit fire simply by diluting the portion of material that can burn (Carraher 2017). Compounds such as sodium carbonate (Na_2CO_3), magnesium carbonate ($MgCO_3 \cdot H_2O$), and microfibrous Dawsonite ($NaAl(OH)_2CO_3$) release carbon dioxide when heated and thus shield the reactants from oxygen (Murphy 1996).

Several flame retardants are linked to human health problems. This includes most of the halogenated organics such as polybrominated diphenyl ethers (PBDEs), hexabromocyclododecane (HBCD), and polybrominated biphenyls (PBBs).

Antioxidants

This group of chemicals is added to restrict oxidative degradation. Plastic products are susceptible to aging processes; this includes deterioration from mechanical abrasion, UV irradiation, thermal effects, chemical oxidation reactions, and biological action (Liu et al. 2020; ter Halle et al. 2017; Zhou et al. 2020). These insults can trigger the formation of free radicals, which promote a chemical chain reaction that breaks polymer chains and increases cross-linking (figure 2.9). Such alterations can adversely affect polymer mechanical strength, morphology, hydrophilicity, and MW, ultimately causing deterioration of plastic properties (Liu et al. 2019; Luo et al. 2021, 2020; Zhang et al. 2018). These degradation reactions can be accelerated by the presence of other components of the polymer chain. Heavy metals, for

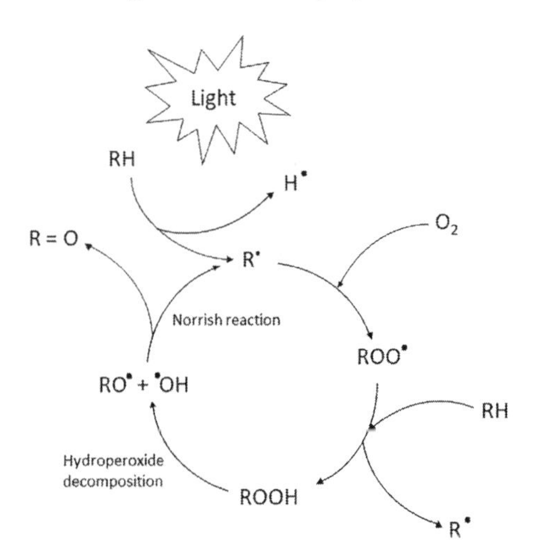

Figure 2.9 Chain Reaction of Polymer Decomposition after Radical Formation
Source: Capiwalter—own work, CC BY-SA 4.0, https://commons.wikimedia.org/w/index.php?curid=79987671.

example, can act as a catalyst in degradation reactions. Cadmium, chromium, lead, and cobalt compounds are commonly added to polymer formulations as pigment compounds.

The presence of minute quantities of antioxidants can slow the rate of the chain reaction.

So-called primary antioxidants work by donating hydrogen atoms to free radicals, thereby quenching or stopping the chain reaction (Carraher 2017). Arylamines are commonly used antioxidants in plastic food packaging. This group includes BHT, BHA, Irganox 1010, BPA (see box below), and Cyanox 2246 (Kattas et al. 2000). Secondary antioxidants are used to reduce hydroperoxides formed during oxidation. Examples include phenolics and organophosphites (e.g., tris-nonylphenyl phosphite) (Bhunia et al. 2013; Hahladakis et al. 2018).

BPA AND HUMAN HEALTH

BPA is one of the most studied additives in the plastics industry. A search query of bisphenol-A on Web of Science returns over 34,000 studies on this compound. BPA has been used as a component of plastics and resins since the 1950s, and billions of pounds are produced annually across the globe (Kadir et al. 2021). One of BPA's primary uses is as a structural component of polycarbonate. It is also used in epoxy resins applied to a range of products, including as a liner in canned food containers.

In 1993, BPA was discovered to leach from polycarbonate flasks and influence estrogen activity (Krishnan et al. 1993). Nearly 30 years later, BPA is recognized by the scientific community as an endocrine disruptor (Bae et al. 2002; Kardas et al. 2015; Rodríguez-Báez et al. 2022; Tiwari et al. 2012). Endocrine disruptors impact the metabolism of hormones in the body and are associated with reproductive disorders, cognitive dysfunction, heart disease, and obesity (Harnett et al. 2021; Thoene et al. 2020).

It is estimated that over 95% of the US populace (adults and children) has been exposed to BPA (Lehmler et al. 2014). Exposure routes are primarily via diet and dust but can also include dental materials, healthcare equipment, thermal paper, and toys (Konieczna et al. 2015). In a cross-sectional study of 170 urine samples from workers in the plastics industry, 90% contained BPA, with concentrations as high as 21.35 μg/L (Rebai et al. 2021).

Regulation of use of BPA in food containers began in 2008 when Canada banned its use in baby bottles; the United States followed in 2012 (Barraza et al. 2013; Jalal 2018). Some countries, however, have avoided issuing a complete ban on the compound, claiming limited evidence that low doses of BPA elicit acute and/or chronic reactions (Jalal et al. 2017).

Petrochemical and plastic manufacturing companies, recognizing the possibility of more stringent regulations in the future, began substituting BPA with chemicals of similar structures such as bisphenol F (BPF), bisphenol S (BPS), bisphenol Z (BPZ), and tetramethyl bisphenol F (TMBF) (Catenza et al. 2020; Harnett et al. 2021). Unfortunately, some of these substitutes have been demonstrated to be as detrimental to human physiology and sometimes more so than the original compound (Harnett et al. 2021).

UV radiation is of sufficient energy to break many covalent chemical bonds; in the case of plastics, UV light causes scission of polymer chains. This is a free-radical process which may be followed by cross-linking and/or reduced chain length. The ultimate result is often weakening of the polymeric material, embrittlement, and yellowing of the polymer (Carraher 2003).

UV susceptibility varies significantly between different polymers. LDPE, PP, and certain polycarbonates, polyesters, and polyurethanes are somewhat susceptible to UV effects (Andrady et al. 2011).

UV stabilizers function by absorbing and dissipating the energy from UV rays as heat. This inhibits the photoinitiation process. Absorption of UV rays by the polymer matrix is thus reduced, which slows the rate of weathering.

Three major classes of light stabilizers are available for polyolefins. UV absorbers are low-cost compounds that function by absorbing UV light. Typical examples are benzophenones and benzotriazoles. Nickel quenchers function by transferring the energy generated from UV absorption from excited state formed during photooxidation to the ground state. These types of stabilizers are not widely used, as they contain a heavy metal, impart color to the final product, and are not as effective as other stabilizers (e.g., HALS). Hindered amine light stabilizers (HALS) function by continuously and cyclically removing free radicals generated by photooxidation of the polymer. HALS are derivatives of tetramethylpiperidine; however, there is a wide range of structural differences.

These three stabilizer types operate via different mechanisms, so they are often used in combination which can offer synergistic benefits. For example, HALS are commonly incorporated with benzotriazoles to minimize color changes in pigmented polymers (Ampacet 2021; Craftech 2021).

Heat Stabilizers

If polyolefins are exposed to excessive heat, free-radical chain degradation occurs. Heat stabilizers provide protection against heat-induced decomposition of plastics. They are used primarily for PVC, as this polymer is quite susceptible to thermal degradation if not stabilized. When heated, PVC loses HCl. Once dehydrochlorination is initiated, the process becomes autocatalytic. Such thermal degradation is accelerated in the presence of iron salts, oxygen, and hydrogen chloride (Carraher 2003).

Some common heat stabilizers include salts of heavy metals (Pb, Ba, Cd), which act as scavengers for HCl. Organotin products are also popular as heat stabilizers in PVC. Some metallic salts of fatty acids are favored, such as calcium and magnesium stearate (Allsopp and Vianello 2012), and certain calcium-zinc formulations (Sastri 2013). Levels in polymers may vary from 2% to 4%.

Curing Agents

Curing is defined as the toughening or hardening of a polymer material by cross-linking of polymer chains during polymerization. A wide range of curing agents (also known as hardeners) is available. Many are dedicated for a specific polymer type. Curing agents are often complex organic molecules which vary in terms of set time, heat resistance, chemical resistance, odor production, degree of hardness of final product, and possible hazards to humans (e.g., aliphatic amines are strong skin irritants) (Murphy 2001).

Curing agent types fall into the following categories (SpecialChem 2021):

- amine
- amide
- isocyanate
- silane
- anhydride
- aziridine
- marcapto

Slip and Antiblocking Agents

In polymer manufacture, the term *slip* (or surface friction) is a measure of the interaction between adjacent polymer surfaces, or between the polymer surface and processing equipment. This effect is expressed as the coefficient of friction or CoF. Variables such as polymer type, process type and temperature, and the presence of other additives affect the development of friction. High friction can cause difficulties with the processing of thin film, winding of film rolls, and other operations.

Slip agents are designed to reduce the CoF of the surface of a polymer. They provide lubrication to the film surface; in addition, slip agents offer antistatic properties, reduce melt viscosity, and provide better mold release and anti-sticking properties (Sablani and Rahman 2007; Bhunia et al. 2013). Over time slip agents float to the plastic surface to act as a lubricant, thereby preventing films from sticking together (Lau and Wong 2000). Common slip agents include fatty acid amides (primary erucamide for PP and oleamide for PE), fatty acid esters, metallic stearates (e.g., zinc stearate), and waxes (Bhunia et al. 2013; Hahladakis et al. 2018). The fatty acid amides are among the most commonly used slip additives in plastics.

A common issue in the manufacture of polyolefin film (PE, PP, PET, and soft PVC) is the unwanted adherence films to each other, commonly referred to as *blocking*. This effect often results during processing and/or storage, when films are stacked or rolled. The blocking tendency of the plastic is determined by measurement of the force required to separate the adhered films under controlled temperatures (Zahedi et al. 2006). Antiblocking agents consist of microparticles that are incorporated into the film and increase the roughness of the surface, thus reducing the degree of blocking that could occur. Common antiblocking agents include silicas (natural and synthetic) and minerals. The agents must match the refractive index of the film to preserve clarity. Antiblocking agents and slip agents are combined to improve the performance of both agents (Nayek et al. 2000).

Antistatic Agents

Most polymers are poor electrical conductors; therefore, they are not capable of dissipating electrical charges, which, as a result, accumulate on plastic surfaces. Charge buildup is especially apparent in cold, dry climates and results in the attraction of dust and sparking. Antistatic agents were developed in the 1950s to dispel static electrical charges. They are typically hygroscopic, thus attracting moisture and dissipating the charge. Antistatic agents often have both a hydrophobic (e.g., a hydrocarbon chain) and a hydrophilic component (salts, quaternary amine, amide). The hydrophobic tail is attracted to the hydrophobic surface of most vinyl polymers and the hydrophilic part interacts with moisture in air.

Antistatic agents can be either internal or external. Internal antistats are incorporated during polymer processing and become a component of the bulk material. Over time, the internal antistat diffuses to the polymer surface, where the hydrophilic portion absorbs moisture and thus provides a conductive layer of water, thus imparting long-term protection to the material.

In contrast, external or topical antistat compounds are applied to a polymer surface by spraying, wiping, or dipping the plastic surface. Given that surface applications are worn away due to attrition, washing and handling, the external antistat may have to be replenished. Examples of antistatic agents are quaternary ammonium salts, long-chain aliphatic amines, phosphate esters, and ethylene glycols.

Biocides

Most synthetic polymers tend to be resistant to microbial attack. However, microbes can colonize plastic surfaces and form biofilms (see chapter 5). Naturally occurring polymeric materials such as rubber, cellulosics, starch, protein, and vegetable oil-based coatings and some polyesters may be susceptible to microbiologic damage as well (Fried 1995; Carraher 2017).

A biocide may be added to a polymer to inhibit bacterial establishment and growth. Biocides prevent microorganisms from feeding on the organic plasticizer in flexible PVC formulations. Biocides can quench odors, prevent color change, and help maintain polymer physical properties. About two-thirds of biocides are applied to PVC formulations, while 20% are used in polyolefins and less than 10% in PU foam (Markarian 2006).

Some biocides are solely organic in nature, while others contain metals such as Ag, Zn and Cu (Markarian 2006). Example biocidal compounds include tributyltin oxide which is used in latex paints, textiles, and plastics (Fried 1995). The biocide should be toxic to the microorganisms but be non-toxic to humans and other organisms. In the United States, all biocides must be registered by the US Environmental Protection Agency. The monomeric organotin-containing compounds are no longer permitted because of the high rate of leaching into the environment (Carraher 2017).

Colorants

Colorants are added to plastics in order to improve appearance. Some also provide shielding against UV light. The colorants used in the plastic industries fall into two major categories: pigments and dyes.

Pigments are finely divided solids which are essentially insoluble in the application medium (i.e., molten thermoplastic). The solids are added by grinding to a fine powder and mixing in.

Pigments are used to provide non-fading color and may be classified as organic and inorganic. Organic pigments contain carbon, hydrogen, oxygen, nitrogen, and sulfur, often in complex arrangements of aromatic hydrocarbons or macrocyclic metal complexes (figure 2.10). They may also be doped with metals; these occur in the pigment formulation as precipitating agents or as pigment particle surface additives. Examples of organic pigment classes

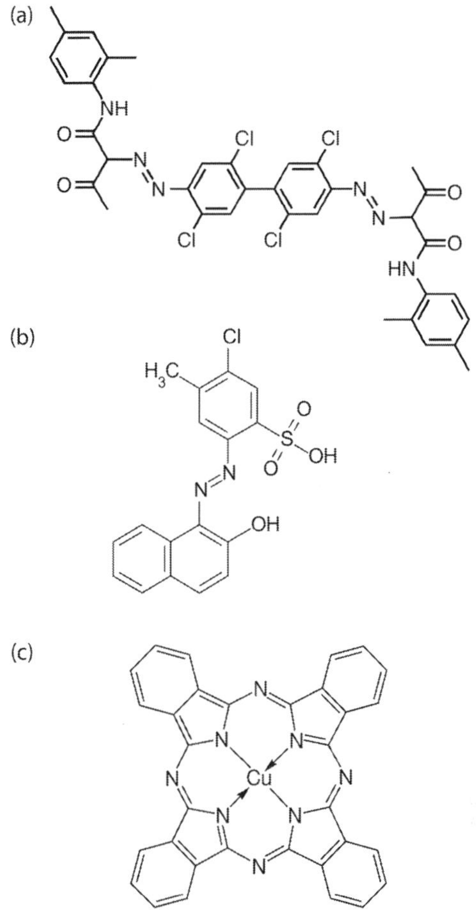

(a)

(b)

(c)

Figure 2.10 Examples of Organic Pigments in Polymers: (a) Benzidine Yellow 10G, (b) Pigment Red 53, and (c) Copper Phthalocyanine

are quinacridone, diazo, azo condensation, mono azo, naphthol, perylene, and pyrrolo-pyrroles (Shrivastava 2018). Unfortunately, several organic pigments are poorly biodegradable if introduced into the environment.

Inorganic pigments contain combinations of metallic or metalloid elements with oxygen, sulfur, or selenium (Charvat 2004). Popular inorganic pigments include titanium dioxide, zinc sulfide, iron oxide, chromate compounds, cadmium compounds, chromium oxide, ultramarines, mixed metal oxides, and carbon black. Certain inorganic pigments are multifunctional; for example, ZnO and Sb_2O_3 both act as white pigments but also act as a fungicide and flame-retardant component, respectively (Turner and Filella 2021). A number of metals occurring in inorganic pigments are potentially hazardous to human health and the environment. These include antimony, barium, cadmium, chromium, cobalt, lead, and manganese. Selected examples of organic and inorganic pigments are listed in table 2.8.

Dyes are soluble colored substances which impart color to plastics via selective absorption of light. Dyes must be transparent for effective use. Most have minimal light-fastness and limited heat stability (Murphy 1996). Dyes are classified as azo, perinon, quinolone, and anthraquinone. Dyes are typically used with polycarbonates, polystyrene, and acrylic polymers because of their color strength and transparency. Dyes are of limited use in the plastics industry as they often react with resins and impart undesirable results.

Unreacted Monomers and Oligomers

The polymeric structure of a plastic product occurs via the chemically induced concatenation of monomers. Not all monomers and oligomers are fully polymerized in plastics, however, resulting in a tendency of free molecules to migrate (Lau and Wong 2000). There are concerns of serious health risks when unreacted monomers migrate from food packaging and

Table 2.8 Examples of Organic and Inorganic Pigments Used to Impart Color to Plastics

Color	*Organic*	*Inorganic*
Red	Anthraquinone BONA Lake Diazo pigments Diketo pyrrolo pyrrole (DPP) Naphthol Lake Quinacridone	Iron oxide Cerium sulfide
Orange	Benzimidazolone diketo pyrrolo pyrrole (DPP) Isoindolinone	Cadmium sulfide Bismuth vanadate orange
Yellow	Anthraquinone Diazo compounds Isoindolinone Mono azo salts	Iron oxide Lead chromates Lead-tin Bismuth vanadates Nickel antimony
Brown	Diazo compounds	Iron oxide Chrome/iron oxide
Green	Phthalocyanine	Chrome green oxide Cobalt-based mixed metal oxides
Blue	Anthraquinone Phthalocyanine	Cobalt oxide/aluminum oxide
Violet	Dioxazine Quinacridone	Manganese (III) pyrophosphate

contaminate food, which is eventually absorbed by the human body (EU 10/2011). For example, it has been found that the styrene monomer may migrate from PS food packaging (Arvanitoyannis and Bosnea 2004). PET contains small quantities of low-MW oligomers which vary in size from dimers to pentamers. Cyclic compounds in the range of 0.06% to 1% were detected, depending on type of PET (Lau and Wong 2000).

CLOSING STATEMENTS

As stated above, very few additives are chemically bound to the polymer material. They can, therefore, leach from plastic as it degrades and pollute aquatic and soil environments. Some plastics contain POPs as additives (e.g., hexabromocyclododecane or HBCD) and/or polybrominated diphenyl ether (PBDE) (Hahladakis et al. 2018). A number of studies have demonstrated that a range of additives leach from plastics when exposed to heat, UV radiation, and other influences. Exceptions to leaching are reactive organic additives such as flame retardants which are polymerized with polymer molecules and become incorporated within the polymer chain.

REFERENCES

Allsopp, M.W., and Vianello, G. (2012). Poly(vinyl chloride). In: *Ullmann's Encyclopedia of Industrial Chemistry*. Weinheim (Germany): Wiley-VCH Verlag GmbH & Co. KGaA.

Ampacet. (2021). Ultraviolet light stabilizers and UV masterbatches [accessed 2021 September 12]. https://www.ampacet.com/faqs/ultraviolet-light-stabilizers.

Ampacet.com. (2021). Flame-retardant masterbatches for thermoplastics [accessed 2021 August 15]. https://www.ampacet.com/faqs/flame-retardants-for-thermoplastics.

Andrady, A.L., and Rajapakse, N. (2016). Additives and chemicals in plastics. In: *Hazardous Chemicals Associated With Plastics in the Marine Environment.* Berlin (Germany): Springer, pp. 1–17.

Arvanitoyannis, I.S., and Bosnea, L. (2004). Migration of substances from food packaging materials to foods. *Critical Reviews in Food Science and Nutrition* 44(2):63–76.

ASTM (American Society for Testing and Materials). (2020). *Standard Terminology Relating to Plastics, ASTM D883-20b.* West Conshohocken (PA): ASTM International. www.astm.org.

Bae, B., Jeong, J.H., and Lee, S.J. (2002). The quantification and characterization of endocrine disruptor bisphenol-A leaching from epoxy resin. *Water Science and Technology: A Journal of the International Association on Water Pollution Research* 46(11–12):381–387.

Barraza, L. (2013). A new approach for regulating bisphenol A for the protection of the public's health. *Journal of Law, Medicine & Ethics* 41(S1):9–12.

Bhunia, K., Sablani, S.S., Tang, J., and Rasco, B. (2013). Migration of chemical compounds from packaging polymers during microwave, conventional heat treatment, and storage. *Comprehensive Reviews in Food Science and Food Safety* 12(5):523–545.

Cadogan, D.F., and Howick, C.J. (2000). Plasticizers. In: *Ullmann's Encyclopedia of Industrial Chemistry*. Weinheim (Germany): Wiley-VCH Verlag GmbH & Co. KGaA.

Cano, J., Marın, M., Sanchez, A., and Hernandis, V. (2002). Determination of adipate plasticizers in poly (vinyl chloride) by microwave-assisted extraction. *Journal of Chromatography A* 963(1–2):401–409.

Carraher, C.E., Jr. (2003). *Polymer Chemistry*, 6th ed. New York: Marcel Dekker.

Carraher, C.E., Jr. (2017). *Introduction to Polymer Chemistry*, 4th ed. Boca Raton (FL): CRC Press.

Catenza, C.J., Farooq, A., Shubear, N.S., and Donkor, K.K. (2020). A targeted review on fate, occurrence, risk and health implications of bisphenol analogues. *Chemosphere* 268:129273.

Charvat, R.A. (2004). *Coloring of Plastics: Fundamentals*, 2nd ed. New York: Wiley-Interscience.

Craftech.com. (2021). The top 3 plastic additives for UV stabilization [accessed 2021 September 19]. https://www.craftechind.com/uv-radiation.

Crawford, M.B., and Quinn, B. (2017). *Microplastic Pollutants*. Amsterdam (Netherlands): Elsevier.

EU (European Union). (2011). Commission Regulation (EU) No. 10/2011 of 14 January 2011 on plastic materials and articles intended to come into contact with food. https://eur-lex.europa.eu/eli/reg/2011/10/2016-09-14.

Fried, J. (1995). *Polymer Science and Technology*. Englewood Cliffs (NJ): PTR Prentice Hall.

Gewert, B., Plassmann, M.M., and MacLeod, M. (2015). Pathways for degradation of plastic polymers floating in the marine environment. *Environmental Sciences: Processes and Impacts* 17:1513–1521.

Gunaalan, K., Fabbri, E., and Capolupo, M. (2020). The hidden threat of plastic leachates: A critical review on their impacts on aquatic organisms. *Water Research* 184:116170.

Hahladakis, J.N., Velis, C.A., Weber, R., Iacovidou, E., and Purnell, P. (2018). An overview of chemical additives present in plastics: Migration, release, fate and environmental impact during their use, disposal and recycling. *Journal of Hazardous Materials* 344:179–199.

Hansen, E. (2013). *Hazardous Substances in Plastic Materials*. TA 3017. Vejle (Denmark): Danish Technological Institute.

Harnett, K.G., Chin, A., and Schuh, S.M. (2021). BPA and BPA alternatives BPS, BPAF, and TMBPF, induce cytotoxicity and apoptosis in rat and human stem cells. *Ecotoxicology and Environmental Safety* 216:112210.

Harper, C.A., and Petrie, E.M. (2003). *Plastics Materials and Processes: A Concise Encyclopedia*. New York: John Wiley & Sons.

Jalal, N., Surendranath, A.R., Pathak, J.L., Yu, S., and Chung, C.Y. (2018). Bisphenol A (BPA) the mighty and the mutagenic. *Toxicology Reports* 5:76–84.

Kadir, E.R., Imam, A., Olajide, O.J., and Ajao, M.S. (2021). Alterations of Kiss 1 receptor, GnRH receptor and nuclear receptors of the hypothalamo-pituitary-ovarian axis following low dose bisphenol-A exposure in Wistar rats. *Anatomy & Cell Biology* 54(2):212–224.

Kardas, F., Bayram, A.K., Demirci, E., Akin, L., Ozmen, S., Kendirci, M., Canpolat, M., Oztop, D.B., Narin, F., Gumus, H., Kumandas, S., and Per, H. (2016). Increased serum phthalates (MEHP, DEHP) and Bisphenol A concentrations in children with autism spectrum disorder: The role of endocrine disruptors in autism etiopathogenesis. *Journal of Child Neurology* 31(5):629–635.

Kattas, L., Gastrock, F., and Cacciatore, A. (2000). Plastic additives. In: Harper, C.A., editor. *Modern Plastics*. New York: McGraw-Hill, pp. 4.1–4.66.

Krishnan, A.V., Stathis, P., Permuth, S.F., Tokes, L., and Feldman, D. (1993). Bisphenol-A: An estrogenic substance is released from polycarbonate flasks during autoclaving. *Endocrinology* 132(6):2279–2286.

Lau, O.-W., and Wong, S.-K. (2000). Contamination in food from packaging material. *Journal of Chromatography A* 882:255–270.

Lehmler, H.J., Liu, B., Gadogbe, M., and Bao, W. (2018). Exposure to Bisphenol A, Bisphenol F, and Bisphenol S in U.S. adults and children: The national health and nutrition examination survey 2013–2014. *ACS Omega* 3(6):6523–6532.

Lithner, D., Larsson, Å., and Dave, G. (2011). Environmental and health hazard ranking and assessment of plastic polymers based on chemical composition. *Science of the Total Environment* 409(18):3309–3324.

Lithner, D., Nordensvan, I., and Dave, G. (2012). Comparative acute toxicity of leachates from plastic products made of polypropylene, polyethylene, PVC, acrylonitrile–butadiene–styrene, and epoxy to *Daphnia magna*. *Environmental Science and Pollution Research* 19:1763–1772 [accessed 2021 August 22]. DOI: 10.1007/s11356-011-0663-5.

Liu, P., Qian, L., Wang, H., Zhan, X., Lu, K., Gu, C., and Gao, S. (2019). New insights into the aging behavior of microplastics accelerated by advanced oxidation processes. *Environmental Science & Technology* 53(7):3579–3588.

Liu, P., Zhan, X., Wu, X., Li, J., Wang, H., and Gao, S. (2020). Effect of weathering on environmental behavior of microplastics: Properties, sorption and potential risks. *Chemosphere* 242:125193.

Luo, H., Zeng, Y., Zhao, Y., Xiang, Y., Li, Y., and Pan, X. (2021). Effects of advanced oxidation processes on leachates and properties of microplastics. *Journal of Hazardous Materials* 413:125342.

Luo, H., Zhao, Y., Li, Y., Xiang, Y., He, D., and Pan, X. (2020). Aging of microplastics affects their surface properties, thermal decomposition, additives leaching and interactions in simulated fluids. *Science of the Total Environment* 714:136862.

Markarian, J. (2006). Steady growth predicted for biocides. *Plastics, Additives and Compounding* 8(1):30–33.
MCpolymers.com. (2020). An introduction to amorphous polymers [accessed 2021 August 2]. https://www.mcpolymers.com/library/introduction-to-amorphous-polymers.
Murphy, J. (1996). *The Additives for Plastics Handbook.* Amsterdam (Netherlands): Elsevier.
Murphy, J. (2001). *Additives for Plastics Handbook*, 2nd ed. Amsterdam (Netherlands): Elsevier.
Nayak, K., and Tollefson, N.M. (2000). An experimental design approach: Effect of slip and antiblocking agents on the performance of an LLDPE polymer. *Journal of Plastic Film and Sheeting* 16(2):84–94.
NOVA Chemicals. (2020). *ESG 2020 Report.* Calgary (Alberta, Canada): Novachem.com.
Peacock, A.J., and Calhoun, A. (2012). *Polymer Chemistry: Properties and Application.* Munich (Germany): Carl Hanser Verlag GmbH Co KG.
PlasticsEurope. (2020). Plastics—The facts 2020. *PlasticsEurope* 1:1–64.
Rebai, I., Fernandes, J.O., Azzouz, M., Benmohammed, K., Bader, G., Benmbarek, K., and Cunha, S.C. (2021). Urinary bisphenol levels in plastic industry workers. *Environmental Research* 202:111666.
Robertson, G. (2006). Modified atmosphere packaging. In: *Food Packaging: Principles and Practice*, pp. 313–329.
Rochman, C.M. (2015). The complex mixture, fate and toxicity of chemicals associated with plastic debris in the marine environment. In: Bergmann, M., Gutow, L., and Klages, M., editors. *Marine Anthropogenic Litter.* Berlin (Germany): Springer, pp. 117–140.
Rodríguez-Báez, A.S., Medellín-Garibay, S.E., Rodríguez-Aguilar, M., Sagahón-Azúa, J., Milán-Segoviaa, R., and Flores-Ramírez, R. (2022). Environmental endocrine disruptor concentrations in urine samples from Mexican Indigenous women. *Environmental Science and Pollution Research International* 29(25):38645–38656.
Sablani, S.S., and Rahman, M.S. (2007). Food packaging interaction. In: *Handbook of Food Preservation.* Boca Raton (FL): CRC Press, pp. 957–974.
Sastri, V.R. (2013). *Plastics in Medical Devices: Properties, Requirements, and Applications.* Cambridge (MA): William Andrew.
Shrivastava, A. (2018). Additives for plastics. In: *Introduction to Plastics Engineering.* Amsterdam (Netherlands): Elsevier.
SpecialChem.com. (2021). Select curing agents for coating formulations. https://coatings.specialchem.com/selection-guide/curing-agents-for-coating-formulations.
Ter Halle, A., Ladirat, L., Martignac, M., Mingotaud, A.F., Boyron, O., and Perez, E. (2017). To what extent are microplastics from the open ocean weathered? *Environmental Pollution* 227:167–174.
Thoene, M., Dzika, E., Gonkowski, S., and Wojtkiewicz, J. (2020). Bisphenol S in food causes hormonal and obesogenic effects comparable to or worse than Bisphenol A: A literature review. *Nutrients* 12(2):532.

Tiwari, D., Kamble, J., Chilgunde, S., Patil, P., Maru, G., Kawle, D., Bhartiya, U., Joseph, L., and Vanage, G. (2012). Clastogenic and mutagenic effects of bisphenol A: An endocrine disruptor. *Mutation Research* 743(1–2):83–90.

Turner, A., and Filella, M. (2021). Hazardous metal additives in plastics and their environmental impacts. *Environment International* 156:106622.

UNEP. (2014). Marine plastic debris and microplastics: Global lessons and research to inspire action and guide policy change.

UNEP (United Nations Environment Programme). (2015). Plastic in cosmetics.

Van Melkebeke, M. (2019). Exploration and Optimization of Separation Techniques for the Removal of Microplastics from Marine Sediments During Dredging Operations [Master's Thesis]. Ghent (Belgium): Ghent University.

Wang, J., Peng, J., Tan, Z., Gao, Y., Zhan, Z., Chen, Q., and Cai, L. (2017). Microplastics in the surface sediments from the Beijiang River littoral zone: Composition, abundance, surface textures and interaction with heavy metals. *Chemosphere* 171:248–258.

Zahedi, A., Asiaban, A.R., and Asiaban, S. (2006). Optimizing COF, blocking force, and printability of low density polyethylene. *Journal of Plastic Film and Sheeting* 22(3):163–176.

Zhang, H., Wang, J., Zhou, B., Zhou, Y., Dai, Z., Zhou, Q., Chriestie, P., and Luo, Y. (2018). Enhanced adsorption of oxytetracycline to weathered microplastic polystyrene: Kinetics, isotherms and influencing factors. *Environmental Pollution* 243:1550–1557.

Zhou, G., Wang, Q., Zhang, J., Li, Q., Wang, Y., Wang, M., and Huang, X. (2020). Distribution and characteristics of microplastics in urban waters of seven cities in the Tuojiang River basin, China. *Environmental Research* 189:109893.

Zhou, L., Wang, T., Qu, G., Jia, H., and Zhu, L. (2020). Probing the aging processes and mechanisms of microplastic under simulated multiple actions generated by discharge plasma. *Journal of Hazardous Materials* 398:122956.

Zimmermann, L., Bartosova, Z., Braun, K., Oehlmann, J., Volker, C., and Wagner, M. (2021). Plastic products leach chemicals that induce in vitro toxicity under realistic use conditions. *Environmental Science & Technology* 55(17):11814–11823.

Chapter 3

Microplastic Types

INTRODUCTION

It is important, from a public health and environmental perspective, to recognize the various resin types of microplastics (MPs) present in water and soil, and to understand how they interact with organisms and the environment. The behavior and fate of MPs are a function of numerous inter-related factors including chemical and physical properties, degree of weathering, and others.

MPs are often classified as primary or secondary. Primary MPs are designed and manufactured in this size range as a feedstock for numerous products, for specific applications, and for ease in transport. Such particles include nurdles, microbeads, and powders. In contrast, secondary MPs originate from degradation of larger plastic objects (table 3.1). Due to differences resulting from aging and other characteristics, primary and secondary MPs often behave differently in the biosphere.

PRIMARY MPs

Primary MPs include plastic feedstock ("nurdles"), that is, the raw material for manufacture of products. Primary MPs are also manufactured for specific commercial uses. Examples include facial cleansers (Zitko and Hanlon 1991), hand cleansers, and cosmetics. Primary MPs can be employed as delivery system for pharmaceuticals *in corpore* (Patel et al. 2009) and in dental tooth polish (Sharma et al. 2020). They are also included in drilling liquids used for oil and gas exploration as well as in industrial abrasives (cleaning products and air-blasting media) (Gregory 1996; Sharma et al. 2020). Examples of primary MPs are discussed below.

Table 3.1 Sources of Primary and Secondary MPs

Type	*Source*	*Description*
Primary microplastics	Virgin pellets (pre-production pellets, nurdles, mermaid tears)	Losses from pellet manufacturing facilities and facilities using pellets. Losses during loading, transfer, and in transport (i.e., truck, rail, ship accidents).
	Personal care items	Release of exfoliating microbeads from cosmetics, facial cleansers, hand cleansers, toothpastes, etc. during consumer use.
	Industrial abrasive media	Used for cleaning of automotive, marine, aerospace, and electronics components. Includes paint stripping, mold removal, and deburring and deflashing metallic surfaces.
	Paints	Synthetic pigments and synthetic binders.
	Rubber granules	Release due to wear of certain artificial turfs, running lanes, playgrounds, etc.
	Pharmaceuticals	Drug delivery systems; allows for controlled release of active ingredients.
Secondary microplastics	Consumer products	Release of microplastic fragments and fibers during use, improper disposal to land and water.
	Industrial and commercial activities	Release of microplastic dust from manufacturing, grinding, cutting, polishing, etc. of plastic products.
	Roads and highways	Release of tire particles, road-marking materials, and polymers in asphaltic materials in road pavement.
	Textiles	Releases of microfibers during manufacture (fiber production, yarn production, fabric production, pre-treatment, dyeing, finishing). Releases of microfibers during wear and washing by consumer.
	Household use	Release of plastic dust from abrasion, wear, or weathering of floor coverings, furniture, roof coverings, and kitchen utensils. Release of plastic fibers from textiles such as carpets, furniture, and clothing.
	Agriculture	Weathering of agricultural plastic films used as mulch and silage wrap.
	Aquaculture	Weathering and abrasion of fishing gear such as nets and ropes.
	Painted surfaces	Release of plastic dusts from paint abrasion and maintenance work.
	Waste management	Plastic particles are released during municipal solid waste loading, mixing and unloading; losses during processing such as shredding, trommel screening, air classification, etc.
	Environmental exposure	Fragmentation of macroplastic litter by physical (e.g., abrasion), chemical (reactions with O_2 and UV light) and biological processes.

Nurdles

Also known as mermaid tears or pre-production resin pellets, these are the raw materials for the manufacture of plastic products (figure 3.1). They occur in the form of granules which are commonly cylinder- or disk-shaped, and may be transparent or colored (white, blue, etc.). Nurdle size is generally about 1–6 mm in diameter; given this size range, they can be classified as MPs (< 5 mm) or mesoplastics (5–25 mm) (Andrady 2011; GESAMP 2015). Nurdles are composed of different polymer types, most commonly PE, PP, and PVC. The processes involved in converting polymer resins to nurdles is shown in figure 3.2.

Nurdles have been unintentionally released into freshwater and marine environments via inadvertent losses from land-based facilities during manufacture and processing. Elevated levels of pellets are sometimes reported in waterways adjacent to plastic production facilities (Lozano and Mouat 2009). In recent years, several companies have faced severe fines and other legal actions as a result of releases from their facilities (California 2011; US EPA 2011, 2016; Fernandez 2019; Stormwater 2020; JDSUPRA 2020; PSF 2020).

Nurdles are also lost during loading and while in transport, both on land and at sea. Shipping containers that store nurdles have fallen from ships during accidents and storms. Of all MPs entering the world's oceans, nurdles make up only a small portion, that is, less than 1% of the total (Boucher and Friot 2017). During a storm in August 2020, the cargo ship *CMA CGM Bianca* broke free from its moorings at a wharf and a large container of nurdles fell into the Mississippi River (Burick 2020). The nurdles broke free from sacks spilling 48,000 pounds of pellets into the river.

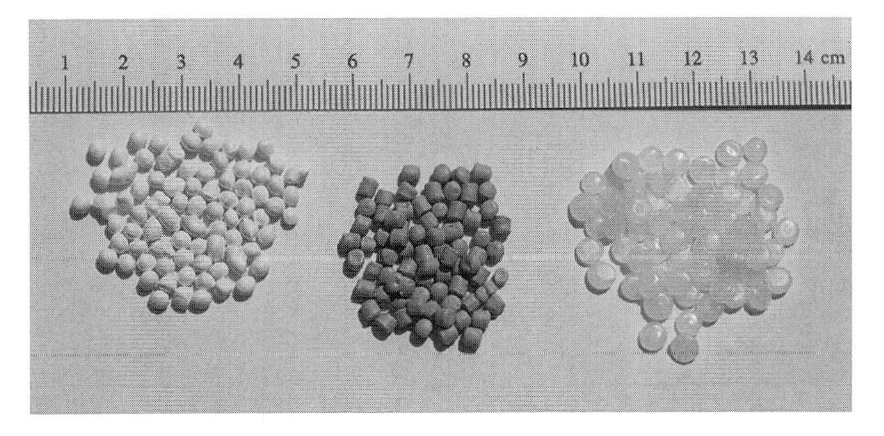

Figure 3.1 Nurdle Types

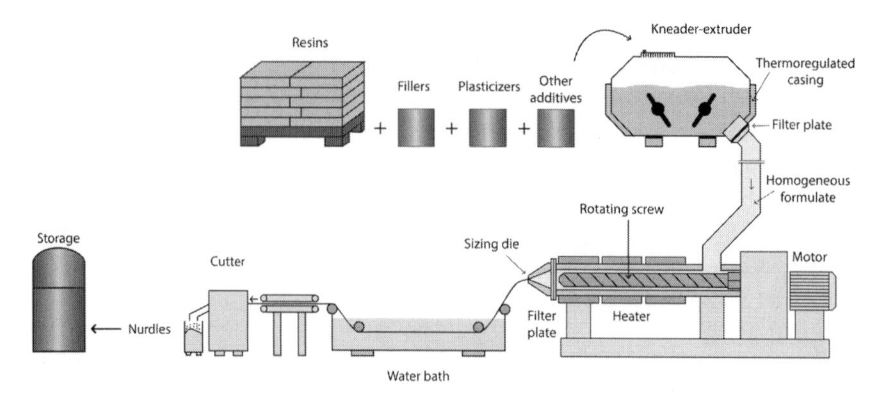

Figure 3.2. Conversion of Resin Feedstock to Nurdles
Source: Adapted from Cjp24 https://commons.wikimedia.org/wiki/File:Compounding-en.png

More than 6,000 container ships operate at sea on a continuous basis, according to the World Shipping Council (WSC). About 226 million containers are transported annually with cargo valued at more than $4 trillion (BIFA 2020). An average of 1,382 containers were lost each year between 2008 and 2019 as estimated by the WSC (Franklin 2021). However, in the span of two months from late 2020 to early 2021 more than 2,675 containers were lost in five separate incidents at sea, according to *American Shipper* magazine (Link-Wills 2021). In 2017 Saudi Basic Industries Corporation agreed to pay for cleanup operations following a massive spill of nurdles in KwaZulu-Natal waters off the coast of South Africa (Kockott 2017). The spill occurred after two container ships collided in the mouth of Durban harbor during a severe storm.

In May 2021 the container ship *X-Press Pearl* suffered a catastrophic accident off the southern coast of Sri Lanka, where it caught fire and sank. Nearly 1,500 containers were aboard the ship, many of which contained nurdles. Many beaches were devastated by thick waves of pellets (figure 3.3). Nurdles have been documented on a stretch of about 250 miles along the coast of Sri Lanka, and were carried by currents to the east.

To date, reliable numbers of nurdles losses to the environment during production do not exist.

Microbeads

This category of MPs includes particles that are spherical or irregular in shape and formulated as ingredients in personal care and cosmetic products (PCCPs), paints, and industrial abrasives. They are also termed microspheres, nanospheres, microcapsules, nanocapsules, as well as various registered

Figure 3.3 Nurdles from the *X-Press Pearl* Spill on a Sri Lankan Beach
Source: Reproduced with kind permission of Oceanswell.

trademark names (Leslie 2015). Although much of the nomenclature implies particles of spherical form, this is not always so. Microbeads often have irregular shapes, but spherical particles are sometimes used to enhance visual attraction to consumers (Kalcíkova et al. 2017).

Microbeads are often solid; however, others are hollow, which enables them to be packed with an active ingredient for consumer use (Lidert 2005a, 2005b). These are sometimes termed microcapsules (Ansaldi 2005; Kvitnitsky et al. 2005). Certain polymers such as cross-linked poly(methyl methylacrylate) (PMMA) may be formulated into fine, sponge-like particles measuring between 1 and 50 μm (Saxena and Nacht 2005). These can sorb active ingredients, especially hydrophobic ones (Lidert 2005a).

Microbeads are designed to be applied, rinsed off, and discarded down the drain. They have been incorporated in cosmetics to replace natural exfoliating materials such as pumice, oatmeal, apricot, or walnut husks. Microspheres are used in toothpaste as they remove plaque and stains due to their abrasive action. Some personal care products that contain microbeads are listed in table 3.2.

Microbeads are manufactured in an array of colors in order to impart visual appeal to cosmetics. Colored microbeads have also been used as visible markers in various techniques of microscopy and biotechnology (Miraj et al. 2019). Microbeads and other plastic ingredients occur in products at varying concentrations, ranging from less than 1% to more than 90% (Leslie 2015).

Microbeads are manufactured in a range of sizes. Some are large enough to be visible to the unaided eye, while others occurring in PCCP formulations are in the nano-size (Leslie 2014).

Table 3.2 Personal Care Products That Contain Microbeads

soap	anti-wrinkle creams	nail polish
facial scrubs	moisturizers	liquid makeup
shower gel	facial masks	deodorant
shampoo	eye shadow	toothpaste
conditioner	mascara	hair spray
bubble bath	lipstick	baby care products
shaving cream	blush powders	sunscreen
hair coloring	makeup foundation	insect repellant
skin creams		

Composition

As with other MPs, the microbeads occurring in PCCPs are solids composed of synthetic homopolymers and/or copolymers which are insoluble in water and nondegradable (table 3.3). Polymer types in PCCPs include thermoplastics such as PE (most common), PP, PS, polytetrafluoroethylene (Teflon), PMMA; and thermosets such as polyurethanes and certain polyesters. Many of these have low-melting temperatures and rapid phase transitions (e.g., melting transition, glass transition) which allow for the creation of a porous structure in cosmetics and other products (Hardesty et al. 2019). Copolymers have been developed to enhance material properties in PCCP applications, such as resistance to degradation (Guerrica-Echevarría and Eguiazábal 2009). Some popular copolymer blends used in PCCPs are ethylene/propylene styrene, butylene/ethylene styrene, acrylate copolymers, and many others (CIR 2002, 2012; Leslie 2015).

Brightly colored microbeads are used in industrial paints. Such formulations commonly contain polyurethanes, polyesters, polyacrylates, polystyrenes, alkyls, and epoxies (Gaylarde et al. 2021). When studying atmospheric deposition of MPs, researchers identified pink microbeads composed of PMMA, which are used in a variety of industrial paint and coating applications (Brahney et al. 2020).

Implications

An analysis of wastewater and MPs by Rochman et al. (2015) estimated that 8 trillion microbeads are emitted into aquatic habitats daily worldwide. The Netherlands was the first country to institute a ban on microbeads in cosmetic products in 2014. Several countries including Australia, Canada, Italy, Korea, New Zealand, Sweden, the UK, and the US have since followed (OECD 2021). In much of the world, however, a significant percentage of households are not connected to wastewater treatment facilities; therefore, MPs are

Table 3.3 Examples of Plastic Ingredients Incorporated in PCCPs

Polymer	*Function*
Acrylates copolymer	Binder, hair fixative, film formation, suspending agent
Allyl stearate/vinyl acetate copolymers	Film formation, hair fixative
Butylene/ethylene/styrene copolymer	Viscosity controlling
Ethylene/acrylate copolymer	Film formation in waterproof sunscreen, gellant (e.g. lipstick, stick products, hand creams)
Ethylene/ methylacrylate copolymer	Film formation
Ethylene/propylene/styrene copolymer	Viscosity controlling
Nylon-12 (polyamide-12)	Bulking, viscosity controlling, opacifying (e.g. wrinkle creams)
Nylon-6	Bulking agent, viscosity controlling
Poly(butylene terephthalate)	Film formation, viscosity controlling
Poly(ethylene isoterephthalate)	Bulking agent
Poly(ethylene terephthalate)	Adhesive, film formation, hair fixative; viscosity controlling, aesthetic agent, (e.g. glitters in bubble bath, makeup)
Poly(methyl methylacrylate)	Sorbent for delivery of active ingredients
Poly(pentaerythrityl terephthalate)	Film formation
Poly(propylene terephthalate)	Emulsion stabilizing, skin conditioning
Polyethylene	Abrasive, film forming, viscosity controlling, binder for powders
Polypropylene	Bulking agent, viscosity increasing agent
Polystyrene	Film formation
Polytetrafluoroethylene (Teflon)	Bulking agent, slip modifier, binding agent, skin conditioner
Polyurethane	Film formation (e.g. facial masks, sunscreen, mascara)
Polyacrylate	Viscosity controlling
Styrene acrylates copolymer	Aesthetic, colored microspheres (e.g. makeup)
Trimethylsiloxysilicate (silicone resin)	Film formation (e.g. color cosmetics, skin care, sun care)

discharged directly to surface waterways. In the United States, 76% of the population is connected to municipal wastewater treatment facilities (ASCE 2017), and that in the EU is about 70% (EuroStat 2012). In less developed countries, the infrastructure and resources needed to construct centralized wastewater treatment plants (WWTPs) is lacking. In developing countries, 80–90% of untreated wastewater is discharged to water bodies and coastal zones (UN Water 2008). Furthermore, PCCP regulation is often non-existent in these countries (Leslie et al. 2014).

Given the potential environmental risks of MPs, a precautionary approach toward MP management has been recommended (Leslie 2015). An eventual phase-out and ban in PCCPs is considered the optimal solution. Greater emphasis is required for designing products that are more environmentally friendly, lower in plastic content, and contain safer additives.

These combined steps can contribute toward reducing potential health and environmental hazards posed by MPs in PCCPs (Leslie 2015).

SECONDARY MPs

Secondary MP particles originate from fragmentation and degradation of primary MPs and also from degradation of larger plastic products such as water bottles and plastic bags. Secondary MPs take on various forms; the major types detected in the environment include fragments, pellets, fibers, films, and granules (see figure 1.5).

Conversion of Macroplastics to Secondary MPs

Plastics present in different environments (marine, freshwater, WWTPs, soil, air) are exposed to myriad processes and stresses including mechanical shear, solar exposure, thermal aging, oxidation, hydrolysis, microbial colonization, and biofilm development (Andrady 2011), all of which result in chemical changes and physical alteration (Guo and Wang 2019; Wang et al. 2020). The major commercial polymers tend to be resistant to degradation; however, they can decompose to some extent by a combination of the above chemical, physical, and biological processes. It is beneficial to identify and quantify the contribution of differing insults on plastic weathering, considering all types of polymers and additive compounds, so as to develop appropriate methods for their removal from water, wastewater, and sediment, and for their ultimate destruction.

When plastic articles are exposed to sunlight, oxygen, natural organic matter, or other substances and stimuli, cracks and craters form, surface chemical groups are oxidized, and particles may become coated with microorganisms and their exudates (*biofilms*). As a result, particle size, density, surface morphology, and crystallinity may be altered. New, reactive surface functional groups appear; in other cases, surface charges are neutralized, reversed, or masked (figure 3.4) (Rincon-Rubio et al. 2001; Fazey and Ryan 2016; Rouillon et al. 2016). The modified surfaces interact with organic matter, minerals, metallic and hydrocarbon-based pollutants, and biota (including pathogenic organisms) (Fotopoulou and Karapanagioti 2012). Weathering (aging) of MPs is typically initiated on the polymer surface, which becomes oxidized, embrittled, and/or crazed. Along with these effects, particles may become discolored. The particle interior is eventually degraded by diffusion-controlled processes (Lithner et al. 2011).

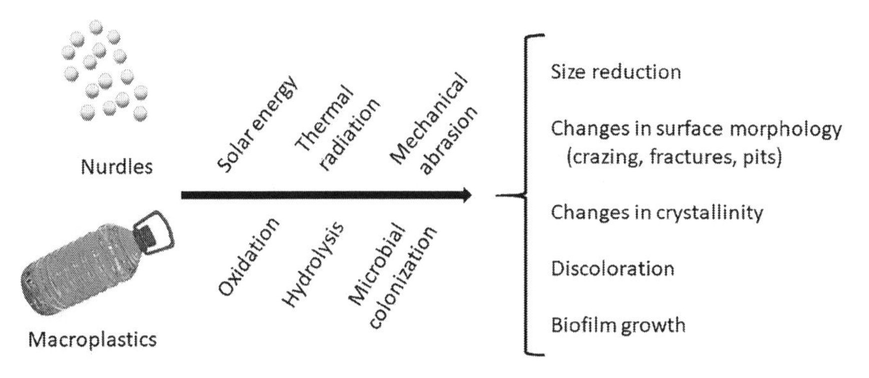

Figure 3.4. Changes in Properties of MPs via Abiotic and Biotic Degradation

Degradation occurs by chain scission, that is, cleavage of chemical bonds within the polymer (Alger 1997; Lithner et al. 2011). For most polymers, chain breakage occurs at random locations; it may also target any monomer repeat unit (for example, a C-O or C-Cl bond) within the polymer (Lithner et al. 2011). For certain polymers, chain scission occurs at the end monomer repeat unit so that monomers are released one by one (i.e., an "unzipping" depolymerization mechanism) (Alger 1997; Braun et al. 2005; Lithner et al. 2011). In other cases, polymer degradation occurs by chain stripping, where side atoms or chemical groups attached to the polymer backbone are detached and released. For example, in the case of PVC, the chloride ion may be stripped from the main carbon chain. Depending on the polymer, these processes lead to progressive depolymerization which includes production and release of carbonyl groups, hydroperoxides, alcohols, and hydrochloric acid, among others. This is often followed by cross-linking (Wilkes et al. 2005; Lithner et al. 2011; Gewert et al. 2015). Overall effects include reduced MW, loss of material properties, particle embrittlement, and fragmentation (Alger 1997; Vasile 2000; Ravve 2000; Lithner et al. 2011).

Abiotic degradation of MPs in the ambient environment occurs via photo-induced damage and to a lesser extent by thermal alteration, hydrolysis, and mechanical fragmentation. Air pollution also exerts an impact. The specific pathways involved in plastic degradation are a function of the physicochemical properties of the polymer and probably embrace a succession of abiotic and/or biotic processes. The final products formed during degradation will therefore vary by polymer type (Ravve 2000). Other factors which control degradation include types of additives in the original polymer, and environmental factors (O_2 level, solution pH and Eh, temperature) (Ravve 2000; Lithner et al. 2011).

Photodegradation

The majority of synthetic plastics are capable of absorbing light at wavelengths which occur in sunlight. The relevant wavelengths have sufficient energy to result in bond cleavage (Gijsman et al. 1999); therefore, photo-induced processes play an important role in abiotic degradation. Degradation of MPs by sunlight is a consequence of both *photolysis* and *photooxidation.*

Photolysis occurs when the absorption of light leads directly to scission of chemical bonds. Subsequent chemical reactions cause advanced deterioration of the polymer. In contrast, during photooxidation, the light which is absorbed causes the formation of radicals that trigger oxidation at various locations within the polymer (Gijsman and Diepens 2009). With either mechanism, particles are ultimately fragmentated by the effects of light.

Photooxidation due to sunlight occurs when the polymer contains chromophores which absorb wavelengths of the sunlight spectrum > 290 nm. Chromophores may be: (1) part of the molecular structure of the polymer; (2) impurities occurring within the chain such as hydroperoxides or carbonyls formed during processing or weathering; and (3) external impurities such as residues from polymerization catalysis, additives (pigments, antioxidants), pollutants from the atmosphere or trace metals from processing equipment (Gijsman et al. 1999; Gijsman and Diepens 2009).

When photons of UV light, particularly those occurring in the UVB region (315–280 nm) of the electromagnetic spectrum contact the polymer surface, reactive oxygen species (ROS) are generated which result in a free-radical chain reaction (Crawford and Quinn 2017). Free radicals include hydroxyl (·OH), alkyl (R·), alkoxyl (RO·) and peroxyl (ROO·). These energetic molecules attack chemical bonds in the polymer, causing chain cleavage (Liu et al. 2019; Bracco et al. 2018; Tian et al. 2019; Wang et al. 2021).

In the initiation step, chemical bonds along the polymer chain are broken by UV light either by hydrogen abstraction or by homolytic cleavage of a carbon-carbon bond to generate a free radical (Eq. 3.1) (Gewert et al. 2015; PPD 2021).

1. Initiation

$$RH \xrightarrow{h\upsilon} R^{\bullet} + H^{\bullet} \tag{3.1}$$

During the propagation step the polymer radical reacts with oxygen with consequent formation of a peroxy radical (Eq. 3.2) (Gewert et al. 2015; Singh and Sharma 2008). In addition to formation of hydroperoxides, other radical reactions occur, resulting in autoxidation (Eqs. 3.3–3.5) (Singh and Sharma 2008). These processes ultimately result in chain scission or cross-linking (Gewert et al. 2015).

2. Propagation

$$R^{\bullet} + O_2 \rightarrow ROO^{\bullet} \tag{3.2}$$

$$ROO^{\bullet} + RH \rightarrow ROOH + R^{\bullet} \tag{3.3}$$

$$ROOH \xrightarrow{h\upsilon} RO^{\bullet} + OH \tag{3.4}$$

$$2\ ROOH \xrightarrow{h\upsilon} ROO^{\bullet} + RO^{\bullet} + H_2O \tag{3.5}$$

The radical-induced reaction is terminated when two radicals combine to form inert products (Eqs. 3.6–3.8) (Gewert et al. 2015).

3. Termination

$$R^{\bullet} + R^{\bullet} \rightarrow \text{stable products} \tag{3.6}$$

$$R^{\bullet} + ROO^{\bullet} \rightarrow ROOR + O_2 \text{ and other stable products} \tag{3.7}$$

$$ROO^{\bullet} + ROO^{\bullet} \rightarrow \text{stable products} \tag{3.8}$$

Products of radical termination include olefins, aldehydes, and ketones, all of which may be prone to photo-induced degradation, as they contain unsaturated double bonds (Gewert et al. 2015).

The above oxidation reactions result in chain breakage, branching, cross-linking, and formation of oxygen-containing functional groups (Gewert et al. 2015). The ultimate result is embrittlement and cracking of the polymer (PPD 2021) which is, consequently, susceptible to further fragmentation. The greater surface area will foster additional reactions (Gewert et al. 2015).

Degradation pathways differ among plastics based on the chemical structure of the backbone. Two key groups are: (1) those with a backbone solely built of carbon atoms (examples include PE, PP, PS, and PVC); and (2) those containing heteroatoms in the main chain (see table 2.5) (PET and PU) (Gewert et al. 2015). In the case of PE, PP, and PS, are all prone to photooxidation; this is likely the most important abiotic degradation pathway in the outdoor environment (Gewert et al. 2015; Gijsman and Diepens 2009).

In the case of opaque plastics, essentially all initial oxidative degradation occurs at surface layers only. Degradation is localized due to the high extinction coefficient of UV-B radiation in plastics (Cunliffe and Davis 1982). The presence of fillers will also restrict oxygen diffusion in the material. Degradation occurs more rapidly in virgin pellets that contain no UV stabilizers.

Air Pollution

In urban and industrial environments, elevated levels of ozone, sulfur dioxide, nitrogen oxide and carbon monoxide are relatively common, and can damage plastic (Ravve 2000; Lithner et al. 2011). The ultimate effects of these gases are embrittlement and fragmentation of the polymer into smaller pieces.

Ozone (O_3) readily attacks certain polymers and causes significant loss of physical properties. In the process of ozonolysis, long carbon chains that normally impart strength to polymers will break. Double bonds react readily with ozone; therefore, those polymers with many unsaturated carbons (i.e., rubbers) are especially subject to ozone-induced degradation. Reaction of a double bond with ozone results in chain cleavage. Molecular weight decreases, consequently reducing the strength of the polymer and causing it to become brittle. Reactions showing ozone degradation of an olefin are shown in Equation 3.9. Ozone reacts with saturated polymers, but at a somewhat slower rate.

$$\mathrm{>C=C<} + O_3 \longrightarrow \text{(O–O–O ring)} \longrightarrow \mathrm{CH{-}O^{+}{-}O^{-}} + \mathrm{>C=O} \qquad (3.9)$$

$$\longrightarrow \text{(O–O, O ring)} + \mathrm{-C-O-O-} + \mathrm{C(O{-}OH)(OR)}$$

When elastomers such as natural rubber, styrene-butadiene, polyurethane elastomers and nitrile rubber are exposed to the atmosphere, ozone induces cracking on surfaces. This effect may be observed on old auto tires, especially in urban areas which experience significant air pollution. Certain elastomers such as polychloroprene (Neoprene®) have good ozone resistance because of the presence of chlorine atoms in the polymer backbone (Crawford and Quinn 2017).

Other atmospheric gases degrade polymers. PE and PP become cross-linked when exposed to SO_2 in the presence of oxygen and light, resulting in a more rigid and less elastic structure (Crawford and Quinn 2017). Nylon 66 is also readily attacked by NO_2 (Jellinek 1978).

Overall, atmospheric pollutants appear to impart little effect on saturated polymers (Davis and Sims 1983).

Hydrolysis

Hydrolysis (splitting of a molecule via reaction with water) is another mechanism which accounts for formation of secondary MPs; however, this effect is not highly significant at reducing the size of plastic particles (Andrady 2011; Wang et al. 2021). When certain plastic articles are placed in water, the water molecules will diffuse into the amorphous regions of the polymer (Crawford and Quinn 2017). Diffusion can subsequently result in addition of water molecules to the polymer via cleavage of chemical bonds. In an early paper (Golike and Lasoski 1960), PET was decomposed via hydrolysis at elevated temperatures, e.g., 60–175°C. In some plastics, such as nylons, the rate of absorption is substantial. Polyurethanes are also subject to damage from the effects of moisture (Crawford and Quinn 2017). In other polymers such as polytetrafluoro-ethylene (Teflon®), absorbance is negligible.

Thermal

Elevated temperatures accelerate the rate of weathering of plastics according to the Arrhenius equation (Geburtig et al. 2019; Wang et al. 2021). Plastic particles on many beaches are exposed to high temperatures: sandy beach surfaces and plastic litter can attain temperatures of 40°C in summer. On dark soil and when plastic debris is pigmented dark, heat build-up due to solar absorption can raise plastic temperatures even higher (Shaw and Day 1994; Andrady 2011).

Under ambient conditions, the steps involved in thermal degradation of plastics are similar to those for photochemical processes (Singh and Sharma 2008). The main difference between these two mechanisms is the sequence of initiation steps leading to the auto-oxidation cycle. Another difference is that while photochemical reactions occur on particle surfaces, thermal reactions occur throughout the plastic article (Tyler 2004).

Thermal degradation consists of two distinct reactions which occur concurrently; one is random cleavage of chemical links along the length of the chain; the other is chain-end scission of C–C bonds (aka *unzipping*) (Singh and Sharma 2008). Random degradation can occur at any point along the polymer chain, that is, an active site is not needed. However, weak links in the structure (for example, a peroxide or ether link) may be particularly susceptible to initiation of depolymerization.

The polymer fragments into lower MW units with little or no liberation of monomers. In chain-end degradation, monomers are released from the end of the chain and is followed by shedding of successive monomer units. Volatile products may be generated. This mechanism occurs when chemical bonds in the backbone are weaker than those of the side groups, and when the polymer molecule carries active chain ends with a free radical or ion. In chain-end degradation, the MW of the polymer decreases gradually (Singh and Sharma 2008).

Polyolefins are documented to be susceptible to thermal oxidation due to the presence of impurities produced during manufacture at high temperatures (Khabbaz et al. 1999). PE thermally decomposes via random degradation: a hydrogen atom migrates from one carbon to another, generating two fragments (Aguado et al. 2007). Vinyl polymers such as PS and poly(acrylonitrile) are also degraded by the random chain scission process, and PET experiences the random thermal degradation route as well (Singh and Sharma 2008). Thermally oxidized polyesters produce a range of products including formaldehyde, acetaldehyde, formic acid, acetic acid, CO_2, and H_2O. In addition, small quantities of other compounds like hydroxyaldehydes, hydroxyacids, aldehyde acids, and so on, have been detected (Boenig 1999).

Mechanical

Plastic articles in the aquatic environment are subjected to a range of mechanical stresses including collisions with and abrasions from mineral particles as well as the shear forces of the water (figure 3.5) (Crawford and Quinn 2017;

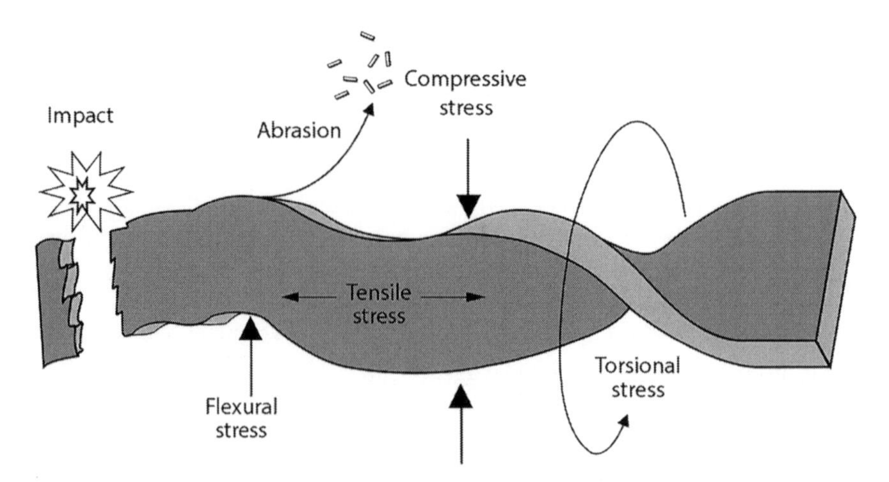

Figure 3.5 The Various Stresses That Result in Formation of MPs
Source: Adapted from *Microplastic Pollutants*, Crawford, M.B., and Quinn, B., Chapter 4, Physiochemical properties and degradation, p. 97 (2017), with permission from Elsevier.

Wang et al. 2021). Fragmentation of plastic pieces occurs in the sea swash zone (i.e., the part of a beach where water washes up after an incoming wave has broken) (Efimova et al. 2018; Wang et al. 2021). In a WWTP, fragmentation of plastic wastes occurs due to the turbulence of water flow and abrasion against mechanical components (Lv et al. 2019; Wang et al. 2021).

The mechanical fragmentation of macroplastic articles is caused primarily by formation of cracks (Enfrin et al. 2020; Julienne et al. 2019). Continued stress results in breakage of carbon-carbon bonds on the polymer backbone followed by fracture. Scanning electron microscopic images of surfaces of mechanically damaged plastics indicate the presence of fractures, horizontal notches, flakes, pits, and grooves. Such mechanically produced textures provide ideal locations for chemical weathering to occur, which further weakens the polymer surface (Cooper and Corcoran 2010).

Microbial

Microbial degradation involves the biochemical transformation of organic compounds. ASTM standard D-5488-94d defines biodegradation as "process which is capable of decomposition of materials into carbon dioxide, methane, water, inorganic compounds, or biomass in which the predominant mechanism is the enzymatic action of microorganisms, that can be measured by standard tests, in a specified period of time, reflecting available disposal conditions" (ASTM 1994). Microorganisms may play a significant role in the long-term weathering of macroplastics to smaller articles. In many cases, simple colonization by bacteria and fungi weakens the polymer. In other cases, the polymer matrix may be used as a source of carbon and nitrogen (Wang et al. 2021).

Biodegradation of polymers, whether synthetic or natural, involve several mechanisms including chain breakage and loss of substituents; reduction in average MW of polymers; alteration of surface properties; loss of mechanical strength; degradation by enzymes; and ultimate assimilation by microorganisms (Singh et al. 2008). Degradation can occur by any of the mechanisms listed above, or in combination. The metabolites released via biodegradation are expected to be nontoxic and eventually redistributed through global carbon, nitrogen, and sulfur cycles. Complete decomposition (mineralization) results in the formation of CO_2 and H_2O under aerobic conditions and CH_4 and CO_2 in anaerobic conditions.

Abiotic effects on polymers such as photooxidation, thermal oxidation, hydrolysis and others likely enhance biodegradation by increasing MP water solubility and surface area.

(Microbial decomposition of MPs will be discussed in detail in chapter 11.)

IMPORTANT SECONDARY MPs

Tire and Road Wear Particles

The road transportation sector is considered a major contributor of MPs to aquatic environments. MPs originate from several sources, including wear of vehicle tire treads, polymers incorporated into bitumen in road pavement, and thermoplastic elastomers used in road marking paints (Roychand and Pramanik 2020; Vogelsgang et al. 2019). These microparticles are sometimes referred to collectively as tire and road wear particles (TRWP).

Particles from tire treads are lost via abrasion on road surfaces (Sommer et al. 2018). When tires are in motion, wear occurs on the tread and friction heat is produced within the tire. The wear leads to liberation of rubber particles, while the increase in temperature may cause evaporation of volatile tire components (Cadle and Williams 1979; Andersson-Sköld et al. 2020).

Road wear particles (e.g., paint, bitumen additives) are generated during vehicle contact with the road surface: particles are dislodged from road surface material (asphalt or concrete) and from road markings. If the bitumen in the asphalt is polymer-modified, release of these polymers may also be a minor source of MPs (Andersson-Sköld et al. 2020)

Both TRWPs accumulate in road dust (Dehgahni et al. 2017; Vogelsang et al. 2019). According to various studies (Wagner et al. 2018; Hann et al. 2018; Unice et al. 2019; Järlskog et al. 2020), the majority of TRWP deposits adjacent to the curb or close to the road settle in nearby soil. The remainder is eventually detached and transported, and enters runoff and the stormwater system during precipitation events.

Tire tread particles may be one of the major sources of MPs to the environment (Kole et al. 2017; Sommer et al. 2018; Järlskog et al. 2020). The tread is defined as the portion of the tire that contacts the road surface and provides grip and traction. Researchers state that accurate data on the transfer of MPs from tires to global oceans is limited (Boucher and Friot 2017).

Siegfried et al. (2017) estimated that TRWPs constitute as much as 42% of the total MPs transported to the ocean, thus comprising one of the major sources of MP pollution in the marine environment (Roychand and Pramanik 2020). Based on results from several models, Sommer et al. (2018) calculated that 30 vol% of all MPs that pollute surface waters are composed of tire fragments (Järlskog et al. 2020).

Particulates from Tire Wear

Tire wear fragments have been measured to range in size from 10 nm to several 100 μm (Kole et al. 2017). This range of particle sizes is likely a function of differing tire compositions, but also of the wear process (e.g., driving behavior and driving conditions) (Wagner et al. 2018), differing road surface properties, temperatures, and meteorological conditions. Recorded variations in types and amounts of tire fragments may also be a result of differences in sampling procedures, sample preparation, analytical methods, and sampled particle sizes (Järlskog et al. 2020). Finally, TRWPs are difficult to analyze; filler materials such as carbon black occurring in tire particles cause disruptive fluorescence phenomena upon irradiation, resulting in near-complete absorption of IR light (Eisentraut et al. 2018; Wagner et al. 2018). Discussion of MP sampling and analysis appears in chapter 6.

Tire treads are formulated to impart various benefits including durability, flexibility, and grip (Camatini et al. 2001; Gieré et al. 2004; Kocher et al. 2010; Apeagyei et al. 2011; Gunawardana et al. 2012; Wu 2017; Sommer et al. 2018). The chemical composition of tires varies markedly according to brand and type (e.g., summer vs. winter; car vs. heavy duty truck). The exact composition of a tire tread is proprietary information, so only estimates can be provided herein. A complex mixture of polymers occurs in tire formulations, which comprises from 40% to 60% of the tire mass. The main component is styrene butadiene rubber, a polymer based on styrene, a precursor of polystyrene. Also present are polybutadiene and natural rubber (polyisoprene, C_5H_8). Other key additives include carbon black (20–35%) as a reinforcing agent, fillers such as SiO_2 and $CaCO_3$ (variable), oils and resins (15–20%) as softeners and extenders, and vulcanization chemicals (e.g., sulfur and zinc oxide) (1–5%). Antioxidants, antiozonants, and other additives may also be present (table 3.4) (U.S. Tire 2021). Other additives in the original rubber core may include Al, Ti, Fe, Zn, Cd, Sb, or Pb (Sommer et al. 2018).

Beyond the principal chemicals used in tire manufacture, impurities may be present in rubber polymers and additives. Polycyclic aromatic hydrocarbons (PAHs) are known to occur in extender oils used in the manufacture of new car tires and tread for new and re-treaded tires.

EU directive 2005/69/EU has restricted the levels of benzo[*a*]pyrene and eight listed PAHs in the manufacture of tires in Europe (Andersson-Sköld et al. 2020). Tire-derived MPs may contain potentially hazardous metals and metalloids contained in particles from brake abrasion (e.g., Al, Fe, Cu, Sb, or Ba). These contaminants increase the potential of public health and environmental impacts from tire-wear particles. (Sommer et al. 2018)

During laboratory-scale tire wear tests (Cadle and Williams 1979), gases containing hydrocarbons and sulfur were released from tires, along with tire particles. Hydrocarbons were identified as 1,3-butadiene, isoprene, vinyl cyclohexene, and others; these are components of the styrene-butadiene rubber copolymer. Sulfur species included sulfur dioxide, carbon disulfide, and carbonyl disulfide. These data suggest that degradation of tire rubber may occur at hot spots on the tire surface as temperature increases (Pohrt 2019).

Table 3.4 Composition and Functions of Selected Tire Tread Chemicals

Constituent	*Function*	*Examples / Chemical Forms*
Natural rubber	Provides elastic properties to tread and sidewalls. Improves traction, friction, and rolling resistance; provides comfort to user. Important for tear and fatigue crack resistance.	Polyisoprene
Synthetic rubber polymers		Butadiene rubber and styrene butadiene rubber (the most common); halogenated polyisobutylene rubber (halobutyl rubber).
Fillers	Improves tensile strength, prevent tearing and abrasion. Improves rolling resistance, wear resistance, and traction, and overall longevity of tires. Carbon black is effective in acting against UV rays to prevent rubber from fissuring and cracking.	Carbon black, amorphous silica, calcium carbonate, calcium and aluminum silicates, barium sulfate, clays, resins.
Vulcanization curing agents	Converts rubber to a solid product through vulcanization (i.e., curing of the rubber) by cross-linking polymer chains	Sulfur
Vulcanization activators	Activates and aids the curing process during vulcanization	Zinc oxide, stearic acid, resins
Vulcanization accelerators	Speeds the vulcanization process	Sulfenamides (N-cyclohexyl-2-benzothiazole-sulfe namide)
Inhibitors (pre-vulcanization inhibitors)	Inhibits premature vulcanization	Nitrosodiphenylamine, salicylic acid, benzoic acids
Antioxidants	Prevents rubber from degrading due to oxygen, ozone pollution, and excessively high temperatures.	Trimethylquinoline, amines (especially paraphenylene diamines), phenols, quinoline, phosphites
Antiozonants	Protects synthetic rubber in the tire tread and sidewalls from ozone cracking; protects natural rubber compounds from flex-cracking and heat degradation.	Mono-crystalline waxes and chemical antiozonants, e.g., *p*-phenylenediamines, alkylaryl-*p*-phenylenediamines

Peptizers	Accelerates the mastication process (i.e., the process where the viscosity of natural rubber is reduced to the desired level). Peptizers remove free radicals formed during mixing of the elastomer; this allows for reduction in compound viscosity and enables incorporation of compounding materials.	Zinc salts of pentachlorophenol, pentachlorothiophenol, phenylhydrazine, certain diphenylsulfides, and xylyl mercaptan.
Plasticizers	Makes rubber softer and more flexible. Reduces viscosity during processing.	Aromatic petroleum process oils, mild extracted solvate oil, rapeseed oil, resins
Other processing agents	Aids mixing and processing (e.g., dispersing agents)	Resin from coniferous trees, synthetic resins, process oils, pine tar, fatty acids, esters of fatty acids.

Plastics in Road Construction

In recent decades polymers have been incorporated into bitumen in road pavement at rates of approximately 5–10% by weight, with resultant improvements in pavement stability, strength, and durability (Sasidharan et al. 2019). Given that waste plastic typically serves as the feedstock as a road-building ingredient, chemical composition will vary with location.

Road-Marking Materials

Road-marking products consist of plastic polymers, pigments, fillers, and various additives. Glass beads are often incorporated to provide reflective properties to the material. Some of the most common road-marking products appear in table 3.5.

Thermoplastic Road-Marking Paint

Thermoplastic road-marking paint, also known as hot melt marking paint, consists of a binding agent of thermoplastic polymers mixed with pigment, filler, glass beads, and other additives (Babić et al. 2015; Chu 2021). The thermoplastic polymers can be either hydrocarbon resins or alkyd resins (Babić et al. 2015) or a mix of these. Depending on climate, ethylene vinyl acetate copolymer (EVA) may be included as part of the thermoplastic binding agent in order to improve road wear resistance and prevent crack formation (Andersson-Sköld et al. 2020) and to make the material more elastic. Styrene block copolymers may also be used (BEC Materials 2019).

An example of a thermoplastic mixture for road marking may be as follows (Chu 2021):

- 16.5% binding agent (pentaerythritol rosin ester, C_5-hydrocarbon resin, or a mixture of these)
- 10% pigment (titanium dioxide)
- 49.5% filler (e.g., 22% calcium carbonate [or calcium magnesium carbonate; dolomite] and 27.5% quartz sand)
- 4% additives (2% phthalate plasticizers [DOP], 1% PE wax, 1% hydrogenated castor oil)
- 20% glass beads

The thermoplastic material occurs in solid powder form. A hot melt kettle heats the thermoplastic to about 200°C (392°F) (Geveko 2019), following which it is applied to the road surface at a thickness of between 2 and 4 mm (Vägmarkeringar 2019); alternatively, it is sprayed on to the road in thin layers (Andersson-Sköld et al. 2020). The marking then cools and hardens.

Table 3.5 Chemical Composition of Some Prominent Road-Marking Materials

Road-Marking Product	*Composition*
Thermoplastic mixtures	Binding agents: hydrocarbon resins (usually C_5 aliphatic hydrocarbon resin) or alkyd resins (i.e., organic polyesters) (such as pentaerythritol rosin ester or maleic acid-modified rosin esters) or a mixture of these. Some mixtures contain ethylene vinyl acetate copolymer or styrene block copolymers Other additives include pigments (TiO_2), fillers ($CaCO_3$, SiO_2), and glass beads.
Water-based paints	Pigments, polymer resin, extenders, fillers, dispersants, film formers, biocides, etc. in a medium composed of emulsion polymers, water, and co-solvents.
Solvent-based paints	Dispersions of primary pigments, extenders and specialty chemicals (driers and skinning agents, etc.) in a dispersion medium of alkyd resins and fast evaporating solvents. The extenders used are calcium carbonate, calcined clays, and silica. Solvents are a combination of aromatic (toluene, xylene) and aliphatic (heptane and naphtha) compounds. Binding agents are thermoplastic acrylic reins and styrene acrylic mixes.
Two-component systems	Formed by mixing a pigmented base resin with a polymer curing agent. One component is composed of acrylic resin (e.g. methyl methacrylate), pigments, fillers, and additives. Second component contains active resins, additives, and curing agents (e.g., dibenzoyl peroxide).
Road-marking tape	Consists of marking tape and thermoplastic (e.g., polyurethane) paint. May include plasticizers, resins, synthetic rubber. Glass beads or micro-crystalline ceramic beads may be included to add reflective properties. Some are pre-coated with self-bonding adhesive.

Common colors of road lines are yellow and white. White pigments are composed primarily of titanium dioxide, zinc oxide, and lithopone, and yellow pigment is mainly heat-resistant yellow lead (Babić et al. 2015; Dubey 1999).

Reflective glass beads are added to thermoplastic paint in order to improve the identification of the lines at night and to improve the anti-skid performance, brightness, and durability of the marking paint.

Water-Based Paints

Water-borne road-marking paints comprise approximately 90% of all traffic paints in the United States. This proportion is 70% and 15% for Australia and Europe, respectively (Fatemi et al. 2006).

These traffic paints are dispersions of pigments, extenders, fillers, and specialty chemicals (dispersants, defoamers, adhesion promoters, film formers, biocides) in a medium composed of emulsion polymers, water, and co-solvents (Dubey 1999). The resin used in water-based road-marking paints is typically a thermoplastic acrylic resin in a water-based emulsion (Babić et al. 2015; Chu 2021). Volatile organic compounds occur at levels of usually less than 2% and originate from the required chemical additives (Babić et al. 2015). The paint is applied at a thickness of between 0.4 and 0.6 mm using a high-pressure machine (Vägmarkeringar 2019).

Solvent-Based Paints

In solvent-based road-marking paints, the resin (binding agent) is dissolved in an organic solvent such as esters or ketones. The binding agent is usually a thermoplastic acrylic resin (Babić et al. 2015), although styrene-acrylic resins are also used.

Two-Component Paints

Two-component systems, also termed cold plastic or cold-applied plastic, consist of two or more components (a polymer resin and a curing agent), which form a thermoset when mixed. There are different types of two-component systems; they are usually either acrylic-based or epoxy-based (Babić et al. 2015).

Road-Marking Tape

Road-marking tape is available either as permanent tape or as a temporary, removable tape. In permanent, prefabricated tape, glass beads or microcrystalline ceramic beads are embedded in the surface (Lopez 2004; 3M 2019). Common binding agents include polyurethane in combination with a flexible polymer (e.g., rubber) (Lopez 2004; 3M 2019). Temporary road-marking tape, used for short-term applications, are thin and consist of a metal foil backing coated with an adhesive on the underside and a pigmented binding agent with glass beads on the upper surface (Lopez 2004).

Hazardous Substances in Road-Marking Products

A number of additives (binding agents, pigments, solvents) in manufacture of road-marking products are potentially hazardous to human health and the environment. The binding agents that pose greatest public health and environmental hazards are polyurethane and epoxy resins and their curing agents (Andersson-Sköld et al. 2020). Lead chromate had historically been used as the pigment for yellow road markings, and is still employed in several countries. The most commonly used pigment for white road markings is titanium dioxide, and sometimes zinc oxide or sulfide, and barium sulfate are used. Some hazardous aromatic solvents are still used in solvent-based paint in certain countries. Glass beads, often formulated from waste glass, can contain elevated levels of lead, arsenic, and antimony (dos Santos et al. 2013).

In many nations, only a small fraction of total stormwater volume is treated in WWTPs or local treatment facilities (Magnusson et al. 2016). This implies that MPs from many urban areas are discharged directly into receiving waterways (Bondelind et al. 2018). It is estimated that road-related MPs, transported with stormwater, account for significant proportions of the total MP input into oceans (Järlskog et al. 2020; Boucher and Friot 2017; Siegfried et al. 2017).

Microfibers

Synthetic microfibers (MFs) are universally composed of polymers including nylon, polyester, rayon, PET, PP, acrylic, or spandex. Disagreement exists regarding the appropriate fiber diameter for purposes of defining MFs—some researchers state that MFs measure less than 5 mm in diameter (Singh et al. 2020; Jerg and Baumann 1990), while others use 10 μm as the cutoff point (Gago et al. 2018).

MFs are considered a major component of aquatic MP pollution (Browne et al. 2011)—a strong correlation has been found between population density of a coastal area and extent of MF pollution. They have been detected in freshwater habitats (Eerkes-Medrano et al. 2018a; Li et al. 2018) and drinking water (Eerkes-Medrano et al. 2018b; Pivokonsky et al. 2018). In marine habitats, MFs are found on the sea surface (Eriksen et al. 2014; Lebreton et al. 2018), along shorelines (Barrows et al. 2018; Browne et al. 2011), in deep-sea sediment (Sanchez-Vidal et al. 2018), and in Arctic sea ice (Obbard et al. 2014; Peeken et al. 2018).

MF release from textiles has been studied at different stages in its life cycle, including manufacture, usage, and disposal.

Manufacture

The degree to which the textile release fibers during manufacturing is a function of variables including yarn type, fabric type (i.e., woven, knit, nonwoven), texture (open or dense), and the nature and number of the different fiber types involved (Hernandez et al. 2017). During the spinning of yarn, a number of rough mechanical actions can cut and break down fibers. In consequence, fiber fragments become embedded in the yarn and are later washed out. During clothing manufacture, higher MF generation was reported in wet processing stages (scouring, bleaching, dyeing, finishing) (Ramasamy and Subramanian 2021). Physical agitation, mechanical stress, and chemical treatments may be significant in this stage. Significant quantities of MFs have been determined in wastewater even following effluent treatment (Zhou et al. 2020; Ramasamy and Subramanian 2021).

Washing

The primary contributor of MFs to water is domestic washing of synthetic clothing (Mishra et al. 2019). Washing is more responsible for fabric damage than is use: as much as 90% of fiber attrition occurs during this stage (Mohamed 1982). It has been estimated that one 6-kg household wash has the potential to release up to 700,000 fibers (Napper and Thompson 2016); however, different fabric types release differing quantities of MFs. For example, a single washing can release more than 1,900 polyester (PET) fibers (Browne et al. 2011). One fleece garment (usually made from PET) can release about 110,000 fibers (Carney Almroth et al. 2018). An estimated 6,000,000 MFs may be lost from polyester fabrics during washing a typical 5-kg wash load, depending on type of detergent used (De Falco et al. 2018).

The quantity of fibers released depends on additional factors including temperature, time, washing speed, and the type of detergents and fabric softeners used (Cristaldi et al. 2020). (Carney Almroth et al. 2018; de Falco et al. 2018). Wash temperature affects fiberglass transition temperature, fiber swelling, and water diffusion rate between fiber strands (Hernandez et al. 2017). Fabric wetting and fiber swelling generally occurs within 30 seconds of immersion in the wash water and immediately places strain across the fabric network. Water hardness and laundering products may affect the rate of water diffusion into the fiber.

Laundry detergents significantly alter the physical effects of mechanical action on both natural and synthetic fibers. The churning motion of the surfactant (detergent) solution produces foam, which protects the fabric against impaction and friction during washing, thereby limiting fabric damage. Secondly, surfactants are deposited on the fiber surface, which reduces frictional forces. This serves to lubricate fibers and mitigate damage to fibers (Hernandez et al. 2017). The relatively hydrophobic synthetic fibers are not markedly affected by alkaline detergent solutions when washed at or below recommended temperatures. Hernandez et al. (2017) did not observe a significant difference between different detergents on shedding of MFs. There was no increase in fiber release over multiple wash cycles; in other words, fabric will release an equivalent quantity of mass during the first wash and in later washes. Such findings imply that the repetitive stress of washing does not mechanically degrade the fibers to the point that they release more fibers over time.

Disposal

Another key source of MF pollution is disposal of domestic waste (Liu et al. 2021). WWTPs are capable of removing MFs to a large extent; however, large quantities nevertheless escape these facilities. The sludge collected from WWTPs is often disposed in landfills. Depending on landfill management practices, MFs can be lost from waste piles as runoff and/or leach to groundwater.

Research collated by the Ellen MacArthur Foundation (2017) suggests that clothing production worldwide has doubled over the past 15 years, with garments being worn much less and discarded more rapidly than ever. This has been driven by a growing middle-class population worldwide, and increased per capita sales in developed nations. So-called fast fashion has motivated consumers to purchase more clothing per capita than during previous generations.

Fast fashion relies on cheap manufacturing, frequent consumption, and short-lived garment use (Niinimäki et al. 2020). Fast fashion textiles tend to be cheaper and trend-specific, thus resulting in more rapid disposal (Ramasamy and Subramanian 2021). The fashion industry overproduces products by about 30–40% each season, and large quantities of unsold or short-lived clothing are landfilled (Song et al. 2013).

Implications

It is estimated that synthetic MFs and nanofibers textiles, transported with wastewater, account for approximately 35% of MPs in freshwater and marine environments (Boucher and Friot 2017). Of all MFs entering the aquatic environment, Asian nations contribute an estimated 65% of the total release (Belzagui et al. 2020), most likely a combined effect of manufacture, use, and disposal. Using conservative estimates, 4 billion fibers·km^{-2} may be present in Indian Ocean sediment (Woodall et al. 2014).

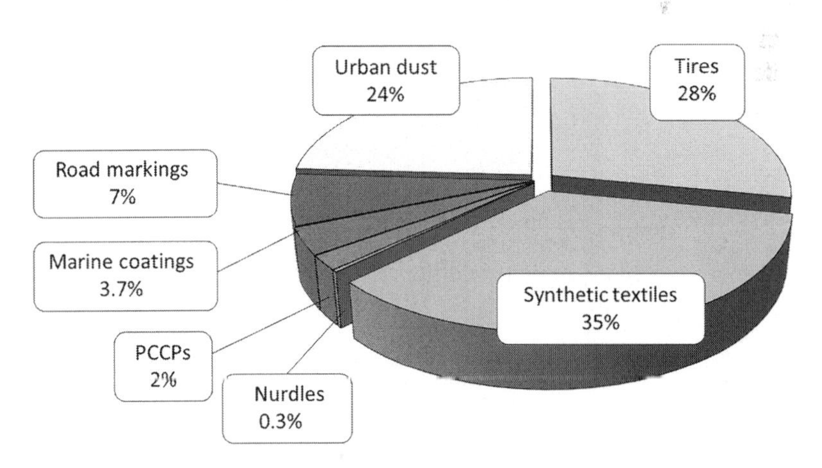

Figure 3.6 Releases of MPs to Oceans
Source: Reproduced with permission from Boucher, J. and Friot D. (2017). *Primary Microplastics in the Oceans: A Global Evaluation of Sources*. Gland, Switzerland: IUCN. https://doi.org/10.2305/IUCN.CH.2017.01.en

Studies under carefully controlled conditions are necessary in order to better understand the mechanisms involved in fiber shedding during use. Factors to consider include fiber production (extruded filaments or wound staples), fabric structure (woven, knitted, non-woven), washing conditions (temperature, detergent type, length of washing, multiple versus single washes), and size distribution of fibers shed during use (Hernandez et al. 2017).

CLOSING STATEMENTS

According to recent studies, about 35% of MP releases are due to laundering of synthetic textiles and about 28% from tire wear. The third important contribution (24.2%) is urban dust (Boucher and Friot 2017). The percentage contribution of MP types to the environment is shown in figure 3.6.

REFERENCES

3M. (2010). Market test specification for durable, retroreflective, preformed, patterned pavement markings with improved pre-coated pressure sensitive adhesive. 3M™ Stamark™ Tape Series 270 ES. May 2010 [accessed 2022 February 2].

3M. (2019). 3M™ Stamark™ pavement marking tape. Series 380 [accessed 2020 September 20]. https://multimedia.3m.com/mws/media/849277O/3m-stamark-pavement-marking-tape-series-380-flyer.pdf.

ACS (American Chemical Society). (2022). What's that stuff? Road markings [accessed 2022 February 3]. https://cen.acs.org/articles/88/i36/Road-Markings.html.

Aguado, J., Serrano, D.P., and Miguel, G.S. (2007). European trends in the feedstock recycling of plastic wastes. *Global Nest Journal* 9(1):12–19.

Alger, M. (1997). *Polymer Science Dictionary*, 2nd ed. London: Chapman & Hall.

Almroth, B.M.C., Åström, L., Roslund, S., Petersson, H., Johansson, M., and Persson, N.-K. (2018). Quantifying shedding of synthetic fibers from textiles; A source of microplastics released into the environment. *Environmental Science and Pollution Research* 25(2):1191–1199.

Andersson-Sköld, Y., Johannesson, M., Gustafsson, M., Järlskog, I., Lithner, D., Polukarova, M., and Strömvall, A.-M. (2020). *Microplastics From Tyre and Road Wear: A Literature Review*. VTI Rapport. 10281. Gothenburg (Sweden): Chalmers University of Technology.

Andrady, A.L. (2011). Microplastics in the marine environment. *Marine Pollution Bulletin* 62(8):1596–1605.

Ansaldi, A. (2005). Porous entrapment spheres as delivery vehicles. In: *Delivery System Handbook for Personal Care and Cosmetic Products*. Amsterdam (Netherlands): Elsevier, pp. 323–332.

Apeagyei, E., Bank, M.S., and Spengler, J.D. (2011). Distribution of heavy metals in road dust along an urban-rural gradient in Massachusetts. *Atmospheric Environment* 45(13):2310–2323.

ASCE (American Society of Civil Engineers). (2017). *Infrastrure Report Card 2017*. Reston (VA): American Society of Civil Engineers.

Babić, D., Burghardt, T.E., and Babić, D. (2015). Application and characteristics of waterborne road marking paint. *International Journal for Traffic and Transport Engineering* 5(2):150–169.

Barnes, D.K., Galgani, F., Thompson, R.C., and Barlaz, M. (2009). Accumulation and fragmentation of plastic debris in global environments. *Philosophical Transactions of the Royal Society B: Biological Sciences* 364(1526):1985–1998.

Barrows, A., Cathey, S.E., and Petersen, C.W. (2018). Marine environment microfiber contamination: Global patterns and the diversity of microparticle origins. *Environmental Pollution* 237:275–284.

BEC Materials. (2019). Global styrenic block copolymers (SBC) market to grow at 5.53% CAGR by 2023. http://www.becmaterials.com/news/183-global sbc market forecast 2017 2023.

Belzagui, F., Gutiérrez-Bouzán, C., Álvarez-Sánchez, A., and Vilaseca, M. (2020). Textile microfibers reaching aquatic environments: A new estimation approach. *Environmental Pollution* 265:114889.

BIFA (British International Freight Association). (2020). Containers lost at sea – 2020 update [accessed 2021 October 10]. https://www.bifa.org/news/articles/2020/jul/containers-lost-at-sea-2020-update.

Boenig, H.V. (1964). *Unsaturated Polyesters: Structure and Properties*. Amsterdam (Netherlands): Elsevier.

Bondelind, M., Nguyen, A., Sokolova, E., and Björklund, K. (2018). Transport of traffic-related microplastic particles in receiving water. In: Mannina, G., editor. *New Trends in Urban Drainage Modelling*. UDM 2018. Green Energy and Technology. Berlin (Germany): Springer. DOI: 10.1007/978-3-319-99867-1_53.

Boucher, J., and Friot, D. (2017). *Primary Microplastics in the Oceans: A Global Evaluation of Sources*. Vol. 10. Switzerland: IUCN Gland.

Bracco, P., Costa, L., Luda, M.P., and Billingham, N. (2018). A review of experimental studies of the role of free-radicals in polyethylene oxidation. *Polymer Degradation and Stability* 155:67–83.

Brahney, J., Hallerud, M., Heim, E., Hahnenberger, M., and Sukumaran, S. (2020). Plastic rain in protected areas of the United States [accessed 2022 March 30]. *Science* 368(6496):1257–1260. DOI: 10.1126/science.aaz5819.

Braun, D., Cherdron, H., Rehahn, M., Ritter, H., and Voit, B. (2005). *Polymer Synthesis: Theory and Practice—Fundamentals, Methods, Experiments*, 4th ed. Berlin/Heidelberg (Germany): Springer-Verlag.

Brenntag.com. (2021). Chemicals in rubber tires [accessed 2021 December 8]. https://www.brenntag.com/en-us/industries/rubber/chemicals-in-rubber-tires.

Browne, M.A., Crump, P., Niven, S.J., Teuten, E., Tonkin, A., Galloway, T., and Thompson, R. (2011). Accumulation of microplastic on shorelines woldwide: Sources and sinks. *Environmental Science & Technology* 45(21):9175–9179.

Burick, T. (2020). No cleanup planned as millions of plastic pellets wash up along Mississippi River and flow to the Gulf [accessed 2020 August 24]. Nola.com. https://www.nola.com/news/environment/article_b4fba760-e18d-11ea-9b0b-b3a2123cf48b.html.

Cadle, S., and Williams, R. (1979). Gas and particle emissions from automobile tires in laboratory and field studies. *Rubber Chemistry and Technology* 52(1):146–158.

California Regional Water Quality Control Board, San Francisco Bay Region. (2011). Cleanup and Abatement Order No. R2-2011-033 [accessed 2020 August 9].

Camatini, M., Crosta, G.F., Dolukhanyan, T., Sung, C., Giuliani, G., Corbetta, G.M., Cencetti, S., and Regazzoni, C. (2001). Microcharacterization and identification of tire debris in heterogeneous laboratory and environmental specimens. *Materials Characterization* 46(4):271–283.

Chu, H.-M. (2019). *Thermoplastic Pavement Marking Technology: TPM-Advanced*, 4th ed. [Location Unknown]: Heng-mo Chu.

Chu, H.-M. (2020). *Thermoplastic Pavement Marking Technology: TPM-Advanced.* [Location Unknown]: Heng-mo Chu.

Cosmetic Ingredient Review. (2011). *Silylates and Surface Modified Siloxysilicates as Used in Cosmetics. Final Safety Assessment.* Washington, D.C.

Cosmetic Ingredient Review. (2012). *Safety Assessment of Modified Terephthalate Polymers as Used in Cosmetics.* Tentative Report for Public Review. Washington, D.C.

Crawford, M.B., and Quinn, B. (2017). *Microplastic Pollutants.* Amsterdam (Netherlands): Elsevier.

Cristaldi, A., Fiore, M., Zuccarello, P., Oliveri Conti, G., Grasso, A., Nicolosi, I., Copat, C., and Ferrante, M. (2020). Efficiency of wastewater treatment plants (WWTPs) for microplastic removal: A systematic review. *International Journal of Environmental Research and Public Health* 17(21):8014.

Cunliffe, A.V., and Davis, A. (1982). Photo-oxidation of thick polymer samples—Part II: The influence of oxygen diffusion on the natural and artificial weathering of polyolefins. *Polymer Degradation and Stability* 4(1):17–37.

Davis, A., and Sims, D. (1983). *Weathering of Polymers.* Berlin/Heidelberg (Germany): Springer Science + Business Media.

De Falco, F., Gullo, M.P., Gentile, G., Di Pace, E., Cocca, M., Gelabert, L., Brouta-Agnesa, M., Rovira, A., Escudero, R., Villalba, R., Mossotti, R., Montarsolo, A., Gavignano, S., Tonin, C., and Villalba, R. (2018). Evaluation of microplastic release caused by textile washing processes of synthetic fabrics. *Environmental Pollution* 236:916–925.

Dehghani, S., Moore, F., and Akhbarizadeh, R. (2017). Microplastic pollution in deposited urban dust, Tehran metropolis, Iran. *Environmental Science and Pollution Research* 24(25):20360–20371.

dos Santos, É.J., Herrmann, A.B., Prado, S.K., Fantin, E.B., dos Santos, V.W., de Oliveira, A.V.M., and Curtius, A.J. (2013). Determination of toxic elements in glass beads used for pavement marking by ICP OES. *Microchemical Journal* 108:233–238.

Eerkes-Medrano, D., Leslie, H.A., and Quinn, B. (2019). Microplastics in drinking water: A review and assessment. *Current Opinion in Environmental Science & Health* 7:69–75.

Eerkes-Medrano, D., Thompson, R.C., and Aldridge, D.C. (2015). Microplastics in freshwater systems: A review of the emerging threats, identification of knowledge gaps and prioritisation of research needs. *Water Research* 75:63–82.

Efimova, I., Bagaev, M., Bagaev, A., Kileso, A., and Chubarenko, I.P. (2018). Secondary microplastics generation in the sea swash zone with coarse bottom sediments: Laboratory experiments. *Frontiers in Marine Science* 5:1–15.

Eisentraut, P., Dümichen, E., Ruhl, A.S., Jekel, M., Albrecht, M., Gehde, M., and Braun, U. (2018). Two birds with one stone—Fast and simultaneous analysis of microplastics: Microparticles derived from thermoplastics and tire wear. *Environmental Science & Technology Letters* 5(10):608–613.

Enfrin, M., Lee, J., Gibert, Y., Basheer, F., Kong, L., and Dumee, L.F. (2020). Release of hazardous nanoplastic contaminants due to microplastics fragmentation under shear stress forces. *Journal of Hazardous Materials* 384:121393.

Eriksen, M., Lebreton, L.C., Carson, H.S., Thiel, M., Moore, C.J., Borerro, J.C., Galgani, F., Ryan, P., and Reisser, J. (2014). Plastic pollution in the world's oceans: More than 5 trillion plastic pieces weighing over 250,000 tons afloat at sea. *PLoS ONE* 9(12):e111913.

Fatemi, S., Khakbaz Varkani, M., Ranjbar, Z., and Bastani, S. (2006). Optimization of the water-based road-marking paint by experimental design, mixture method. *Progress in Organic Coatings* 55:337–344.

Fazey, F.M.C., and Ryan, P.G. (2016). Biofouling on buoyant marine plastics: An experimental study into the effect of size on surface longevity. *Environmental Pollution* 210:354–360.

Fendall, L.S., and Sewell, M.A. (2009). Contributing to marine pollution by washing your face: Microplastics in facial cleansers. *Marine Pollution Bulletin* 58(8):1225–1228.

Fernández, S. (2019). Plastic company set to pay $50 million settlement in water pollution suit brought on by Texas residents. *The Texas Tribune*, October 15, 2019.

Fiume, M.Z. (2002). Final report on the safety assessment of acrylates copolymer and 33 related cosmetic ingredients. *International Journal of Toxicology* 21:1–50.

Fotopoulou, K.N., and Karapanagioti, H.K. (2012). Surface properties of beached plastic pellets. *Marine Environmental Research* 81:70–77.

Franklin, I. (2021). Shipping firms experience a sharp rise in containers lost at sea. https://global.lockton.com/gb/en/news-insights/shipping-firms-experience-a-sharp-rise-in-containers-lost-at-sea.

Gago, J., Carretero, O., Filgueiras, A., and Viñas, L. (2018). Synthetic microfibers in the marine environment: A review on their occurrence in seawater and sediments. *Marine Pollution Bulletin* 127:365–376.

Geburtig, A., Wachtendorf, V., and Trubiroha, P. (2019). Exposure response function for a quantitative prediction of weathering caused aging of polyethylene. *Materials Processing* 61:517–526.

Geveco Markings. (2019). Hot applied thermoplastics [accessed 2021 October 4]. https://www.geveko-markings.com/products/hot-applied-thermoplastics.

Gewert, B., Plassmann, M.M., and MacLeod, M. (2015). Pathways for degradation of plastic polymers floating in the marine environment. *Environmental Sciences: Processes and Impacts* 17:1513–1521.

Gieré, R., LaFree, S.T., Carleton, L.E., and Tishmack, J.K. (2004). Environmental impact of energy recovery from waste tyres. *Geological Society, London, Special Publications* 236(1):475–498.

Gijsman, P., and Diepens, M. (2009). Photolysis and photooxidation in engineering plastics. In: Celine, M.C., Wiggins, J.S., and Billingham, N.C., editors. *Polymer Degradation and Performance.* ACS Symposium Series 1004. Washington, DC: American Chemical Society.

Golike, R.C., and Lasoski, S.W. (1960). Kinetics of hydrolysis of polyethylene terephthalate films. *Journal of Physical Chemistry* 64:895–898.

Grand View Research. (2019). *Traffic Road Marking Coatings: Market Estimates and Forecasts to 2025.* San Francisco (CA): Grand View Research, Inc.

Grady, D. (2015). The unifying role of dissipative action in the dynamic failure of solids. *Procedia Engineering* 103:143–150.

Gregory, M.R. (1996). Plastic 'scrubbers' in hand cleansers: A further (and minor) source for marine pollution identified. *Marine Pollution Bulletin* 32(12):867–871.

Gregory, M.R., and Andrady, A.L. (2003). Plastics in the environment. In: Andrady, A.L., editor. *Plastics and the Environment.* New York: John Wiley & Sons, pp. 379–401.

Guerrica-Echevarría, G., and Eguiazábal, J. (2009). Structure and mechanical properties of impact modified poly (butylene terephthalate)/poly (ethylene terephthalate) blends. *Polymer Engineering & Science* 49(5):1013–1021.

Gunawardana, C., Goonetilleke, A., Egodawatta, P., Dawes, L., and Kokot, S. (2012). Source characterisation of road dust based on chemical and mineralogical composition. *Chemosphere* 87(2):163–170.

Guo, X., and Wang, J. (2019). The chemical behaviors of microplastics in marine environment: A review. *Marine Pollution Bulletin* 142:1–14.

Habib, R.Z., Thiemann, T., and Al Kendi, R. (2020). Microplastics and wastewater treatment plants—A review. *Journal of Water Resource and Protection* 12(1):1.

Hann, S., Sherrington, C., Jamieson, O., Hickman, M., Kershaw, P., Bapasola, A., and Cole, G. (2018). Investigating options for reducing releases in the aquatic environment of microplastics emitted by (but not intentionally added in) products. *Report for DG Environment of the European Commission* 335.

Hardesty, B.D., Polidoro, B., Compa, M., Shim, W.J., Widianarko, B., and Wilcox, C. (2019). Multiple approaches to assessing the risk posed by anthropogenic plastic debris. *Marine Pollution Bulletin* 141:188–193.

Hernandez, E., Nowack, B., and Mitrano, D.M. (2017). Polyester textiles as a source of microplastics from households: A mechanistic study to understand microfiber release during washing. *Environmental Science & Technology* 51(12):7036–7046.

Hubbs, A.F., Mercer, R.R., Benkovic, S.A., Harkema, J., Sriram, K., Schwegler-Berry, D., and Sargent, L.M. (2011). Nanotoxicology—A pathologist's perspective. *Toxicologic Pathology* 39(2):301–324.

Hüffer, T., Wagner, S., Reemtsma, T., and Hofmann, T. (2019). Sorption of organic substances to tire wear materials: Similarities and differences with other types of microplastic. *TrAC Trends in Analytical Chemistry* 113:392–401.

Järlskog, I., Strömvall, A.-M., Magnusson, K., Gustafsson, M., Polukarova, M., Galfi, H., Aronsson, M., and Andersson-Sköld, Y. (2020). Occurrence of tire and bitumen wear microplastics on urban streets and in sweeps and and washwater. *Science of the Total Environment* 729:138950.

JDSUPRA. (2020). Resource Conservation and Recovery Act/Clean Water Act citizen suit action: Southern Environmental Law center Alleges violations by Charleston county, South Carolina, Plastic-Pellet Packager [accessed 2020 August 11]. https://www.jdsupra.com/legalnews/resource-conservation-and-recovery-act-41749.

Jellinek, H.H.G. (1978). *Aspects of Degradation and Stabilization of Polymers*. New York: Elsevier.

Jerg, G., and Baumann, J. (1990). Polyester microfibers – A new generation of fabrics. *Textile Chemist and Colorist* 22(12):12–14.

Johnson, S. (2019). Two years on, multiple investigations have shed little light on Warrnambool's nurdle spill [accessed 2020 August 24]. https://www.abc.net.au/news/2019-10-04/two-years-no-no-light-shed-on-warrnambool-plastic-spill/11555394.

Julienne, F., Lagarde, F., and Delorme, N. (2019). Influence of the crystalline structure on the fragmentation of weathered polyolefines. *Polymer Degradation and Stability* 170:109012.

Kalčíková, G., Alič, B., Skalar, T., Bundschuh, M., and Gotvajn, A.Ž. (2017). Wastewater treatment plant effluents as source of cosmetic polyethylene microbeads to freshwater. *Chemosphere* 188:25–31.

Kershaw, P. (2015). Sources, fate and effects of microplastics in the marine environment: A global assessment. IMO/FAO/UNESCO-IOC/UNIDO/WMO/IAEA/UN/UNEP/UNDP Joint Group of Experts on the Scientific Aspects of Marine Environmental Protection. Rep. Stud. GESAMP No. 90, p. 96.

Khabbaz, F., Albertsson, A.C., and Karlsson, S. (1999). Chemical and morphological changes of environmentally degradable polyethylene films exposed to thermo-oxidation. *Polymer Degradation and Stability* 63:127–138.

Kocher, B. (2010). Stoffeinträge in den Straßenseitenraum-Reifenabrieb.

Kockott, F. (2017). Government probes KZN cargo spill [accessed 2020 August 11]. GroundUp. https://www.groundup.org.za/article/government-probes-kzn-cargo-spill.

Kole, P.J., Löhr, A.J., Van Belleghem, F.G., and Ragas, A.M. (2017). Wear and tear of tyres: A stealthy source of microplastics in the environment. *International Journal of Environmental Research and Public Health* 14(10):1265.

Kreider, M.L., Panko, J.M., McAtee, B.L., Sweet, L.I., and Finley, B.L. (2010). Physical and chemical characterization of tire-related particles: Comparison of particles generated using different methodologies. *Science of the Total Environment* 408(3):652–659.

Kvitnitsky, E., Lerner, N., and Shapiro, Y.E. (2005). Tagravit™ microcapsules as controlled drug delivery devices and their formulations. In: *Delivery System Handbook for Personal Care and Cosmetic Products*. Amsterdam (Netherlands): Elsevier, pp. 215–258.

La Mantia, F. (2002). *Handbook of Plastics Recycling*. Shrewsbury (UK): Smithers Rapra Publishing.

Lebreton, L., Slat, B., Ferrari, F., Sainte-Rose, B., Aitken, J., Marthouse, R., Hajbane, S., Cunsolo, S., Schwarz, A., Levivier, A., Noble, P., Debeljak., P., Maral, H., Schoeneich-Argent, R., Brambini, R., and Reisser, J. (2018). Evidence that the Great Pacific Garbage Patch is rapidly accumulating plastic. *Scientific Reports* 8(1):1–15.

Leslie, H. (2014). *Review of Microplastics in Cosmetics; Scientific Background on a Potential Source of Plastic Particulate Marine Litter to Support Decision-Making*, Vol. 33. Amsterdam (Netherlands): Institute for Environmental Studies.

Leslie, H. (2015). *Plastic in Cosmetics: Are We Polluting the Environment Through Our Personal Care? Plastic Ingredients That Contribute to Marine Microplastic Litter*. United Nations Environment Programme. ISBN: 978-92-807-3466-9.

Li, J., Liu, H., and Chen, J.P. (2018). Microplastics in freshwater systems: A review on occurrence, environmental effects, and methods for microplastics detection. *Water Research* 137:362–374.

Lidert, Z. (2005a). Microencapsulation: An overview of the technology landscape. In: *Delivery System Handbook for Personal Care and Cosmetic Products*. Amsterdam (Netherlands): Elsevier, pp. 181–190.

Lidert, Z. (2005b). Microencapsulation: An overview of the technology landscape. In: *Delivery System Handbook for Personal Care and Cosmetic Products*. Amsterdam (Netherlands): Elsevier, pp. 215–258.

Link-Wills, K. (2021). Ocean container losses topple annual average in 2 months [accessed 2021 October 10]. https://www.freightwaves.com/news/ocean-container-losses-topple-annual-average-in-2-months.

Lithner, D., Larrson, Å., and Dave, G. (2011). Environmental and health hazard ranking and assessment of plastic polymers based on chemical composition. *Science of the Total Environment* 409(18):3309–3324.

Liu, J., Liang, J., Ding, J., Zhang, G., Zeng, X., Yang, Q., Zhu, B., and Gao, W. (2021). Microfiber pollution: An ongoing major environmental issue related to the sustainable development of textile and clothing industry. *Environment, Development and Sustainability* 23:11240–11256.

Liu, X., Yuan, W., Di, M., Li, Z., and Wang, J. (2019). Transfer and fate of microplastics during the conventional activated sludge process in one wastewater treatment plant of China. *Chemical Engineering Journal* 362:176–182.

Lopez, C.A. (2004). *Pavement Marking Handbook.* Austin (TX): Texas Department of Transportation.

Lozano, R., and Mouat, J. (2009). Marine litter in the North-East Atlantic Region: Assessment and priorities for response. KIMO International. 127.

Lv, X., Dong, Q., Zuo, Z., Liu, Y., Huang, X., and Wu, W.-M. (2019). Microplastics in a municipal wastewater treatment plant: Fate, dynamic distribution, removal efficiencies, and control strategies. *Journal of Cleaner Production* 225:579–586.

MacArthur, F.E. (2017). *A New Textiles Economy: Redesigning Fashion's Future*. Ellen MacArthur Foundation, pp. 1–150.

Magnusson, K., Eliaeson, K., Fråne, A., Haikonen, K., Olshammar, M., Stadmark, J., and Hultén, J. (2016). Swedish sources and pathways for microplastics to the marine environment. In: *IVL Svenska Miljöinstitutet.*

Miraj, S.S., Parveen, N., and Zedan, H.S. (2019). Plastic microbeads: Small yet mighty concerning. *International Journal of Environmental Health Research* 31:788–804.

Mishra, S., Charan Rath, C., and Das, A.P. (2019). Marine microfiber pollution: A review on present status and future challenges. *Marine Pollution Bulletin* 140:188–197.

Mohamed, S.S. (1982). Comparison of phosphate and carbonate built detergents for laundering polyester/cotton. *Textile Chemist & Colorist* 14(3).

Napper, I.E., and Thompson, R.C. (2016). Release of synthetic microplastic plastic fibres from domestic washing machines: Effects of fabric type and washing conditions. *Marine Pollution Bulletin* 112(1–2):39–45.

Niinimäki, K., Peters, G., Dahlbo, H., Perry, P., Rissanen, T., and Gwilt, A. (2020). The environmental price of fast fashion. *Nature Reviews Earth & Environment* 1(4):189–200.

Obbard, R.W., Sadri, S., Wong, Y.Q., Khitun, A.A., Baker, I., and Thompson, R.C. (2014). Global warming releases microplastic legacy frozen in Arctic Sea ice. *Earth's Future* 2(6):315–320.

OECD (Organization for Economic Cooperation and Development). (2021). Microbeads in cosmetics [accessed 2021 September 9]. https://www.oecd.org/stories/ocean/microbeads-in-cosmetics-609ea0bf/.

Patel, M.M., Goyal, B.R., Bhadada, S.V., Bhatt, J.S., and Amin, A.F. (2009). Getting into the brain. *CNS Drugs* 23(1):35–58.

Peeken, I., Primpke, S., Beyer, B., Gütermann, J., Katlein, C., Krumpen, T., Bergman, M., Hehemann, L., and Gerdts, G. (2018). Arctic sea ice is an important temporal sink and means of transport for microplastic. *Nature Communications* 9(1):1–12.

Perry, R.J. (2005). "Pro-fragrant" silicone delivery polymers. In: *Delivery System Handbook for Personal Care and Cosmetic Products.* Amsterdam (Netherlands): Elsevier, pp. 667–682.

Pivokonsky, M., Cermakova, L., Novotna, K., Peer, P., Cajthaml, T., and Janda, V. (2018). Occurrence of microplastics in raw and treated drinking water. *Science of the Total Environment* 643:1644–1651.

PPD (Polymer Properties Database). (2021). Ozonation reactions and ozone cracking in elastomers [accessed 2021 October 18]. https://polymerdatabase.com/polymer%20chemistry/Ozone.html.

PSF (PlasticSoupFoundation). (2020). Ducor petrochemicals to be held responsible for plastic nurdle pollution [accessed 2020 August 11]. https://www.plasticsoupfoundation.org/en/2020/03/ducor-petrochemicals-to-be-held-responsible-for-plastic-nurdle-pollution/.

Ramasamy, R., and Subramanian, R.B. (2021). Synthetic textile and microfiber pollution: A review on mitigation strategies. *Environmental Science and Pollution Research* 28:41596–41611.

Rao, J.P., and Geckeler, K.E. (2011). Polymer nanoparticles: Preparation techniques and size-control parameters. *Progress in Polymer Science* 36(7):887–913.

Ravve, A. (2000). Degradation of polymers. In: *Principles of Polymer Chemistry*. New York: Kluwer Academic/Plenum Publishers.

Rincon-Rubio, L., Fayolle, B., Audouin, L., and Verdu, J. (2001). A general solution of the closed-loop kinetic scheme for the thermal oxidation of polypropylene. *Polymer Degradation and Stability* 74(1):177–188.

Rochman, C.M., Kross, S.M., Armstrong, J.B., Bogan, M.T., Darling, E.S., Green, S.J., and Veríssimo, D. (2015). *Scientific Evidence Supports a Ban on Microbeads*. ACS Publications.

Rodgers, B., and Waddell, W. (2013). The science of rubber compounding. In: *The Science and Technology of Rubber*, 4th ed. New York: Academic Press.

Rouillon, C., Bussiere, P.O., Desnoux, E., Collin, S., and Vial, C. (2016). Is carbonyl index a quantitative probe to monitor polypropylene photodegradation? *Polymer Degradation and Stability* 128:200–208.

Roychand, R., and Pramanik, B.K. (2020). Identification of micro-plastics in Australian road dust. *Journal of Environmental Chemical Engineering* 8(1):103647.

Sasidharan, M., Eskandari Torbaghan, M., and Burrow, M. (2019). *Using Waste Plastics in Road Construction*. Birmingham (UK): University of Birmingham. K4D research helpdesk.

Saxena, S., and Nacht, S. (2005). Polymeric porous delivery systems: Polytrap® and Microsponge®. In: *Delivery System Handbook for Personal Care and Cosmetic Products*. Amsterdam (Netherlands): Elsevier, pp. 333–351.

Sharma, S., Basu, S., Shetti, N.P., Nadagouda, M.N., and Aminabhavi, T.M. (2020). Microplastics in the environment: Occurrence, perils, and eradication. *Chemical Engineering Journal* 408:127317.

Shaw, D.G., and Day, R.H. (1994). Colour-and form-dependent loss of plastic microdebris from the North Pacific Ocean. *Marine Pollution Bulletin* 28(1):39–43.

Siegfried, M., Koelmans, A.A., Besseling, E., and Kroeze, C. (2017). Export of microplastics from land to sea. A modelling approach. *Water Research* 127:249–257.

Singh, B., and Sharma, N. (2008). Mechanistic implications of plastic degradation. *Polymer Degradation and Stability* 93:561–584.

Singh, R.P., Mishra, S., and Das, A.P. (2020). Synthetic microfibers: Pollution toxicity and remediation. *Chemosphere* 257:127199.

Sneath, S. (2021). Resolution aimed at preventing plastic pollution in state waterways passes committee [accessed 2021 October 10]. *Louisiana Illuminator.* https://lailluminator.com/briefs/resolution-aimed-at-preventing-plastic-pollution-in-state-waterways-passes-committee.

Sommer, F., Dietze, V., Baum, A., Sauer, J., Gilge, S., Maschowski, C., and Gieré, R. (2018). Tire abrasion as a major source of microplastics in the environment. *Aerosol and Air Quality Research* 18(8):2014–2028.

Song, H.K., and Lewis, V.D. (2013). Development of a system for sustainable fashion from recycled clothes – Based on US fashion brands. *The Research Journal of the Costume Culture* 21(1):139–150.

Stormwater Solutions. (2020). EPA requires plastic manufacturers to protect waterways [accessed 2020 August 11]. https://www.estormwater.com/commercialindustrial-storm-water/epa-requires-plastic-manufacturers-protect-waterways.

Sundt, P., Schulze, P.-E., and Syversen, F. (2014/2015). *Sources of Microplastic Pollution to the Marine Environment.* Report No. M-321/2015. Mepex for the Norwegian Environment Agency (Miljødirektoratet, 86).

Tian, L., Chen, Q., Jiang, W., Wang, L., Xie, H., Kalogerakis, N., Ma, Y., and Ji, R. (2019). A carbon-14 radiotracer-based study on the phototransformation of polystyrene nanoplastics in water: Versus in air. *Environmental Science: Nano* 6:2907–2917.

Tyler, D.R. (2004). Mechanistic aspects of the effects of stress on the rates of photochemical degradation reactions in polymers. *Journal of Macromolecular Science, Part C: Polymer Reviews* 44(4):351–388.

Unice, K., Weeber, M., Abramson, M., Reid, R., van Gils, J., Markus, A., Vethaak, A., and Panko, J. (2019). Characterizing export of land-based microplastics to the estuary-part I: Application of integrated geospatial microplastic transport models to assess tire and road wear particles in the Seine watershed. *Science of the Total Environment* 646:1639–1649.

US EPA. (2011). Government ordered cleanup of illegally discharged plastic pellets will protect San Francisco Bay, endangered species [accessed 2020 August 11]. https://archive.epa.gov/region9/mediacenter/web/html/index-33.html.

US EPA. (2016). US EPA, Canyon plastics come to agreement about clean water act violations [accessed 2020 August 11]. https://www.hydrocarbonprocessing.com/news/2016/12/us-epa-canyon-plastics-come-to-agreement-about-clean-water-act-violations.

US Tire Manufacturers Association. (2019). What's in a tire [accessed 2021 October 4]. https://www.ustires.org/whats-tire-0.

Vägmarkering, A.B. (2019). Tjänster [services] [accessed 2021 October 4]. http://www.vagmarkering.se/vara-tjanster.

Vasile, C. (ed.). (2000). *Handbook of Polyolefins*. New York: Marcel Dekker.

Vidal, A.S., Thompson, R.C., Artigas, M.C., and De Haan, W.P. (2018). The imprint of microplastics from textiles in southern European deep seas. Paper presented at the MICRO 2018. Fate and Impact of Microplastics: Knowledge, Actions and Solutions.

Vogelsang, C., Lusher, A., Dadkhah, M.E., Sundvor, I., Umar, M., Ranneklev, S.B., Eidsvoll, D., and Meland, S. (2019). *Microplastics in Road Dust–Characteristics, Pathways and Measures*. NIVA-Rapport.

Wagner, S., Hüffer, T., Klöckner, P., Wehrhahn, M., Hofmann, T., and Reemtsma, T. (2018). Tire wear particles in the aquatic environment – A review on generation, analysis, occurrence, fate and effects. *Water Research* 139:83–100.

Wang, L., Wu, W.-M., Bolan, N.S., Tsang, D.C., Li, Y., Qin, M., and Hou, D. (2021). Environmental fate, toxicity and risk management strategies of nanoplastics in the environment: Current status and future perspectives. *Journal of Hazardous Materials* 401:123415.

Water, U. (2008). *Tackling a Global Crisis*. International Year of Sanitation.

Wilkes, C.E., Summers, J.W., and Daniels, C.A. (2005). *PVC Handbook*. München: Hanser Verlag.

Woodall, L., Sanchez-Vidal, A., Canals, M., Paterson, G., Coppock, R., Sleight, V., Calafat, A., Rogers, A.D., Narayanaswamy, B.E., and Thompson, R.C. (2014) The deep sea is a major sink for microplastic debris. *Royal Society of Open Science* 1:140317.

Wu, G. (2017). The mechanisms of rubber abrasion [accessed 2021 October 18]. Queen Mary University of London. http://qmro.qmul.ac.uk/xmlui/handle/123456789/25986.

Zhou, H., Zhou, L., and Ma, K. (2020). Microfiber from textile dyeing and printing wastewater of a typical industrial park in China: Occurrence, removal and release. *Science of the Total Environment* 739:140329.

Zitko, V., and Hanlon, M. (1991). Another source of pollution by plastics: Skin cleaners with plastic scrubbers. *Marine Pollution Bulletin* 22(1):41–42.

Chapter 4

Distribution of Microplastics in the Biosphere

INTRODUCTION

Microplastics (MPs) are widely distributed throughout global marine and freshwater ecosystems, from the surface microlayer to great depths. Additional, though less comprehensive, data reveals MP pollution of terrestrial environments as well.

Accumulation and distribution of MPs in the environment vary as a function of both human activity and environmental factors. Human factors include population density, economic status, types and amounts of industrial facilities, quality of waste management practices, and other factors. Environmental factors include rates of precipitation and runoff, river hydrodynamics, wave currents, tides, wind action, time, and other variables. It has been proposed that environmental variables play a greater role in the distribution of MPs than do anthropogenic factors (Hamid et al. 2018).

Although a substantial body of data has been compiled regarding the occurrence of MPs in the biosphere, comparing published data is challenging, due to lack of standardization in both sampling methods and units of measure (Claessens et al. 2011; Lassen et al. 2015). Different sampling tools and methods in the field can produce conflicting results, some of which will not accurately represent the MP distribution in the environment. (This is discussed further in chapter 6.) In addition, researchers have employed multiple size ranges for defining MP particles, which yields discrepancies across studies (Nerland et al. 2014). This issue is especially relevant given that small MPs particles (i.e., < 1 mm) are recognized as being more abundant than larger particles (McDermid and McMullen 2004; Browne et al. 2010; Eriksen et al. 2013; Song et al. 2014; Zhao et al. 2014).

MPs IN MARINE ECOSYSTEMS

Of all environmental compartments affected by MPs, the global oceans are considered by many to be the most contaminated (Van Melkebeke 2019). Using model results, Eriksen et al. (2014) estimated that at least 5.25 trillion plastic particles weighing 268,940 tons are currently floating at sea. In general, MPs enter deep oceans from ships and sea platforms (offshore littering), accidents and mismanagement during shipping, and from fishing (recreational and commercial). MPs in near-shore areas originate mostly from the mainland via wastewater discharges, surface runoff, river transport, and airflow. It is estimated that about 80% of marine plastic debris originates from terrestrial sources (Jambeck et al. 2015; Mani et al. 2015; Wagner et al. 2014) (figure 4.1).

Because of their relatively low density (table 2.1), many MPs are buoyant. Due to currents, winds and the vast areas involved, once MPs reach the oceans they can be rapidly and widely dispersed, travelling significant distances from the source (Van Sebille et al. 2012). Several global surveys have been conducted in recent years to evaluate the presence of floating MPs in different water bodies (tables 4.1 and 4.2) (Cózar et al. 2014; Eriksen et al. 2014; Reisser et al. 2015). MPs are now widespread in oceans, including pelagic and coastal zones from the arctic to the equator, as well as in semi-enclosed water bodies (e.g., Caribbean Sea, Mediterranean Sea) (Cózar et al. 2014; Law et al. 2010; Ory et al. 2018; Welden and Lusher 2017). MPs have been reported to occur from the sea surface microlayer, extending downward to ocean abyssal zones (Barnes et al. 2009).

The reported data on MPs in the oceans is markedly dependent on geographic location. Cole et al. (2014) measured a concentration of 0.27

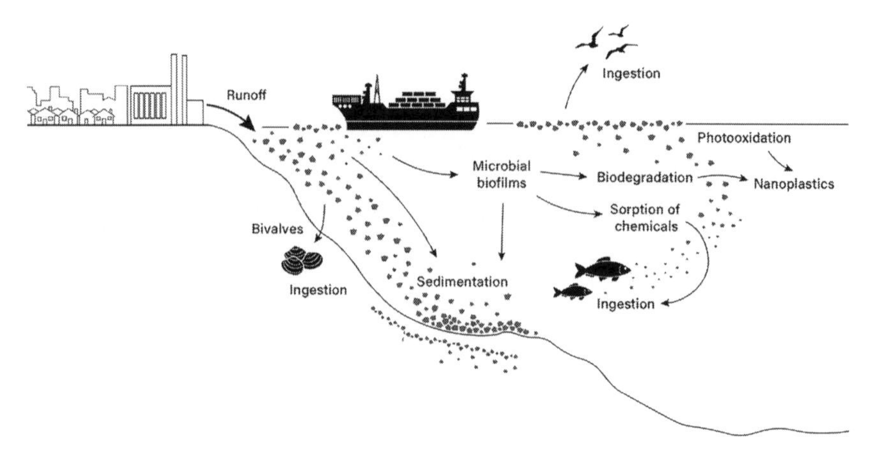

Figure 4.1 Migration and Fate of MPs in the Marine Environment

Table 4.1 Distribution of MPs in Marine Ecosystems, Recent Data

Location	*Media*	*Abundance (Counts or Weight)*	*Particle Size (Diameter or Length)*	*Polymer Type(s)*	*Comments*	*Reference*
Indian Ocean and Bay of Bengal	Seawater (36 stations)	Ocean: 0.01–4.53 items·m^{-2}. Mean 0.34 items·m^{-2}. Bay of Bengal: mean 2.04 items·m^{-2}	Particles < 2 mm comprise more than 70% of all MPs in the eastern Indian Ocean. Of these, 46.71% < 1 mm	PP 51.11%; PE 20.07%.	MP pollution in Eastern Indian Ocean among highest of world's oceans. Bay of Bengal likely to become MP hot spot due to vast input of land-based plastics and presence of recirculation gyres.	Li et al. 2021
Pearl River Estuary, China	Mangrove wetland sediment	100–7900 items·kg^{-1} dry weight (Mean 851 items·kg^{-1})	< 500 μm (69.4% of samples)	Predominantly PP-PE copolymer	Mostly green/black, and fibers/fragments.	Zuo et al. 2020
Tuscany, Italy	Coastal waters	Water column 0.26 items·m^{-3}; Floating 69,161.3 items·km^{-2}	1–2.5 mm	PE; PP	Most abundant size class 1–2.5 mm as fragments and sheets.	Baini et al. 2018
Xiangshan Bay, China	Seawater	Surface water 8.9 items·m^{-3}; sediment 1739 items·kg^{-1}	seawater 1.54 mm; sediment 1.33 mm	PE foams	Porous structure of PE foams led to high fragmentation. Mariculture-derived MPs in Xiangshan Bay transported to open sea.	Chen et al. 2018

(*continued*)

Table 4.1 (Continued)

Location	*Media*	*Abundance (Counts or Weight)*	*Particle Size (Diameter or Length)*	*Polymer Type(s)*	*Comments*	*Reference*
South Korea	20 beaches	Large MPs (1–5 mm) 0–2088 items·m^{-2}; Small MPs (0.02–1 mm) 1400–62,800 items·m^{-2}	Large MPs 1–5 mm; small MPs 0.02–1 mm	EPS (95% of large MPs); PE 49%; PP 38% (small MPs).	Maximum MP abundance in size range 100–150 μm; items < 300 μm account for 81% of total abundance. Input pathways of macroplastics and MPs may differ.	Eo et al. 2018
Ross Sea (Antarctica)	Seawater	0.0032–1.18 items·m^{-3}	> 60 μm	PE and PP 57.1%; polyester 28.6%; PTFE 5.7%; PMMA 5.7%; PA 2.9%.	MPs include fragments (mean 71.9%), fibers (12.7%), and others (15.4%). MP concentrations (items m^{-3}) lower than for other oceans.	Cincinelli et al. 2017
Bay of Bengal	Seawater	Few hundreds to 20,000 items·km^{-2}. (One location > 100,000 items·km^{-2}).	Order of abundance: 1.0–4.75 mm; 0.335–0.999 mm; > 4.75 mm.	n/a	AVANI, manta, and DiSalvo neuston trawl designs yield comparable data.	Eriksen et al. 2018
Wind Farm Yellow Sea, China	Seawater and sediment	Water 0.330 m^{-3}; sediment 2.58 items·g^{-1} (dry)	0.05 to 5 mm	PET; cellophane; PE. Others included polyphthalamide, polyvinyl acetate fibers, PVC, alkyd resin, linear low-density PE, and monoelaidin.	Distribution is affected by hydrodynamic effect based on human activities.	Wang et al. 2018

Northern Gulf of Mexico	Seawater	4.8–18.4 items·m^{-3} (Bongo sampler); 5.0–18.4 items·m^{-3} (Neuston sampler)	86.0% between 0.001 and 0.01 mm^2	n/a	MPs collected on inner continental shelf during this study among highest reported globally.	Di Mauro et al. 2017
South China Sea	Seawater	n/a	Net 1 mean length 125 μm; Net 2 mean length 167 μm.	Polyester most common.	This study focused on ingestion of MPs by natural zooplankton groups.	Sun et al. 2017
Mediterranean Sea, Spain	Seawater sediment	100.78–897.35 items·kg^{-1} dry sediment	Two size fractions: 1–2 mm and 0.5–1 mm	Filaments comprised > 60% of MPs.	High proportion of MP filaments close to populated areas; MP fragments are more common in Marine Protected Areas.	Alomar et al. 2016
Northern Adriatic Sea	Seawater	Range 14.1 × 10^3–3.1 × 10^6 items·km^{-2} ; mean 472 × 10^3–1 × 10^3 items·km^{-2}	Highest numbers in size range 2.2–2.71 mm	PE 80%. Remainder PP, PS, PVC, polyolefin, ABS, and unidentified.	This area experiences significant locational and temporal variation in MP pollution, linked to surface currents.	Gajšt et al. 2016
Hong Kong coastline	Beach sediment	5595 items·m^{-2}	0.315–5 mm	EPS 92%; fragments 5%; pellets 3%.	Quantities of macroplastics and MPs of same types were positively correlated, suggesting MP origins.	Fok and Cheung 2015

(continued)

Table 4.1 (Continued)

Location	*Media*	*Abundance (Counts or Weight)*	*Particle Size (Diameter or Length)*	*Polymer Type(s)*	*Comments*	*Reference*
Jinhae Bay, South Korea	Seawater (sea surface microlayer)	33–247 items·l^{-1}	Fiber lengths 100–2000 μm; diameters ranged from 10–35 μm. Spherules had a peak abundance at < 50 μm; no spherules > 100 μm detected. Sheets measured 50–500 μm; EPS < 500 μm (peak abundance 100–200 μm).	Paint resin items 75%; spherules 14%; fibers 5.8%; PS 4.6%; sheets 1.6%	Abundance of floating MPs in surface water among the highest worldwide.	Song et al. 2015
Halifax Harbor, Nova Scotia	Sediment	20-80 fibers·10 g^{-1} sediment	Grain diameters (median) 9–220 μm.	Microfibers the predominant MP type.	MP concentrations similar between three beaches analyzed. All located on outskirts of urbanized harbor. Polychaete worm fecal casts and mussel tissue contained MP fibers.	Mathalon and Hill 2014

Table 4.2 Presence of MPs in Freshwater Ecosystems, Recent Data

Location	*Media*	*Abundance (Counts or Weight)*	*Particle Size (Diameter or Length)*	*Polymer Type(s)*	*Comments*	*Reference*
Anchar Lake (Kashmir Valley, India)	Lake sediment	Mean 606 items·kg^{-1}	0.3–5 mm	Fibers (91%), fragments/films (8%), pellets (1%). PA (96%), PET (1.4%), PS (1.4%), PVC (0.9%), PP (0.7%).	Sources likely automobile, textile, and packaging industries.	Neelavannan et al. 2022
Veeranam Lake (Tamil Nadu, India)	Surface water and lake sediments	Water 13–54 items·km^{-2} (mean 28 items·km^{-2}) Sediment 92–604 items·kg^{-1} (mean 309 items·kg^{-1})	0.3–2 mm > 60% of MPs were < 1 mm	nylons > polythene > fibers/PVC > fragments > foam > pellets. Dominant colors white > red > black > green > blue and yellow	Inflowing rivers are major sources of the MPs.	Bharath et al. 2021
Lake Guaíba (southern Brazil)	Surface water	11.9–61.2 items ·m^{-3}	Collected with manta net (60 µm mesh size).	PP and PE comprised 98% of polymers. Predominant colors were white/transparent and red.	Most frequent morphology and size were fragments 100–250 µm.	Bertoldi et al. 2021

(*continued*)

Table 4.2 (Continued)

Location	*Media*	*Abundance (Counts or Weight)*	*Particle Size (Diameter or Length)*	*Polymer Type(s)*	*Comments*	*Reference*
Red Hills Lake, India	Surface water and sediment	Water 5.9 items·l^{-1}; sediment 27 items·kg^{-1}	0.3–2 mm	In both sediment and water, the most MPs types were fibers (37.9%), fragments (27%), films (24%), pellets (11.1%).	Lake is located in densely populated region; also surrounded by industries situated in Chennai.	Gopinath et al., 2020
River Elbe, Germany	Surface water and sediment	Water (mean): 5.57 items·m^{-3}, (150–5000 μm). Sediment (mean): 3,350,000 items·m^{-3} (125–5000 μm)	150–5000 μm	Suspended MPs predominantly PE and PP. Sediment contained greater diversity of polymer types.	MP concentrations correspond well with previous results from other European rivers.	Scherer et al. 2020
Tuojiang River, China	Surface water	912–3395 items·m^{-3}	0.5–1 mm predominant	PP, PE, PS, PVC	Microplastics found in all seven cities of the Tuojiang River basin. MP surfaces had numerous fissures; substantial particulate matter adsorbed.	Zhou et al. 2020
Scotland	River water downstream from WWTP	Mean discharge of 2.2×10^7 items·day^{-1} to recipient river	60–2800 μm	PP fibers most abundant.	Secondary MPs were predominant; few primary MPs. Fibers (67%), films (18%), fragments (15%). At least 1.2×10^7 items may be discharged daily from WWTP even during low flow.	Blair et al. 2019

Poyang Lake, China	Water, sediment	Surface water 5–34 items·l^{-1}; Sediment 54–506 items·kg^{-1}	Majority < 0.5 mm	Fibrous and colored were predominant characteristics. PP and PE were the major polymer types.	Domestic sewage and fishing activities might be the main sources of MPs in the lake.	Yuan et al., 2019
15 Flemish rivers, Belgium	Surface water	–	–	EVA, PET, PVA, PP, PVC	Microscopic and spectroscopic analysis showed MPs to be from varied sources.	Slootmaekers, Catarci Carteny et al. 2019
Three Gorges Dam, China	Surface water	1597–12,611 items·m^{-3} (surface water); 25–300 items·kg^{-1} (sediment)	< 0.5 mm size range most abundant, followed by 0.5–1 mm.	PS (38.5%), PP (29.4%), PE (21%)	Fibers were the most abundant MP. MP numbers higher in water in urban areas; higher in sediment in rural areas. MP abundance in surface water and sediment not directly proportional.	Di and Wang 2018
Italian Subalpine Lakes	Surface water	4000 to 57,000·km^{-2}	1–5 mm	PE, EPS, PP	PE (45%), EPS (18%), and PP (15%). Fragments were dominant shape (73.7%). Significant relationship regarding contribution of direct and diffuse sources to the quantity of MPs.	Sighicelli et al. 2018

(*continued*)

Table 4.2 (Continued)

Location	*Media*	*Abundance (Counts or Weight)*	*Particle Size (Diameter or Length)*	*Polymer Type(s)*	*Comments*	*Reference*
Vembanad Lake, India	Lake sediment	96 to 496 items·m^{-2}; mean 252.8 items·m^{-2}	< 5 mm	PE	Morphology of the MPs suggests origin from fragmentation of larger plastic debris.	Sruthy and Ramasamy 2017
Beijiang River, China	Surface sediments in littoral zone	178–544 items·kg^{-1}.	< 5 mm	Predominantly PE (41.7–65.5%)	SEM images: pits, fractures, flakes, and adhering particles. Chemical weathering confirmed by μ-FTIR.	Wang et al. 2017
Lower Saxony, Germany	Rivers receiving effluents from WWTPs	0-5 × 10^1 items·m^{-3} (>500 mm size); 1 × 10^1–9 × 10^3 items·m^{-3} (<500 mm)	20 μm	PE	Quantities of synthetic fibers ranged from 9 × 10^1 to 1 × 10^3 m^{-3} ; predominantly polyester.	Mintenig et al. 2017
Ljubljana, Slovenia	Surface water (effluent from WWTP)	13.9 mg·m^{-3}	<100 μm	PE	Daily release of approx. 112,500,000 items into receiving river is possible, resulting in a microbeads concentration of 21 items·m^{-3}.	Kalcikova, Alic et al. 2017
River Clyde, Glasgow, Scotland	River receiving WWTP effluent	0.25 items·m^{-3}	< 5 mm	PE, PET, PS, PV, PVA, PVE, PVC	Despite treatment operations, WWTP may release 65 million MPs·day^{-1} into receiving water.	Murphy, Ewins et al. 2016

Jurujuba Cove, Niterói, RJ., Brazil	Surface water	16.4 items·m^{-3}	Classified into four size ranges: ≥ 5 mm; 3–5 mm; 1–3 mm; < 1 mm. The <1 mm size range was most abundant.	72% of samples were PE and 26% PP.	MP source probably domestic effluents and mussel farming.	Castro, Silva et al. 2016
Rhine River	Surface water	Mean 892,777 items·km^{-2} at 11 sampling locations	< 1 g cm^{-3}	PS, PP, acrylate, polyester, PVC	Spherules almost 60% of total. Majority were opaque, size range of 300–1000 μm. Almost 70% consisted of cross-linked PS, and 15% PE. In Rhine-Ruhr metropolitan area, a peak concentration of 3.9 million items·km^{-2} measured.	Mani, Hauk et al. 2015
River Seine, urban area, France	Surface water	3–108 items·m^{-3}	100–5000 μm		Fibers predominated in plankton net sampling. MPs observed in atmospheric fallout; more than 90% were fibers.	Dris 2015

(*continued*)

Table 4.2 (Continued)

Location	*Media*	*Abundance (Counts or Weight)*	*Particle Size (Diameter or Length)*	*Polymer Type(s)*	*Comments*	*Reference*
Goiana Estuary, Brazil	Surface water	26 items·100 m^{-3}	Mean MP size 2.23 mm.		Highest amounts of MPs observed during late rainy season, when under the influence of the highest river flow.	Lima et al. 2014
Great Lakes, US	Surface water	466,000 items·km^{-2}	Three size classes: 0.355–0.999 mm, 1.00–4.749 mm, and > 4.75 mm	PE, PP, PS	MP pellets and fragments more abundant than other particle types. Size class 0.355–0.999 mm accounted for 81% of total particle count.	Eriksen et al. 2013
Lake Geneva, Switzerland	Surface water and beaches	1–7 items·l^{-1}	< 5 mm	PS	Macroplastics and MPs identified in large numbers in surface layer and beaches of Lake Geneva. Fibers, pellets, hard plastics, PS.	Alencastro 2012
Lake Huron (US and Canada)	Beach sediment	Beach at Sarnia contained 408 items·m^{-2}; other beaches had low concentrations.	< 5 mm	PE, PP, PET	Majority of pellets near industrial sector along southeastern margin of the lake; abundance decreased northward, following dominant lake current patterns.	Zbyszewski and Corcoran 2011

PE—polyethylene; PS—polystyrene; PP—polypropylene; PET—polyethylene terephthalate; PVC—polyvinylchloride particles·m^{-3} in the English Channel; in contrast, Desforges et al. (2014) reported a concentration of 7630 particles·m^{-3} off the South Island of New Zealand. Over 400 MP items·m^{-2} were detected in estuaries off the South Carolina coast (Gray et al. 2018), and tens of thousands of particles·m^{-2} were measured in the Mediterranean Sea (Schmidt et al. 2018) and off the coast of South Korea (Eo et al. 2018).

The varied MPs numbers in ocean waters are also a function of sampling depth. A concentration of 1.36 kg·m^{-2} total free-floating MPs in the upper 5 m layer of the water column was estimated by Everaert et al. (2018). In Singapore's coastal environment, MP residues consisting of PE, PP, and PS fragments were detected both in the surface microlayer (50–60 μm) and a subsurface layer (1 m) of water bodies (Ng and Obbard 2006). In the North Atlantic Gyre, MPs of size 0.5–1.0 mm were more plentiful in subsurface water than surface water (Reisser et al. 2015). Along the western Irish continental shelf, deeper waters were contaminated by MPs almost everywhere, and 66% of recovered MPs were found at the water-sediment interface (Martin et al. 2017). MPs are apparently widespread throughout the water column (Alomar et al. 2016; Martin et al. 2017).

A substantial fraction of MPs released to waterways eventually settles and ends up on the seabed.

Several factors direct the deposition of MP particles to the sea sediment layer. The most obvious variable is particle density, which is a function of polymer type (see table 2.2). When MP intrinsic density exceeds that of seawater (i.e., 1.02–1.03 g·m^{-3}), the particle will naturally settle. It is possible, however, that low-density MPs can sediment due to density modification. One common example of such modification is biofouling, where microbial biomass accumulates on the particle surface. Such accumulation induces settling of formerly floating particles. This mechanism is recognized as key to the presence of MPs in marine sediment (Andrady 2011; Reisser et al. 2013; Zettler et al. 2013; Kooi et al. 2017).

Concentrations of MPs on the deep-sea floor have been estimated to be as high as 4×10^9 fibers·km^{-2}, with an average of approximately 1×10^9 km^{-2} (Woodall et al. 2014). Courtene-Jones et al. (2020) reported that MPs were ubiquitous in a sediment core obtained from the Rockall Trough (in the North Atlantic) (>2000 m water depth) and a statistically significant historical accumulation was found. Some scientists (Waters et al. 2014) have argued that MPs may comprise a stratigraphic indication of the Anthropocene epoch as a result of accumulation in sediments.

Koelmans et al. (2017) formulated a "whole ocean" plastic waste mass balance model that integrated emissions data, plastic fragmentation rates, estimated concentrations in the ocean surface layer, and removal from the surface by sedimentation. Their simulation showed that 99.8% of all plastic that had entered the ocean since 1950 had settled below the surface layer by 2016 with an additional 9.4 million tons settling per year. The vast majority (83.7%) of the sedimented plastic was macroplastic, while 13.8% was MP, and 2.5% was nanoplastic (measuring < 0.335 mm).

A significant transport route for particulate organic matter to the benthos that is not considered in many mathematical models is the formation of *marine snows*. Marine snows are defined as organic-rich aggregates (particles measuring > 200 μm) composed of fecal pellets, larvacean houses, phytoplankton, microbial biomass, particulate organic matter, and various inorganics (Porter et al. 2018). Marine snows have much higher settling rates than their individual particle components (via Stokes's Law) (Long et al. 2015; Doyle et al. 2014) and are primarily responsible for the mass flux of organic material from surface water to the deep ocean (Porter et al. 2018). Marine snow concentrations are documented to range from < 1 to 100 aggregates·L^{-1} (Simon et al. 2002); however, values can be as high as 5300 aggregates·L^{-1} (Riebesell 1991).

Other transport mechanisms may include fast-sinking fecal pellets, but these tend to remain in the upper few hundred meters of the water column. Much of the global oceans vertical flux of particulate material in global oceans may be directed by the movement of marine snow (Porter et al. 2018).

Beaches

Beaches are significantly affected by MPs pollution—significant quantities of MPs released to waterways eventually wash up on shore. Fok and Cheung (2015) reported that of all plastic debris collected at 25 beaches along the coastline of Hong Kong, more than 90% occurred as MPs (0.315–5 mm) consisting of PS (92%), fragments (5%), and pellets (3%). The mean MPs abundance (5595 items·m^{-2}) is higher for beaches in Hong Kong than international averages. Along the Belgian coast, the average MP concentration in beach sediment was 92.8 particles·kg^{-1} dry sediment. Fibers were the major component (82.1 particles·kg^{-1}).

Lee et al. (2013) reported that abundance of beached MPs in the rainy season was higher because natural phenomena like heavy rainstorms, wind currents, and hurricanes caused plastic debris to accumulate on shores. Ivar do Sul et al. (2009) studied the distribution of MPs on the beaches of Fernando de Noronha, an island off the coast of Brazil. The concentration of plastics was higher on windward beaches compared to the leeward side due to surface currents (Debrot et al. 1999; Ivar do Sul et al. 2009). The beaches of Guanabara Bay on the Brazilian coastline have been noted as among the most polluted beaches, due in part to the extensive presence of MPs. Of the total plastic debris, MPs fragments contributed 56%, along with polystyrene foam

(26.7%), pellets (9.9%), and fibers (7.2%). The concentration of MP ranged from 12 to 1,300 particles•m^{-2} on these beaches (de Carvalho and Neto 2016).

In a study on the island of Malta, a high abundance of MP pellets was reported (>1000 m^{-2} at the surface) on the backshores of beaches (Turner and Holmes 2011). On the coast of Singapore, MPs were detected in four out of seven beaches (Ng and Obbard 2006). The presence of MPs (max. 3 particles•kg^{-1}) on the coastline is believed due to establishment of industries and recreational practices as well as from shipping discharge. Polymer types identified in beach sediments included PE, PP, PS, nylon, and others. Browne et al. (2011) reported MP contamination at 18 beaches (1 cm depth) in six continents from the poles to the equator and reported concentrations in sediments ranging from 2 fibers•L^{-1} in Australia to 160 fibers•L^{-1} in the UK and Portugal. Microbeads/pellets, fragments, and fibers were the major MPs in sediments and were composed mainly of PP, PE, and PS.

In the Great Lakes region, plastic pellets were common in the debris in several surface water and beach studies, with very minor amounts reported in benthic zones, as a result of their low density. Zbyszewski and Corcoran (2011) and Corcoran et al. (2015) found that pellets were the most common plastic types along shorelines of the Great Lakes, with values as high as 33 pellets•m^{-2} (Baxter Beach, Lake Huron) and 21 pellets•m^{-2} (Humber Bay), respectively (Corcoran et al. 2020).

MPs IN FRESHWATER ECOSYSTEMS

MPs enter rivers and are often eventually transported to oceans. Depending on hydrodynamics, however, freshwater ecosystems can also act as sinks, and retain a substantial proportion of MP inputs. Reported values of MPs in freshwater ecosystems vary greatly, from near-zero to several millions of particles per cubic meter (Li et al. 2018). These differences are attributed to geography (i.e., sampling location), types of human activities, and sampling methods used (Eerkes-Medrano et al. 2015). Hydrologic conditions also affect concentrations of MPs (Baldwin et al. 2016; Mani et al. 2015; Yonkos et al. 2014). In urban and non-urban watersheds, the concentrations of MPs are believed to be higher during runoff events than those under low-flow conditions.

Relatively less data is available for MPs pollution of freshwater and brackish environments compared to marine ecosystems. MPs data in freshwater has addressed lakes, reservoirs, and some major rivers; the spatial distribution of MPs in river and lakeshore sediments has also been studied (Peng et al. 2018; Driedger et al. 2015). Some authors consider the level of MP contamination in freshwater to be as severe as that in oceans (Wagner et al. 2014).

An expedition of the US Great Lakes determined that all samples except one contained plastics; the frequency of occurrence among Lake Superior, Lake Huron, and Lake Erie were 100%, 87.5%, and 100%, respectively (Eriksen et al. 2013). MP numbers in surface water were 105,500 particles·km^{-2} (Lake Erie), 5,390 particles·km^{-2} (Lake Superior), and 2,779 particles·km^{-2} (Lake Huron) (Eriksen et al. 2013). Among the top six largest lakes in Switzerland, all samples contained MPs, which were found in beach sediments, lake, and river surfaces (Faure et al. 2015; Wu et al. 2019).

Researchers have reported widespread occurrence of MPs in major rivers (Mani et al. 2015), urban rivers (Moore et al. 2011), and estuarine rivers worldwide (Yonkos et al. 2014). In surface water samples collected from four estuarine tributaries of the Chesapeake Bay, MPs were found in 59 of 60 samples (Yonkos et al. 2014). Baldwin et al. (2016) collected samples from 29 Great Lakes tributaries and measured MP concentrations from 0.05 to 32 items·m^{-3}. Moore et al. (2011) found that MP numbers ranged from 0.01 to 12.9 particles·L^{-1} in the Los Angeles River, San Gabriel River, and tributary Coyote Creek. In the Chicago River, concentrations of MPs in surface water reached 730,000–6.7 million items·km^{-2}, which are comparable or even higher than the values of oceans and the Great Lakes (McCormick et al. 2014). MPs were detected in all samples at 11 locations along the Rhine River in Germany (Mani et al. 2015). In the Yangtze River of China, MP concentrations were 3.4 million–13.6 million·km^{-2} in the main stream and 192,500–11.9 million·km^{-2} in the estuarine areas of four tributaries (Zhang et al. 2015).

As shown in table 4.2, concentrations of MPs in freshwater systems are highly variable, likely a consequence of several factors including particle size, human population density, economic and urban development, waste management practices, and hydrologic conditions (Eriksen et al. 2013; Free et al. 2014; Moore et al. 2011; Zbyszewski and Corcoran 2011).

An inverse relationship between MP concentration and particle size has been observed in many studies of rivers, lakes, and oceans. Such data may also be explained as a consequence of the sampling method (see chapter 6).

Greater numbers of MPs tend to be found in zones of high population density (Eerkes-Medrano et al. 2015). Eriksen et al. (2013) reported greater concentrations of plastics from surface trawls in Lake Erie near population centers (i.e., Buffalo, NY; Erie, PA; Cleveland, OH) compared to significantly less populous regions of the Lakes Huron and Superior. Mani et al. (2015) reported a maximum concentration of 3.9 million particles·km^{-2} in the metropolitan region of Rhine-Ruhr in Germany. In 29 Great Lakes tributaries (Baldwin et al. 2016) and four estuarine tributaries within the Chesapeake Bay, a significant correlation with population was found by Yonkos et al. (2014). However, MP pollution has also been documented in remote regions having low population densities. In a mountain area of Lake Hovsgol in Mongolia, MPs were observed in all nine pelagic survey transects. MP concentrations were as high as 44,435 particles·km^{-2}, apparently a result of improper waste management (Free et al. 2014).

Wastewater treatment plants (WWTPs) are recognized to be a significant source of MPs to freshwater (Radityaningrum et al. 2021; Xu et al. 2021; Akarsu et al. 2020; Alvim et al 2020; Magnusson and Nor´en 2014; Talvitie and Heinonen 2014; Murphy et al. 2016; Mintenig et al. 2017). Globally, approximately 60% of municipal wastewater is treated (Mateo-Sagasta et al. 2015); many studies have determined that WWTPs have remarkable capabilities in removing substantial volumes of MPs from waterways (see chapter 7) (Talvitie and Heinonen 2014; Talvitie et al. 2017); however, massive quantities still escape with effluent. One study (Rochman et al. 2015) estimated that 8 billion MP particles are discharged daily from WWTPs in the United States alone.

MPs IN TERRESTRIAL ECOSYSTEMS

MPs have been documented to occur extensively in terrestrial ecosystems (table 4.3). MPs enter soil from varied sources including land application of sewage sludge, organic fertilizers from composting, improper waste disposal, use of plastic mulch film and greenhouse covering in agricultural applications, irrigation with MP-enriched polluted water, and from atmospheric deposition (Sol et al. 2020).

The annual input of MPs from agricultural lands has been estimated to range from 44,000 to 300,000 tons in North America and 63,000 to 430,000 tons in Europe (Nizzetto et al. 2016a, 2016b). Both values exceed the calculated annual emissions of MPs to the oceans. In croplands in Shanghai, China, 78.0 MP items·kg^{-1} were estimated in shallow soil and 62.5 items·kg^{-1} in deep soil (Liu et al. 2018). In southeast Mexico, MPs were detected in home garden soils at a mean concentration of 0.87 ± 1.9 particles·g^{-1} (Huerta Lwanga et al. 2017). Soils sampled from an industrial area in Australia were reported to contain 0.03–6.7% MPs; concentrations ranged from 300 to 67,500 mg·kg^{-1} (Fuller and Gautam 2016).

MPs were detected in about 90% of Swiss floodplain soils at depths between 0 and 5 cm. The mean MPs concentration to estimated at 5 mg·kg^{-1}, with a maximum value of 55.5 mg·kg^{-1} (Scheurer and Bigalke 2018). All study sites occurred in designated nature reserves, which should not be expected to be directly affected by MP inputs from sludge application or improper waste disposal. It is suggested that flooding may have caused deposition of MP from the river.

Table 4.3 Distribution of MPs in Terrestrial Ecosystems, Recent Data

Location	*Land Use*	*Abundance (Counts or Weight)*	*Particle Size (Diameter or Length)*	*Polymer Type(s)*	*Comments*	*Reference*
Wuxi area, Taihu Lake, China	Agriculture	56.67–180.33 items·kg^{-1} (spring sampling) 206.15–890.49 items·kg^{-1} (winter sampling)	< 5 mm	PP	MPs presumably from plastic mulch. Mainly as films, fibers, debris. Mainly transparent and black in color. PP most abundant polymer (67.26% spring, 54.65% winter)	Jiang et al. 2022
Fuerteventura, Canary Islands, Spain	Agriculture	46–159 items·kg^{-1}	310–4069 μm	Cellulosic, PES, acrylic, PP	No MPs extracted from non-irrigated/non-cultivated soils.	Pérez-Reverón et al. 2022
Arusha, Tanzania	Irrigated farms	0.21–1.5 items·g^{-1}	0.05–5 mm	PP, PET, and HDPE	Numbers of MPs in irrigated farm soils was higher than in surface water and river sediments.	Kundu et al. 2022
Republic of Korea	Urban	49.5 items·kg^{-1} 5.8 items·kg^{-1}	< 1 mm 1–5 mm	EPS, PE, PET, PP, SBR	Soils of Yeoju city contained mean 700 items·kg^{-1} MPs, but varied greatly with land use. Roadside soils had more MPs (1108 items·kg^{-1}), mostly black styrene-butadiene rubber (SBR) fragments associated with tire dust. Largest amount of MPs detected from upland soil (3440 items·kg^{-1}).	Choi et al. 2021

Yeoju City, Republic of Korea	Forest	7.3 items·kg^{-1} 0.8 items·kg^{-1}	< 1 mm 1–5 mm	SBR, SIS		
	Residence	21.0 items·kg^{-1} 4.0 items·kg^{-1}	< 1 mm 1–5 mm	PS, SBR		
Hebei, Shandong, Shaanxi, Hubei, and Jilin provinces, China	Farmland	MP abundances in range 2783–6366 items kg^{-1}, mean 4496 items·kg^{-1}	< 5 mm	PVC, PA (nylon), PP, PS, PE, polyester, acrylic	Six land use types.	Wang et al. 2021
Central Valley, Chile	Cropland, Pasture	306 items·kg^{-1} (cropland); 184 items·kg^{-1} (pasture)	Median length for fibers 1.6 mm; shortest length 0.3 mm	PP, PS, PE, CPE, EVA, acrylates, polyurethane, varnish, nitrile rubber, Polyester	MPs identified in soils under four different land uses (cropland, pasture, rangeland, natural grassland); all somewhat prone to accumulate MPs from different sources.	Corradini et al. 2021
Xinjiang, China	Agricultural	161.5 items·100 g^{-1} (depth 0–300 mm); 11.2 items·100 g^{-1} (depth 400–800 mm)	0.0310–4.9000 mm; 0.0011–0.4080 mm	PE	Various crops	Hu et al. 2021
Ilia County, Greece	Agricultural. Watermelons and tomatoes grown.	69–301 items· kg^{-1}	< 5 mm	PE	SEM images show that mulch films and MPs in soil were oxidized by exposure to sunlight, which led to fragmentation. Number of MPs in watermelon fields > four times higher than MPs in fields with tomatoes.	Isari et al. 2021

(continued)

Table 4.3 (Continued)

Location	*Land Use*	*Abundance (Counts or Weight)*	*Particle Size (Diameter or Length)*	*Polymer Type(s)*	*Comments*	*Reference*
Spanish Mediterranean coast, Spain	Agricultural (Agua Amarga, Roquetas) Urban (Santa Maria)	3819 items·kg^{-1} (Agua Amarga) 2173 items·kg^{-1} (Roquetas) 68–362 items·kg^{-1} (Santa Maria)	100–5000 μm	PU, PS, PVC, PE, PTFE	Fibers and fragments. Direct linkage between intense anthropogenic activity, extensive use of plastics, and plastic contamination in coastal ecosystems such as seagrass meadows.	Dahl et al. 2021
San Juan Cotzocón Municipality, Mexico	Rainforest, savanna, pine plantation, pasture	1.49–1.53 items·g^{-1}	150–500 μm size group most common	–	MPs (fibers and fragments) present in soils of all studied ecosystems. Films and pellets not found.	Álvarez-Lopeztello et al. 2021
Murcia region, Spain	Six vegetable fields	2116 items·kg^{-1}	–	PE	PE mulch. Concern raised about ingestion of plastics by livestock and eventual dispersion in feces.	Beriot et al. 2021
Shaanxi Province, China	Agriculture	1430–3410 items·kg^{-1}	< 5 mm	PS, PE, PP, HDPE, PVC, PET	Fibers and small-sized items (0–0.49 mm) predominated. MP abundance in agricultural soil closely related to planting type and climatic factors.	Ding et al. 2020

Valencia, Spain	16 agricultural fields	Soils without sewage sludge: 930 items·kg^{-1} (light-density plastic); 1100 items·kg^{-1} (heavy density plastic). Soils with addition of sewage sludge: 2130 items·kg^{-1} (light-density plastic); 3060 items·kg^{-1} (heavy density plastic)	Greatest proportion of MPs measured 150–250 µm	PP, PVC	11 fields had history of sewage sludge application	van den Berg et al. 2020
Central Valley, Chile	31 agricultural fields with different sludge application records	18–41 items·g^{-1}	> 2 mm	Acrylic, PES, nylon fibers, LDPE, PVC	Fiber (>97%), film, fragment, and pellet	Corradini et al. 2019
Shanghai, China	Soil from rice-fish co-culture ecosystem	10.3 ± 2.2 items·kg	< 5 mm	PE (61.4%), PP (35.1%), PVC (3.5%)	Fibers (majority), granules, fragments, films	Lv et al. 2019
Switzerland	29 floodplains in Swiss nature reserves	55–593 items·kg^{-1}	125–500 µm	PE (major), PS, latex, PVC, SBR	90% of Swiss floodplain soils contained MPs. MPs associated with mesoplastics (5 mm–2.5 cm diameter).	Scheurer and Bigalke 2018

(continued)

Table 4.3 (Continued)

Location	*Land Use*	*Abundance (Counts or Weight)*	*Particle Size (Diameter or Length)*	*Polymer Type(s)*	*Comments*	*Reference*
Shanghai, China	Vegetable soil	Shallow soil (0–3 cm): 78.0 items·kg^{-1}; deep soil (3–6 cm): 62.50 items·kg^{-1}.	20 μm ~ 5 mm	PP (50.5%), PE (43.43%), PET (6.1%)	Fiber, fragment, film, pellet.	Liu et al. 2018
Yunnan, China		7,100–42,960 items·kg^{-1} (Mean 18,760 items·kg^{-1})	10–0.05 mm	–	72% of plastic items associated with soil aggregates, 28% are dispersed.	Zhang and Liu 2018
Sydney, Australia	Industrial zone	300–67,500 mg·kg^{-1}		PVC, PE, PS		Fuller and Gautam 2016

CPE—chlorinated PE; EPS—Expanded polystyrene; EVA—ethylene vinyl acetate; PE—polyethylene; PET—polyethylene terephthalate; PS—polystyrene; PP—polypropylene; PTFE—Polytetrafluoroethylene; PVC—polyvinylchloride; PU—Polyurethane; SBR—Styrene-butadiene rubber; SIS—Styrene-isoprene-styrene.

Agricultural soils may be a significant source of MPs to freshwater systems via application of sewage sludge and composted biosolids as soil amendments. In some cases, however, a high proportion of applied MPs is retained. A study on MP retention within soils (Zubris et al. 2005) found that synthetic fibers derived from sewage sludge were retained within treated agricultural soil up to 15 years after the last sludge application. The study also suggested that fibers may aggregate within the soil profile; in some cases, fibers were found at depth along preferential flow paths. Fibers were found deeper than 25 cm in areas experiencing significant water infiltration and percolation through the soil (Zubris et al. 2005).

Retention of MPs within soils is promoted by processes such as bioturbation (for example, by root growth, earthworms), which draws particles downward from the surface and into deeper soil layers (Gabet et al. 2003). In one study (Huerta Lwanga et al. 2018) the earthworm *Lumbricus terrestris* introduced substantial proportions (as much as 73.5%) of surface MPs into the walls of their burrows. It is probable that agricultural and forest soils retain more MP particles than does urban land, due to having permeable soils and lower rates of overland flow and runoff (Horton and Dixon 2017).

CLOSING STATEMENTS

Global utilization of plastic products will continue at an increasing rate; unfortunately, mismanagement of waste will persist, with the consequent excessive losses of macro- and MPs to all environmental compartments. In order to better understand the distribution and impacts of MP pollution it is important to correlate sources, particle behavior, and transport mechanisms.

REFERENCES

Akarsu, C., Kumbur, H., Gökdağ, K., Kıdeyş, A.E., and Sanchez-Vidal, A. (2020). MPs composition and load from three wastewater treatment plants discharging into Mersin Bay, north eastern Mediterranean Sea. *Marine Pollution Bulletin* 150:110776.

Alomar, C., Estarellas, F., and Deudero, S. (2016). Microplastics in the Mediterranean Sea: Deposition in coastal shallow sediments, spatial variation and preferential grain size. *Marine Environmental Research* 115:1–10.

Alvim, C.B., Mendoza-Roca, J.A., and Bes-Piá, A. (2020). Wastewater treatment plant as microplastics release source–quantification and identification techniques. *Journal of Environmental Management* 255:109739.

Andrady, A.L. (2011). Microplastics in the marine environment. *Marine Pollution Bulletin* 62(8):1596–1605.

Baini, M., Fossi, M.C., Galli, M., Caliani, I., Campani, T., Finoia, M.G., and Panti, C. (2018). Abundance and characterization of microplastics in the coastal waters of Tuscany (Italy): The application of the MSFD monitoring protocol in the Mediterranean Sea. *Marine Pollution Bulletin* 133:543–552.

Baldwin, A.K., Corsi, S.R., and Mason, S.A. (2016). Plastic debris in 29 Great Lakes tributaries: Relations to watershed attributes and hydrology. *Environmental Science & Technology* 50(19):10377–10385.

Barnes, D.K., Galgani, F., Thompson, R.C., and Barlaz, M. (2009). Accumulation and fragmentation of plastic debris in global environments. *Philosophical Transactions of the Royal Society B: Biological Sciences* 364(1526):1985–1998.

Browne, M.A., Crump, P., Niven, S.J., Teuten, E., Tonkin, A., Galloway, T., and Thompson, R. (2011). Accumulation of microplastic on shorelines woldwide: Sources and sinks. *Environmental Science & Technology* 45(21):9175–9179.

Browne, M.A., Galloway, T.S., and Thompson, R.C. (2010). Spatial patterns of plastic debris along estuarine shorelines. *Environmental Science & Technology* 44(9):3404–3409.

Chen, M., Jin, M., Tao, P., Wang, Z., Xie, W., Yu, X., and Wang, K. (2018). Assessment of microplastics derived from mariculture in Xiangshan Bay, China. *Environmental Pollution* 242:1146–1156.

Cincinelli, A., Scopetani, C., Chelazzi, D., Lombardini, E., Martellini, T., Katsoyiannis, A., Fossi, M.C., and Corsolini, S. (2017). Microplastic in the surface waters of the Ross Sea (Antarctica): Occurrence, distribution and characterization by FTIR. *Chemosphere* 175:391–400.

Claessens, M., De Meester, S., Van Landuyt, L., De Clerck, K., and Janssen, C.R. (2011). Occurrence and distribution of microplastics in marine sediments along the Belgian coast. *Marine Pollution Bulletin* 62(10):2199–2204.

Claessens, M., Van Cauwenberghe, L., Vandegehuchte, M.B., and Janssen, C.R. (2013). New techniques for the detection of microplastics in sediments and field collected organisms. *Marine Pollution Bulletin* 70(1–2):227–233.

Cole, M., Webb, H., Lindeque, P.K., Fileman, E.S., Halsband, C., and Galloway, T.S. (2014). Isolation of microplastics in biota-rich seawater samples and marine organisms. *Scientific Reports* 4(1):1–8.

Corcoran, P.L., de Haan Ward, J., Arturo, I.A., Belontz, S.L., Moore, T., Hill-Svehla, C.M., Robertson, K., Wood, K., and Jazvac, K. (2020). A comprehensive investigation of industrial plastic pellets on beaches across the Laurentian Great Lakes and the factors governing their distribution. *Science of the Total Environment* 747:141227.

Corcoran, P.L., Norris, T., Ceccanese, T., Walzak, M.J., Helm, P.A., and Marvin, C.H. (2015). Hidden plastics of Lake Ontario, Canada, and their potential preservation in the sediment record. *Environmental Pollution* 204:17–25.

Courtene-Jones, W., Quinn, B., Ewins, C., Gary, S.F., and Narayanaswamy, B.E. (2020). Microplastic accumulation in deep-sea sediments from the Rockall Trough. *Marine Pollution Bulletin* 154:111092.

Cózar, A., Echevarría, F., González-Gordillo, J.I., Irigoien, X., Úbeda, B., Hernández-León, S., and Ruiz, A. (2014). Plastic debris in the open ocean. *Proceedings of the National Academy of Sciences* 111(28):10239–10244.

de Carvalho, D.G., and Neto, J.A.B. (2016). Microplastic pollution of the beaches of Guanabara Bay, Southeast Brazil. *Ocean & Coastal Management* 128:10–17.

Debrot, A.O., Tiel, A.B., and Bradshaw, J.E. (1999). Beach debris in Curacao. *Marine Pollution Bulletin* 38(9):795–801.

Desforges, J.-P.W., Galbraith, M., Dangerfield, N., and Ross, P.S. (2014). Widespread distribution of microplastics in subsurface seawater in the NE Pacific Ocean. *Marine Pollution Bulletin* 79(1–2):94–99.

Di Mauro, R., Kupchik, M.J., and Benfield, M.C. (2017). Abundant plankton-sized microplastic particles in shelf waters of the northern Gulf of Mexico. *Environmental Pollution* 230:798–809.

do Sul, J.A.I., and Costa, M.F. (2014). The present and future of microplastic pollution in the marine environment. *Environmental Pollution* 185:352–364.

Doyle, J.J., Palumbo, V., Huey, B.D., and Ward, J.E. (2014). Behavior of titanium dioxide nanoparticles in three aqueous media samples: Agglomeration and implications for benthic deposition. *Water, Air, and Soil Pollution* 225(9):1–13.

Driedger, A.G., Dürr, H.H., Mitchell, K., and Van Cappellen, P. (2015). Plastic debris in the Laurentian Great Lakes: A review. *Journal of Great Lakes Research* 41(1):9–19.

Eerkes-Medrano, D., Thompson, R.C., and Aldridge, D.C. (2015). Microplastics in freshwater systems: A review of the emerging threats, identification of knowledge gaps and prioritisation of research needs. *Water Research* 75:63–82.

Eo, S., Hong, S.H., Song, Y.K., Lee, J., Lee, J., and Shim, W.J. (2018). Abundance, composition, and distribution of microplastics larger than 20 μm in sand beaches of South Korea. *Environmental Pollution* 238:894–902.

Eriksen, M., Lebreton, L.C.M., Carson, H.S., Thiel, M., Moore, C.J., Borerro, J.C., Galgani, F., Ryan, P.G., and Reisser, J. (2014). Plastic pollution in the world's oceans: More than 5 trillion plastic 500 pieces weighing over 250,000 tons afloat at sea. *PLoS ONE* 9(12):e111913.

Eriksen, M., Liboiron, M., Kiessling, T., Charron, L., Alling, A., Lebreton, L., Richards, H., Roth, B., Ory, N.C., Hidalgo-Ruz, V., Meerhoff, E., Box, C., Cummins, A., and Hidalgo-Ruz, V. (2018). Microplastic sampling with the AVANI trawl compared to two neuston trawls in the Bay of Bengal and South Pacific. *Environmental Pollution* 232:430–439.

Eriksen, M., Mason, S., Wilson, S., Box, C., Zellers, A., Edwards, W., Faley, H., and Amato, S. (2013). Microplastic pollution in the surface waters of the Laurentian Great Lakes. *Marine Pollution Bulletin* 77(1–2):177–182.

Everaert, G., Van Cauwenberghe, L., De Rijcke, M., Koelmans, A.A., Mees, J., Vandegehuchte, M., and Janssen, C.R. (2018). Risk assessment of microplastics in the ocean: Modelling approach and first conclusions. *Environmental Pollution* 242:1930–1938.

Faure, F., Demars, C., Wieser, O., Kunz, M., and De Alencastro, L.F. (2015). Plastic pollution in Swiss surface waters: Nature and concentrations, interaction with pollutants. *Environmental Chemistry* 12(5):582–591.

Fok, L., and Cheung, P.K. (2015). Hong Kong at the Pearl River Estuary: A hotspot of microplastic pollution. *Marine Pollution Bulletin* 99(1–2):112–118.

Free, C.M., Jensen, O.P., Mason, S.A., Eriksen, M., Williamson, N.J., and Boldgiv, B. (2014). High-levels of microplastic pollution in a large, remote, mountain lake. *Marine Pollution Bulletin* 85(1):156–163.

Fuller, S., and Gautam, A. (2016). A procedure for measuring microplastics using pressurized fluid extraction. *Environmental Science & Technology* 50(11):5774–5780.

Gabet, E.J., Reichman, O., and Seabloom, E.W. (2003). The effects of bioturbation on soil processes and sediment transport. *Annual Review of Earth and Planetary Sciences* 31(1):249–273.

Gajšt, T., Bizjak, T., Palatinus, A., Liubartseva, S., and Kržan, A. (2016). Sea surface microplastics in Slovenian part of the Northern Adriatic. *Marine Pollution Bulletin* 113(1–2):392–399.

Gray, A.D., Wertz, H., Leads, R.R., and Weinstein, J.E. (2018). Microplastic in two South Carolina Estuaries: Occurrence, distribution, and composition. *Marine Pollution Bulletin* 128:223–233.

Guo, J.-J., Huang, X.-P., Xiang, L., Wang, Y.-Z., Li, Y.-W., Li, H., Cai, Q., Mo, C.E., and Wong, M.-H. (2020). Source, migration and toxicology of microplastics in soil. *Environment International* 137:105263.

Horton, A., and Dixon, S. Microplastics: An introduction to environmental transport processes [accessed 2021 December 4]. WiresWater. https://wires.onlinelibrary.wiley.com/doi/epdf/10.1002/wat2.1268.

Huerta Lwanga, E., Mendoza Vega, J., Ku Quej, V., Chi, J.d.l.A., Sanchez del Cid, L., Chi, C., Segura, G.E., Gertsen.H., Salanki, T., Ploeg, M.V.D., Koelmans, A., and Geissen, V. (2017). Field evidence for transfer of plastic debris along a terrestrial food chain. *Scientific Reports* 7(1):1–7.

Jambeck, J.R., Geyer, R., Wilcox, C., Siegler, T.R., Perryman, M., Andrady, A., Narayan, Ramani, and Law, K.L. (2015). Plastic waste inputs from land into the ocean. *Science* 347(6223):768–771.

Koelmans, A.A., Kooi, M., Law, K.L., and Van Sebille, E. (2017). All is not lost: Deriving a top-down mass budget of plastic at sea. *Environmental Research Letters* 12(11):114028.

Kooi, M., Nes, E.H.V., Scheffer, M., and Koelmans, A.A. (2017). Ups and downs in the ocean: Effects of biofouling on vertical transport of microplastics. *Environmental Science & Technology* 51(14):7963–7971.

Lassen, C., Hansen, S.F., Magnusson, K., Norén, F., Hartmann, N.I.B., Jensen, P.R., Nielsen, and Brinch, A.T.G. (2015). Microplastics-occurrence, effects and sources of releases to the environment in Denmark. Environmental Project

No. 1793. Danish Ministry of the Environment–Environmental Protection Agency (Denmark), 204.

Law, K.L., Morét-Ferguson, S., Maximenko, N.A., Proskurowski, G., Peacock, E.E., Hafner, J., and Reddy, C.M. (2010). Plastic accumulation in the North Atlantic subtropical gyre. *Science* 329(5996):1185–1188.

Lee, J., Hong, S., Song, Y.K., Hong, S.H., Jang, Y.C., Jang, M., Heo, N.W., Han, G.M., Lee, M.J., Kang, D., and Shim, W.J. (2013). Relationships among the abundances of plastic debris in different size classes on beaches in South Korea. *Marine Pollution Bulletin* 77(1–2):349–354.

Li, C., Wang, X., Liu, K., Zhu, L., Wei, N., Zong, C., and Li, D. (2021). Pelagic microplastics in surface water of the Eastern Indian Ocean during monsoon transition period: Abundance, distribution, and characteristics. *Science of the Total Environment* 755:142629.

Li, J., Liu, H., and Chen, J.P. (2018). Microplastics in freshwater systems: A review on occurrence, environmental effects, and methods for microplastics detection. *Water Research* 137:362–374.

Liu, M., Lu, S., Song, Y., Lei, L., Hu, J., Lv, W., Wenzong, Z., Cao, C., Shi, H., Yang, X., and He, D. (2018). Microplastic and mesoplastic pollution in farmland soils in suburbs of Shanghai, China. *Environmental Pollution* 242:855–862.

Long, M., Moriceau, B., Gallinari, M., Lambert, C., Huvet, A., Raffray, J., and Soudant, P. (2015). Interactions between microplastics and phytoplankton aggregates: Impact on their respective fates. *Marine Chemistry* 175:39–46.

Lwanga, E.H., Gertsen, H., Gooren, H., Peters, P., Salánki, T., van der Ploeg, M., Besseling, E., Koelmans, A., and Geissen, V. (2017). Incorporation of microplastics from litter into burrows of *Lumbricus terrestris*. *Environmental Pollution* 220:523–531.

Mani, T., Hauk, A., Walter, U., and Burkhardt-Holm, P. (2015). Microplastics profile along the Rhine River. *Scientific Reports* 5(1):1–7.

Martin, J., Lusher, A., Thompson, R.C., and Morley, A. (2017). The deposition and accumulation of microplastics in marine sediments and bottom water from the Irish continental shelf. *Scientific Reports* 7(1):1–9.

Mateo-Sagasta, J., Raschid-Sally, L., and Thebo, A. (2015). Global wastewater and sludge production, treatment and use. In: *Wastewater.* Berlin (Germany): Springer, pp. 15–38.

Mathalon, A., and Hill, P. (2014). Microplastic fibers in the intertidal ecosystem surrounding Halifax Harbor, Nova Scotia. *Marine Pollution Bulletin* 81(1):69–79.

McDermid, K.J., and McMullen, T.L. (2004). Quantitative analysis of small-plastic debris on beaches in the Hawaiian archipelago. *Marine Pollution Bulletin* 48(7–8):790–794.

Moore, C.J., Lattin, G., and Zellers, A. (2011). Quantity and type of plastic debris flowing from two urban rivers to coastal waters and beaches of Southern California. *Revista de Gestão Costeira Integrada: Journal of Integrated Coastal Zone Management* 11(1):65–73.

Nerland, I.L., Halsband, C., Allan, I., and Thomas, K.V. (2014). *Microplastics in Marine Environments: Occurrence, Distribution and Effects*. Report No. 6754-2014. Oslo: Norwegian Institute for Water Research.

Ng, K., and Obbard, J. (2006). Prevalence of microplastics in Singapore's coastal marine environment. *Marine Pollution Bulletin* 52(7):761–767.

Nizzetto, L., Bussi, G., Futter, M.N., Butterfield, D., and Whitehead, P.G. (2016a). A theoretical assessment of microplastic transport in river catchments and their retention by soils and river sediments. *Environmental Science: Processes & Impacts* 18(8):1050–1059.

Nizzetto, L., Futter, M., and Langaas, S. (2016b). Are agricultural soils dumps for microplastics of urban origin? *Environmental Science & Technology* 50(20):10777–10779.

Ory, N., Chagnon, C., Felix, F., Fernández, C., Ferreira, J.L., Gallardo, C., Ordonez, O.G., Henostroza, A., Laaz, E., Mizraji, R., Mojica, H., Haro, V.M., Medina, L.O., Preciado, M., Sobral, P., Urbina, M., and Thiel, M. (2018). Low prevalence of microplastic contamination in planktivorous fish species from the southeast Pacific Ocean. *Marine Pollution Bulletin* 127:211–216.

Peng, G., Xu, P., Zhu, B., Bai, M., and Li, D. (2018). Microplastics in freshwater river sediments in Shanghai, China: A case study of risk assessment in megacities. *Environmental Pollution* 234:448–456.

Porter, A., Lyons, B.P., Galloway, T.S., and Lewis, C. (2018). Role of marine snows in microplastic fate and bioavailability. *Environmental Science & Technology* 52(12):7111–7119.

Radityaningrum, A.D., Trihadiningrum, Y., Soedjono, E.S., and Herumurti, W. (2021). Microplastic contamination in water supply and the removal efficiencies of the treatment plants: A case of Surabaya City, Indonesia. *Journal of Water Process Engineering* 43:102195.

Reisser, J., Shaw, J., Wilcox, C., Hardesty, B.D., Proietti, M., Thums, M., and Pattiaratchi, C. (2013). Marine plastic pollution in waters around Australia: Characteristics, concentrations, and pathways. *PLoS ONE*. 8(11):e80466.

Reisser, J., Slat, B., Noble, K., Du Plessis, K., Epp, M., Proietti, M., de Sonneville, J., Becker, T., and Pattiaratchi, C. (2015). The vertical distribution of buoyant plastics at sea: An observational study in the North Atlantic Gyre. *Biogeosciences* 12(4):1249–1256.

Riebesell, U. (1991). Particle aggregation during a diatom bloom I. Physical aspects. *Marine Ecology Progress Series* 69:273–280.

Rochman, C.M., Kross, S.M., Armstrong, J.B., Bogan, M.T., Darling, E.S., Green, S.J., Smyth, A.R., and Veríssimo, D. (2015). Scientific evidence supports a ban on microbeads. *Environmental Science & Technology* 49(18):10759–10761.

Schmidt, N., Thibault, D., Galgani, F., Paluselli, A., and Sempéré, R. (2018). Occurrence of microplastics in surface waters of the Gulf of Lion (NW Mediterranean Sea). *Progress in Oceanography* 163:214–220.

Shahul Hamid, F., Bhatti, M.S., Anuar, N., Anuar, N., Mohan, P., and Periathamby, A. (2018). Worldwide distribution and abundance of microplastic: How dire is the situation? *Waste Management & Research* 36(10):873–897.

Simon, M., Grossart, H.-P., Schweitzer, B., and Ploug, H. (2002). Microbial ecology of organic aggregates in aquatic ecosystems. *Aquatic Microbial Ecology* 28(2):175–211.

Sol, D., Laca, A., Laca, A., and Díaz, M. (2020). Approaching the environmental problem of microplastics: Importance of WWTP treatments. *Science of the Total Environment* 740:140016.

Song, Y.K., Hong, S.H., Jang, M., Han, G.M., and Shim, W.J. (2015). Occurrence and distribution of microplastics in the sea surface microlayer in Jinhae Bay, South Korea. *Archives of Environmental Contamination and Toxicology* 69(3):279–287.

Song, Y.K., Hong, S.H., Jang, M., Kang, J.-H., Kwon, O.Y., Han, G.M., and Shim, W.J. (2014). Large accumulation of micro-sized synthetic polymer particles in the sea surface microlayer. *Environmental Science & Technology* 48(16):9014–9021.

Sun, X., Li, Q., Zhu, M., Liang, J., Zheng, S., and Zhao, Y. (2017). Ingestion of microplastics by natural zooplankton groups in the northern South China Sea. *Marine Pollution Bulletin* 115(1–2):217–224.

Talvitie, J., and Heinonen, M. (2014). Preliminary study on synthetic microfibers and particles at a municipal waste water treatment plant. *Baltic Marine Environment Protection Commission* HELCOM, Helsinki, 1–14.

Talvitie, J., Mikola, A., Setälä, O., Heinonen, M., and Koistinen, A. (2017). How well is microlitter purified from wastewater?–A detailed study on the stepwise removal of microlitter in a tertiary level wastewater treatment plant. *Water Research* 109:164–172.

Turner, A., and Holmes, L. (2011). Occurrence, distribution and characteristics of beached plastic production pellets on the island of Malta (central Mediterranean). *Marine Pollution Bulletin* 62(2):377–381.

Van Melkebeke, M. (2019). *Exploration and Optimization of Separation Techniques for the Removal of Microplastics from Marine Sediments During Dredging Operations* [Master's Thesis]. Ghent (Belgium): Ghent University.

Van Sebille, E., England, M.H., and Froyland, G. (2012). Origin, dynamics and evolution of ocean garbage patches from observed surface drifters. *Environmental Research Letters* 7(4):044040.

Wagner, M., Scherer, C., Alvarez-Muñoz, D., Brennholt, N., Bourrain, X., Buchinger, S., Fries, E., Grosbois, C., Klameier, J., Marti, T., Rodriquez-Mozaz, S., Urbatzka, R., Vethaaak, D., Winther-Nielsen, M., and Reifferscheid, G. (2014). Microplastics in freshwater ecosystems: What we know and what we need to know. *Environmental Sciences Europe* 26(1):1–9.

Wang, T., Zou, X., Li, B., Yao, Y., Li, J., Hui, H., Wenwen, Y., and Wang, C. (2018). Microplastics in a wind farm area: A case study at the Rudong Offshore Wind Farm, Yellow Sea, China. *Marine Pollution Bulletin* 128:466–474.

Waters, C.N., Zalasiewicz, J.A., Williams, M., Ellis, M.A., and Snelling, A.M. (2014). A stratigraphical basis for the Anthropocene? *Geological Society* 395(1):1–21.

Welden, N.A., and Lusher, A.L. (2017). Impacts of changing ocean circulation on the distribution of marine microplastic litter. *Integrated Environmental Assessment and Management* 13(3):483–487.

Woodall, L.C., Sanchez-Vidal, A., Canals, M., Paterson, G.L., Coppock, R., Sleight, V., Calafat, A., Rogers, A.D., Narayanaswamy, B.E., and Thompson, R.C. (2014). The deep sea is a major sink for microplastic debris. *Royal Society Open Science* 1(4):140317.

Wu, P., Huang, J., Zheng, Y., Yang, Y., Zhang, Y., He, F., Chen, H., Quan, G., Yan, J., Li, T., and Gao, B. (2019). Environmental occurrences, fate, and impacts of microplastics. *Ecotoxicology and Environmental Safety* 184:109612.

Xu, Z., Bai, X., and Ye, Z. (2021). Removal and generation of microplastics in wastewater treatment plants: A review. *Journal of Cleaner Production* 291:125982.

Yonkos, L.T., Friedel, E.A., Perez-Reyes, A.C., Ghosal, S., and Arthur, C.D. (2014). Microplastics in four estuarine rivers in the Chesapeake Bay, USA. *Environmental Science & Technology* 48(24):14195–14202.

Zbyszewski, M., and Corcoran, P.L. (2011). Distribution and degradation of fresh water plastic particles along the beaches of Lake Huron, Canada. *Water, Air, & Soil Pollution* 220(1):365–372.

Zettler, E.R., Mincer, T.J., and Amaral-Zettler, L.A. (2013). Life in the "plastisphere": Microbial communities on plastic marine debris. *Environmental Science & Technology* 47(13):7137–7146.

Zhang, K., Gong, W., Lv, J., Xiong, X., and Wu, C. (2015). Accumulation of floating microplastics behind the Three Gorges Dam. *Environmental Pollution* 204:117–123.

Zhang, S., Wang, J., Liu, X., Qu, F., Wang, X., Wang, X., Li, Y., and Sun, Y. (2019). Microplastics in the environment: A review of analytical methods, distribution, and biological effects. *Trends in Analytical Chemistry* 111:62–72.

Zhao, J., Liu, L., Zhang, Y., Wang, X., and Wu, F. (2018). A novel way to rapidly monitor microplastics in soil by hyperspectral imaging technology and chemometrics. *Environmental Pollution* 238:121–129.

Zhao, S., Zhu, L., Wang, T., and Li, D. (2014). Suspended microplastics in the surface water of the Yangtze Estuary System, China: First observations on occurrence, distribution. *Marine Pollution Bulletin* 86(1–2):562–568.

Zubris, K.A.V., and Richards, B.K. (2005). Synthetic fibers as an indicator of land application of sludge. *Environmental Pollution* 138(2):201–211.

Zuo, L., Sun, Y., Li, H., Hu, Y., Lin, L., Peng, J., and Xu, X. (2020). Microplastics in mangrove sediments of the Pearl River Estuary, South China: Correlation with halogenated flame retardants' levels. *Science of the Total Environment* 725:138344.

Chapter 5

Interactions of Contaminants with Microplastics

INTRODUCTION

A significant hazard associated with microplastics (MPs) in marine and freshwater ecosystems is possible ingestion of pollutant-bound particles by organisms whose food sources are of similar dimensions and appearance. Recent findings indicate that MP particles in marine and freshwater environments act as carriers of trace metals, hazardous organic compounds, antibiotics, radioactive elements, and pathogenic microorganisms due to adsorption from surrounding water (Holmes et al. 2012; Rochman et al. 2014; Turner and Holmes 2015; Brennecke et al. 2016; Munier and Bendell 2018; Maršić-Lučić et al. 2018; Li et al. 2018; Guo et al. 2019; Zettler et al. 2013; Kirstein et al. 2018). A wide range of suspension-, filter-, and deposit-feeders, detritivores, planktivorous fish, mammals and birds are, therefore, at risk of MP ingestion and related toxicity from adsorbed pollutants. Efforts to manage and recover MPs from aquatic and terrestrial ecosystems must include an understanding of the complex nature of contaminants associated with this debris.

SORPTION OF HEAVY METALS TO MPs

MPs have been documented to concentrate harmful metals at levels many times higher than those of surrounding seawater. Concentrations of metals on MPs from marine and coastal environments vary from 10^{-1} to 10^{4} $\mu g \cdot g^{-1}$ at different locations (Guo et al. 2019). Consumption of particles by marine organisms may therefore result in adsorbed trace metals entering food chains, which presents a potential human health risk from consumption of seafood

(Maršić-Lučić et al. 2018). Selected examples of heavy metal sorption to MPs appears in table 5.1.

Several laboratory and field studies have investigated the mechanisms of metal sorption to MPs (Holmes et al. 2012, 2014; Turner and Holmes 2015; Brennecke et al. 2016; Hodson et al. 2017). According to one study (Turner and Holmes 2015), adsorption of metals from the aqueous phase to pellets is a two-phase process, that is, a period of relatively rapid adsorption followed by a slower, more protracted period in which equilibrium is attained. Various researchers (Holmes et al. 2012, 2014) state that monolayer sorption on active sites of MPs occur; however, others (Hodson et al. 2017) indicate that multilayer sorption takes place.

Trace metal (Cd, Cr, Cu, Fe, Mn, Ni, Pb, and Zn) concentrations in pellets collected from beaches off the coast of Croatia were greater than concentrations reported for seawater in the area, suggesting sorption of metals by pellets (Maršić-Lučić et al. 2018). Copper and Zn leached from an antifouling paint were adsorbed to MPs in seawater; concentrations of Cu and Zn were as high as 3000 and 270 $\mu g \cdot g^{-1}$, respectively (Brennecke et al. 2016). Metal enrichment of MPs has also been detected in river water (Turner and Holmes 2015). Median concentrations of trace elements (Al, Cd, Co, Cr, Cu, Hg, Ni, Pb, and Zn) in beached pellets ranged from as low < 3 $ng \cdot g^{-1}$ for Ag and Hg to as high as 34400 $ng \cdot g^{-1}$ for Fe (table 5.1).

Concentrations of metals on MPs reveal geographic variability; local factors such as the presence of industry near sampling locations and other anthropogenic activities affect concentrations of metals (Vedolin et al. 2017). Zinc accumulation on MPs reached up to 14815 $\mu g \cdot g^{-1}$ in sediment of Beijiang River, China (Wang et al. 2017); in contrast, Zn concentrations on MPs from remote Santubong and Trombol in Malaysia were < 10 $\mu g \cdot g^{-1}$ (Noik et al. 2015). Nurdles were sampled from four beaches along a coastline in southwest England (Ashton et al. 2010), and pellets were enriched with Cd and Pb (up to 10 $ng \cdot g^{-1}$ and 1.1 $ng \cdot g^{-1}$, respectively) in pellets at two sites. In another study of beaches on the English coast (Holmes et al. 2012), highest mean concentrations of Al, Fe, Mn, Co, and Ni (55.8, 97.8, 20.5, 0.1 and 0.1 $mg \cdot kg^{-1}$, respectively) were detected at one site, while a second location showed highest mean concentrations for Cu, Zn, and Pb (1.32, 23.3, and 1.64 $ng \cdot g^{-1}$, respectively). At a third site, Cr and Cd concentrations were 751 and 76.7 $ng \cdot g^{-1}$ (Holmes et al. 2012). Certain trace element concentrations were greater than concentrations of local estuarine sediments. This data is particularly unusual, as sediment minerals have significantly higher surface area and presence of charged sites per unit weight.

Given the capabilities for MP migration, it is likely that trace metals associated with MPs can be transported considerable distances in marine and

Table 5.1 Metals Bound to Microplastics in Freshwater and Marine Environments, Recent Data

Location	*Media*	*Particle Type*	*Polymer Type*	*Ag*	*Cd*	*Co*	*Cr*	*Cu*	*Hg*	*Ni*	*Pb*	*Zn*	*Mn*	*Fe*	*Reference*
China	Seawater	Fragments	PS		0.27[b]		14.9[b]	15.0[b]		17.2[b]	24.8[b]		730[b]		Xie et al. 2021
India	Freshwater sediment		PE, PET		0.65–5.78[b]		26.26–342.28[b]	0.2–119.59[b]		12.43–75.77[b]	0.04–104.63[b]	1.8–1191.52[b]			Sarkar et al. 2021
Norway	Freshwater sediment	Fragments, fibers	PE, PP, PA		0.05–0.58[b]			16.0–41.3[b]			14.0–30.1[b]			3451.6–5349.8[b]	Patterson et al. 2020
Thailand	Freshwater	Fragments, pellets, films, fibers	PP, PS, PE				2.95 [b]	13.0[b]		0.78[b]	17.61[b]				Ta and Babel 2020
England	Beach	Pellets	PE								0.037–0.095[b]				Turner et al. 2020
China	Seawater		PP, PVC		BDL[d]-0.023[b]		≤ 0.003[b]	0.08–0.22[b]			0.03–0.14[b]	BDL[d]	3.0–11.8[b]		Gao et al. 2019
Spain	Seawater	Pellets	PE, PET, PP, PS, PVC		0.81[c]		0.47–4.7[c]	0.26–2.95[c]			1.9–4.9[c]	0.50–0.63[c]			Godoy et al. 2019
France	Seawater	Pellets	PE				10[b]								Prunier et al. 2019
Iran	Seawater	Various debris			0.035[b]		0.915[b]	3.6[b]		2.03[b]	4.59[b]		32.2[b]	531[b]	Dobaradaran et al. 2018
Adriatic Sea, Croatia	Beach sediment	Beached pellets			2.9[a]		0.21[b]	0.21[b]		0.14[b]	0.26[b]	2.08[b]	1.78[b]	40.3[b]	Maršić-Lučić et al. 2018

(*continued*)

Table 5.1 (Continued)

Location	*Media*	*Particle Type*	*Polymer Type*	*Ag*	*Cd*	*Co*	*Cr*	*Cu*	*Hg*	*Ni*	*Pb*	*Zn*	*Mn*	*Fe*	*Reference*
Urban intertidal regions, British Columbia, Canada	Beach	Various	PVC		0.42[b]			3.81[b]			2.67[b]	4.3[b]			Munier and Bendell 2018
			LDPE		1.77[b]			47.53[b]			52.16[b]	604.24[b]			
Brazil	Seawater	Pellets	PE, PP				0.37–0.49[b]					1.35[b]		2.3–7.6[b]	Vedolin et al. 2018
UK			HDPE									236–4505[b]			Hodson et al. 2017
China	Freshwater	Pellets	PE, PP					0.28[b]			1.72[b]	0.25[b]			Wang et al. 2017
Portugal	Seawater	PS Beads (0.7–0.9μm)	PS					11.70[b]				29.33[b]			Brennecke et al. 2016
		PVC Frags (0.8–1.6mm)	PVC					3.11[b]				6.12[b]			
UK	River water	Beached pellets	Mix	< 3[a]	5.0[a]	13.8[a]	42.5[a]	47.0[a]	< 3[a]	29.3[a]	109[a]	196[a]	712[a]	34,400[a]	Turner and Holmes 2015
SW England	Seawater	Pellets	PE	N/A	0.52–14.9[a]	13.8–44.9[a]	42.5–413[a]	47–652[a]		29.3–91.1[a]	0.109–0.74[b]	0.196–1.99[b]	0.669–5.01[b]	34.4–67.5[b]	Holmes et al. 2012
Soar Mill Cove, England	Coastline	Nurdles (3–5 mm)	Mix	2,4[a]	1.7[a]	27[a]	19[a]	0.06[b]			0.15[b]	0.55[b]	1.58[b]	25.85[b]	Ashton et al. 2010

Thurlestone, England	Coastline	Nurdles (3–5 mm)	Mix	4.9[a]	10[a]	25[a]	63[a]	0.14[b]		0.41[b]	0.42[b]	1.28[b]	35.67[b]	Ashton et al. 2010
Bovisan, England	Coastline	Nurdles (3–5 mm)	Mix	6.0[a]	3.6[a]	54[a]	69[a]	0.29[b]		0.73[b]	0.94[b]	2.01[b]	54.6[b]	Ashton et al. 2010
Saltram, England	Coastline	Nurdles (3–5 mm)	Mix	30[a]	5.0[a]	101[a]	151[a]	0.61[b]		1.08[b]	2.34[b]	8.31[b]	64.97[b]	Ashton et al. 2010

a = $ng \cdot g^{-1}$
b = $\mu g \cdot g^{-1}$
c = $mg \cdot g^{-1}$
d = below detectable limits

freshwater environments (Ren et al. 2021; Brennecke et al. 2016; Holmes et al. 2012; Tourinho et al. 2010; Robards et al. 1997).

Mechanisms: Factors Affecting Metal Sorption by MPs

Metals occurring in water and soil are almost exclusively positively charged, yet the polymers occurring in MPs are inherently neutral in charge. It would seem intuitive, therefore, that metals should not sorb to MPs at all; however, this is clearly not the case. Several phenomena are responsible for the attraction between metals and MPs and are discussed below.

Polymer Type

As shown in chapter 3, different resin types possess unique chain configurations and substituents, which may ultimately affect sorption of metals and other contaminants. At three locations in San Diego Bay, PET, PVC, LDPE, and PP typically accumulated greater concentrations of metals (Al, Cr, Mn, Fe, Co, Ni, Zn, Cd, Pb) than did HDPE (Rochman et al. 2014). In another study (Gao et al. 2019), Cd, Cu, and Pb had higher adsorbance on PVC and PP particles compared with PA, PE, and POM. In a laboratory study (Zou et al. 2020), virgin plastic particles including chlorinated polyethylene (CPE), PVC, LDPE, and HDPE could load substantial quantities of Pb^{2+}, Cu^{2+}, and Cd^{2+}. Metal sorption was a function of the chemical structure and electronegativity of the sorbents (pellets). Sorption affinity of metals to MPs followed the sequence: CPE > PVC > HDPE > LDPE. Based on data from adsorption isotherms, sorbent surface area had little effect on sorption affinity. Surface area of the four MP types followed the order: PVC > HPE > CPE > LPE (Zou et al. 2020). In nine urban intertidal regions in Vancouver, British Columbia, Canada, macro- and MP polymers identified in order of abundance were: PVC, LDPE, PS, LDPE, PP, nylon, HDPE, polycarbonate, PET, polyurethane (PUR), and polyoxymethylene. It was found that (Munier and Bedell 2018):

- PVC, HDPE, and LDPE sorbed significantly greater amounts of copper;
- HDPE, LDPE, and PUR had significantly greater amounts of zinc;
- PVC and LDPE sorbed significantly greater amounts of cadmium; and
- PVC tended to have greater levels of lead.

Metals Involved

Different metals sorb to the same polymer to differing degrees, in part due to its oxidation state, electron configuration (i.e., number of occupied *d* orbitals), and other variables. CPE, PVC, LDPE, and HDPE MPs could load substantial amounts of Cu^{2+}, Cd^{2+}, and Pb^{2+} (Zou et al. 2020). The Pb^{2+} ion exhibited significantly stronger sorption than did Cu^{2+} and Cd^{2+}, however, due to strong electrostatic interaction. Chromium interacted relatively weakly with charged regions of beached pellets (Turner and Holmes 2015). The Cr occurred as the Cr^{6+} species, which exists in oxyanionic forms ($HCrO_4^-$ and CrO_4^{2-}).

Concentrations of metals in surrounding water play a significant role in the sorption process. Several studies report that high concentrations of major and trace elements can adhere to and become entrapped within MPs (Ashton et al. 2010; Brennecke et al. 2016; Rochman et al. 2014). At low initial concentrations (<20 $\mu g \cdot l^{-1}$), sorption of Ag, Cd, Co, Cr, Cu, Hg, Ni, Pb, and Zn by MPs ranged from 0.0004 to 2.78 $\mu g \cdot g^{-1}$ (Holmes et al. 2012; Holmes et al. 2014; Turner and Holmes 2015). At high initial Zn concentrations of Zn (up to 100 $mg \cdot L^{-1}$), sorption ranged as high as 236–7171 $\mu g \cdot g^{-1}$ (Hodson et al. 2017).

Solution pH

For many metals, pH strongly influences sorption to particulate plastic as well as other solids. Adsorption of Cd, Co, Ni, and Pb to both virgin and aged pellets was greater in river water, where pH was relatively low, whereas Cr sorption was higher in seawater (higher pH) (Turner and Holmes 2015). As pH increased, so too did adsorption of Cd, Co, Ni, and Pb. No significant effects on Cu or Hg sorption occurred with increasing pH.

Dissolved Organic Matter

In water bodies, dissolved organic matter (DOM) includes both *humic* and *non-humic* material (see box below). Particulate plastics are known to have a high affinity for DOM—organic matter occurs as a common coating on MPs in freshwater and marine environments (Bradney et al. 2019). Greater interaction and retention of metals is expected to occur on DOM-coated particles than would occur on pristine particles.

BIOFILMS AND ORGANIC MATTER

Biofilm. Mixture of microbial consortia which are enclosed within self-produced extracellular polymeric substances comprising a meshwork-like structure of carbohydrates, proteins, and DNA. In natural environments, biofilms serve as a protected growth system that allows microorganisms to survive in hostile environments.

Humic Material. High-MW polymers of soil organic matter composed of compounds of biological origin and synthesized by microbial reactions in soil.

Non-Humic Material. Organic molecules released directly from cells of fresh plant and animal residues, such as proteins, amino acids, sugars, and starches.

Aged PVC fragments have shown a high affinity to trace elements. A study (Brennecke et al. 2016) found that aged PVC adsorbed Cu, and that Zn sorption increased with time, possibly caused by accumulation of DOM which conferred greater reactivity to the surface. In another study (Wijesekara et al. 2018), the quantity of adsorbed Cu in particulate plastics reacted with different types of organic matter was significantly higher than that adsorbed to pristine particles.

Contribution of Biofilms

When MPs are suspended in marine and freshwater environments, their hydrophobic surfaces often stimulate rapid formation of biofilms (Ma et al. 2020). Biofilms are highly heterogeneous microenvironments which provide ecological advantages to microorganisms such as a protective barrier, a site for accumulation of nutrients, and a location for consortia of different species that degrade complex substrates. Biofilms typically consist of multiple species of microorganisms from one or more kingdoms such as bacteria, fungi, algae, and archaea. Members vary in terms of environmental requirements such as nutrient types and concentrations. Microbial numbers are often substantial; populations may range between 10^9 and 10^{11} cells·mL^{-1} biofilm mass (Pichtel 2017).

Rapid formation of biofilms has been observed on other hydrophobic substrates immersed in water; for example, organic matter and mineral deposits were identified on surfaces of glass slides after submersion in an urban lake for two weeks (Hua et al. 2012). The extent of biofilm-related sorption depends on the chemical nature of the biofilm, its rate of accretion, and residence time of the plastic substrate in the water column (Turner and Holmes 2015). As a result of biofilm formation, therefore, MPs acquire charge and a greater surface area (Ashton et al. 2010). These surface changes allow for charged metal ions and other contaminants to sorb to MPs.

Extent of Particle Weathering

MPs tend to acquire a surface charge and greater surface area through photo-oxidation, thermal processes, friction, hydrolysis, and other abiotic processes which result in the formation of various charged functional groups (see chapter 3). Polar functional groups such as hydroxyl, carbonyl, and others have been detected on surfaces of weathered plastics. These electron-rich groups readily bond with metals (Bradney et al. 2019; Wang et al. 2017). In a study by Rochman et al. (2014), concentrations of metals (Al, Cr, Mn, Fe, Co, Ni, Zn, Cd, Pb) sorbed to plastic particles increased over a 12-month period. Turner and Holmes (2015) determined that beached (i.e., aged) pellets adsorbed more metals than did virgin pellets. Holmes et al. (2014) showed sorption of Cd, Co, Cr, Cu, Ni, and Pb to MPs under estuarine conditions. Adsorption of all metals was considerably greater on beached pellets than on virgin pellets, presumably due to the weathering and attrition of the former. Vedolin et al. (2017) collected pellets from 19 beaches along the coast of São Paulo State in southeastern Brazil. Metal adsorption on beached pellets was greater than that on virgin pellets. Highest levels were for Fe (227.8 $mg{\cdot}kg^{-1}$) and Al (45.3 $mg{\cdot}kg^{-1}$). Kedzierski et al. (2018) determined that aged PVC adsorbed both Cu and Ag. PVC degradation occurred via the formation of chips from the polymer surface, possibly from desorption of plasticizers. This degradation resulted in increased surface area which may have contributed to greater metal sorption. Acosta-Coley et al. (2019) found that aged particulate plastics contained higher detectable concentrations of 28 metals than pristine particulate plastics. Their study demonstrated that weathering and degradation are important factors in trace element redistribution and concentration.

SORPTION OF PERSISTENT ORGANIC POLLUTANTS TO MPs

Seawater typically contains low levels of hazardous organic chemical species including insecticides, herbicides, and industrial chemicals; however, sites having elevated concentrations have been documented. These pollutants enter oceans via surface runoff and wastewater effluents (Wurl and Obbard 2004).

MPs are known to act as scavengers and transporters of persistent organic pollutants (POPs) (Bakir et al. 2012). Because of their small size, great surface-to-volume ratio, and hydrophobic properties, MPs are capable of adsorbing and accumulating organic pollutants from surrounding water (table 5.2) (Holmes et al. 2012, 2014; Hodson et al. 2017; Brennecke et al. 2016; Karapanagioti et al. 2011; Fries and Zarfl 2012; Bakir et al. 2014; Li et al. 2018; Guo et al. 2019a). In a study of plastic particles on Hawaiian, Mexican, and Californian beaches it was found that significant levels of POPs were sorbed to particles (Rios et al. 2007). Reported values ranged from PAHs 39–1200 $ng{\cdot}g^{-1}$; PCBs 27–980 $ng{\cdot}g^{-1}$; and DDT 22–7100 $ng{\cdot}g^{-1}$. POP levels on pellets collected near industrial facilities were much higher. Highest values reported were for PAHs (12000 $ng{\cdot}g^{-1}$) and DDT (7100 $ng{\cdot}g^{-1}$). A 2009 study reported data for eight US beaches (of which six were in California)

Table 5.2 Persistent Organic Pollutants Bound to Microplastics in Freshwater and Marine Environments, Recent Data

Location	*Media*	*MP Type*	*Pollutant*	*Concentration*	*Comments*	*Reference*
Taiwan	Six sandy beaches	Pellets	PCDD/Fs, PBDD/Fs, PBDEs, PCBs, PBBs, and their congeners	PCDD/Fs on MP surfaces from 1.9 to 14.6 $pg \cdot g^{-1}$. PBDD/Fs n.d.–29,343 $pg \cdot g^{-1}$ PBDEs 0.228–3888.6 $ng \cdot g^{-1}$ PCBs 0.04–5.7 $ng \cdot g^{-1}$	POPs could reach inner parts of pellets via diffusional transport.	Wang et al. 2021
Japan	Surface water	Various	PCBs and PBDEs in MPs and zooplankton	PCBs n.d. - 429.3 PBDEs n.d. - 2504.4 $ng \cdot g^{-1}$	PCBs higher in urban bay areas compared to off shore samples. Sporadic high PBDEs observed in off-shore samples (mostly higher BDE congeners).	Yeo et al. 2020
Brazil South Atlantic coastline	Beach stand line	Pellets	PAHs, PCBs	PAHs 1454 to 6002 $ng \cdot g^{-1}$ PCBs 0.8 to 104.6 $ng \cdot g^{-1}$	Older pellets, having undergone greater weathering, may be more prone to accumulate organic contaminants and thus pose a greater risk.	Gorman et al. 2019
Chile (San Vicente Bay)	Beach	Pellets	PBDEs, PCBs, DDT	PBDEs 10–133 $ng \cdot g^{-1}$ PCBs 3–60 $ng \cdot g^{-1}$ DDT 0.1–7 $ng \cdot g^{-1}$	Pellet weathering may influence sorption of one or more contaminants.	Pozo et al. 2019
Laboratory batch studies	Water	PE, PP, PA, PS 4 mm granules	PBDE	PE 3.78 $ng \cdot g^{-1}$ PP 5.52 $ng \cdot g^{-1}$ PA 5.69 $ng \cdot g^{-1}$ PS 6.41 $ng \cdot g^{-1}$	Sorption capacity in the order of PS > PA > PP > PE due to different crystallinity, specific surface area, and surface structure.	Xu et al. 2019

Cartagena, Spain	Treated urban effluent from WWTP	PE microbeads included in four commercial facial cleansers	PCBs	120–380 $\mu g \cdot g^{-1}$ (est.)	Maximum PCB sorption in facial cleanser containing TiO_2.	Bayo et al. 2018
Pacific Ocean	Seawater	PE, PP, PS	Deca-BDE		Flame-retardant deca-BDE tentatively identified on some PP trawl particles.	Ghosal et al. 2018
China (Three Gorges Reservoir)	Surface water and sediment	PS, PP, PE, PC, PVC	Pharmaceutical Intermediates or solvents (nonanoic acid, 4-aminobenzoic acid, *p*-tolualdehyde, pth-methionine		PS the most common (38.5%); PP (29.4%); PE (21%).	Di and Wang 2018
Santos Bay, Brazil	Sandy beaches	PE, PP	PAHs	386–1996 $ng \cdot g^{-1}$	A predictable variation in PAH concentrations and compositions of pellets of different ages and resin types. This directly influences potential for toxicological effects.	Fisner et al. 2017
Norway	Seabirds	Various particles	PCBs, DDT, PBDEs	PCBS 2.21–10.94 $ng \cdot g^{-1}$ DDT 17,2–139 $ng \cdot g^{-1}$ PBDE 0.62–1669 $ng \cdot g^{-1}$	Correlations among POP concentrations in plastic and bird tissues do not necessarily imply that plastic acts as a substantial carrier for POPs.	Herzke et al. 2016

(*continued*)

Table 5.2 (Continued)

Location	*Media*	*MP Type*	*Pollutant*	*Concentration*	*Comments*	*Reference*
Brazil	41 beaches (15 cities) along coastline of Sao Paulo	Pellets	PAHs, PCBs, organochlorine pesticides such as DDTs, PBDEs	PAHs 192 to 13708 $ng \cdot g^{-1}$ PCBs 3.41 to 7554 $ng \cdot g^{-1}$ DDTs < 0.11 to 840 $ng \cdot g^{-1}$	Lower concentrations on the less urbanized and industrialized southern coast, highest values in central coastline, which is affected by both waste disposal, large port, and industrial complex.	Taniguchi et al. 2016
Switzerland	Surface water, beach sediment, birds, fish	PE, PP, PS (fragments, pellets, cosmetic beads, lines, fibers, films, foams)	PCBs, OCPs, PAHs, PBDEs, phthalates, BPA	PCBs 97.6 $ng \cdot g^{-1}$ OCPs 187.3 $ng \cdot g^{-1}$ PAHs 1202 $ng \cdot g^{-1}$ PBDEs 125.9 $ng \cdot g^{-1}$ Phthalates 18039 $ng \cdot g^{-1}$ BPA 16.6 $ng \cdot g^{-1}$	Particles sampled after large rain events generally contain lower concentrations of adsorbed compounds. PE has greater affinity for pesticides and PAHs, and less with phthalates.	Faure et al. 2015
North Pacific Ocean	Seawater	PP, PE	PAHs PCBs	1–846 $ng \cdot g^{-1}$ 1–223 $ng \cdot g^{-1}$	Transfer of these toxic chemicals from plastic to oily membranes in marine creatures should be expected.	Rios and Jones 2015
China	Two beaches	Pellets	PAHs, PCBs, HCHs, DDTs, chlordane, heptachlor, endosulfan, aldrin, dieldrin, endrin	PAHs 136.3–2384.2 $ng \cdot g^{-1}$ PCBs 21.5–323.2 $ng \cdot g^{-1}$ DDTs 1.2–127.0 $ng \cdot g^{-1}$	MPs can serve as an effective media for monitoring local concentrations of organic pollutants.	Zhang et al. 2015
	In seawater and under simulated gut conditions.	PVC, PE	Phenanthrene, DDT, PFOA, DEHP	–	Gut surfactants present a significant effect on enhancing desorption rates of sorbed contaminants.	Bakir et al. 2014

Portugal	Shoreline	Pellets	PAHs PCBs DDT	PAH: 53–44800 $ng \cdot g^{-1}$ PCBs: 2–223 $ng \cdot g^{-1}$ DDT: 0.42–41 $ng \cdot g^{-1}$	Efforts needed to assess points of entry of plastic pellets in order to minimize impacts on ecosystems.	Antunes et al. 2013
17 countries	30 beaches	PE pellets	PCBs DDTs HCHs	PCBs 5–605 $ng \cdot g^{-1}$ DDTs 1.69–267 $ng \cdot g^{-1}$ HCHs 0.15–37.1 $ng \cdot g^{-1}$	High concentrations of DDTs on the US west coast and in Vietnam. High concentrations of pesticide HCHs detected in pellets from southern Africa.	Ogata et al. 2009
Japan	Beaches	PE, PP pellets	PCBs	< 28–2300 $ng \cdot g^{-1}$	Sporadic high concentrations of PCBs found in pellets from remote islands, suggesting that pellets could be dominant route of exposure to the contaminants at remote sites.	Endo et al. 2005

BDE—decabrominated diphenyl ether; BPA—bisphenol A; DDT—dichlorodiphenyltrichloroethane; DEHP—di(2-ethylhexyl) phthalate; HCB—hexachlorobenzene; HCH—hexachlorocyclohexane; OCP—organochlorine pesticides; PAH—polycyclic aromatic hydrocarbons; PBB—polybrominated biphenyls; PBDD/F—polybrominated dibenzodioxin/furan; PBDE—polybrominated diphenyl ethers; PCB—polychlorinated biphenyl; PFOA—perfluorooctanoic acid.

as follows (Ogata et al. 2009): PCBs 32–605 $ng \cdot g^{-1}$; DDT 2–106 $ng \cdot g^{-1}$; and HCH (4 isomers) 0–0.94 $ng \cdot g^{-1}$.

Concentrations of PAHs on MPs have been found to be very high (>5000 $ng \cdot g^{-1}$) in East Asia, such as on beaches on the Yellow Sea, China, and Ookushi, Japan. MPs from the coast of São Paulo State, Brazil, and southwest England also had substantial concentrations of PAHs (Guo et al. 2019). High concentrations of PCBs (>2000 $ng \cdot g^{-1}$) on MPs were found on beaches in Brazil and Japan (Guo et al. 2019). In rather remote areas such as the Caribbean Sea, the Hawaiian Islands, and the east coast of Australia, concentrations of PAHs and PCBs were much lower (<500 $ng \cdot g^{-1}$). Concentrations of POPs on MPs also demonstrate spatial variability. It has been concluded that concentrations of PCBs, PAHs, and HCHs on MPs from big cities were about an order of magnitude greater than those from small cities and remote areas (Wang et al. 2018).

Mechanisms of Hydrocarbon Sorption

The hydrocarbon sorption behavior of MPs is related to a suite of factors including chemical composition of particles, types and concentrations of pollutants, and environmental conditions in surrounding water (Holmes et al. 2012; Guo et al. 2012; Turner and Holmes, 2015; Zhang et al. 2018a; Hu et al. 2017). Different MP types have distinct sorption affinities for the same pollutants, which are associated with differences in rubbery domains, polarity, and functional groups on the polymer (Guo et al. 2012; Wang et al. 2015). Polyethylene (PE) has been reported to sorb more organic pollutants than other kinds of MPs (Bakir et al. 2012; Wang and Wang 2018a, 2018b). Various researchers (Bakir et al. 2012; Wang and Wang 2018a, 2018b; Wang et al. 2015, 2018) have determined that the sorption capacities of pyrene, phenanthrene (PHE), PCBs, PFOS, DDT, and lubricating oils on PE were higher than for other types of MP polymers.

Aged MPs are inclined to sorb more hydrocarbon pollutants than do virgin ones (Brennecke et al. 2016; Müller et al. 2018).

POPs such as PCBs, PBDEs, and PFOA have large polymer-water distribution coefficients, indicated as $K_{P/W}$ (see box below); in other words, these pollutants prefer to sorb to a solid surface rather than remain in the water phase. Researchers have determined high values for $K_{P/W}$ on common polymers (Mato et al. 2001). One study (Teuten et al. 2007) reported sorption of PHE by PE, PP, and PVC. The distribution coefficients were ranked as: PE = PP > PVC. The authors found that desorption of the PHE was very slow, and that sediment tended to desorb the PHE more rapidly than did the plastics fragments.

POLYMER-WATER DISTRIBUTION COEFFICIENTS

A linear isotherm model relates the mass of the chemical sorbed per unit mass of solid polymer (q_e) [µg·kg^{-1}] to the equilibrium solute concentration (C_e) [µg·L^{-1}] by the following equation:

$$q_e = KP/W \times C_e$$

where KP/W (L·kg^{-1}) is the equilibrium distribution coefficient for the system (Friedman et al. 2009).

Many POPs are complex compounds which include different congeners or isomers. MPs may adsorb certain isomers of POPs preferentially (Wang et al. 2018).

SORPTION OF RADIONUCLIDES TO MPs

As discussed earlier in this chapter, a range of heavy metals has been documented to sorb to MP surfaces. Published data regarding sorption of radionuclides to MPs, is, however, severely lacking. A number of nuclides occur naturally in water and soil. Biota have evolved to tolerate and even thrive in the presence of radioactive elements such as ^{40}K, ^{87}Rb, ^{235}U, ^{238}U, and ^{232}Th. These are widely distributed throughout global oceans and have extremely long half-lives.

Given the likelihood of MPs acting as vectors of environmental contaminants, concern exists regarding possible sorption, transport, and biotic uptake of anthropogenic radionuclides.

The greatest inputs of nuclides from anthropogenic sources were a consequence of the above-ground nuclear weapons tests that began in 1945 and continued up through the early 1960s. A total of 528 above-ground nuclear weapons tests occurred across the globe between 1945 and 1980 (Arms Control Association 2020).

The largest detonations of the 1960s punched radionuclides into the stratosphere; as a result, so-called *fallout* has been distributed worldwide. The range of radionuclides released from weapons testing is vast; they are sometimes grouped into those of greatest radiological concern—^{90}Sr, ^{137}Cs, ^{131}I, ^{238}Pu, ^{239}Pu, ^{240}Pu, and ^{241}Am. The second group is represented by radionuclides such as ^{3}H, ^{14}C, ^{99}Tc, and ^{129}I which are of lower health concern (Livingston and Povinec 2000). The largest source of ^{137}Cs is global fallout from nuclear weapons testing; cumulative deposition from

1945 through 1980 is estimated at approximately 950 PBq (Buesseler et al. 2014).

Much of the fallout was inevitably delivered to the world's oceans, given their preponderance on the earth's surface. Most of these radionuclides are highly soluble in seawater (Buesseler 2014); their mobility may, therefore, be significant, depending on ocean currents and mixing processes. In contrast, isotopes of Pu and Am are of low solubility. In coastal regions, these isotopes tend to be removed rapidly from the water column via sinking as particulate matter (Livingston and Povinec 2000).

Other events have released significant quantities of radionuclides to aquatic ecosystems. In 1957, a fire occurred at the Windscale (UK) nuclear reprocessing site on the Irish Sea, which released vast quantities of nuclides into the atmosphere. Fallout spread across the United Kingdom and the rest of Europe. Nuclides of greatest concern were ^{131}I, ^{137}Cs, and ^{133}Xe, among others. The Chernobyl accident of 1986 (Ukraine, USSR) is considered by many to be the largest accidental release of ^{137}Cs to the environment. This event also released Pu, Ru, and Ce isotopes (Schwantes et al. 2012; Yoshida and Kanda 2012). A preponderance of the fallout touched down on European soil. Due to prevailing winds and the great distance between Chernobyl and the Baltic, North, Black, and Mediterranean Seas, the input of ^{137}Cs from Chernobyl to global oceans was about 15% to 20% of the total released (Aarkrog 2003). In 2011, the Tohoku earthquake and resulting tsunami led to meltdowns at the Fukushima Daiichi Nuclear Power Plant with consequent massive release of radionuclides to the Pacific Ocean. Parts of the Japanese east coast, in particular Fukushima Prefecture, were heavily contaminated by radionuclides, especially Cs (^{134}Cs, ^{137}Cs), ^{40}K, Sr (^{89}Sr, ^{90}Sr), and Zr (^{91}Zr, ^{95}Zr) (Tazaki 2015; Yu et al. 2015).

Many other significant releases of nuclides to the world's oceans have occurred. In the early 1990s the Soviet Union admitted to dumping large volumes of radioactive waste into rivers and seas, notably into the Arctic Ocean, between 1959 and 1992. The locations of the containers are not well documented, and concerns exist that some containers have breached with consequent release of radionuclides. From 1946 through the end of 1962, the United States dumped about 86,000 containers of radioactive waste into the oceans (Hamblin 2008).

To date, few studies have addressed sorption of radionuclides to MPs. Elevated ^{137}Cs, with putative origin from the Fukushima accident, were detected on the surfaces of plastics collected from a contaminated lake and was associated with adsorbed clay and diatoms (Tazaki et al. 2015). The adsorption of ^{232}U by MPs (PE, polyamide nylon [PN6] and PVC) was evaluated in a laboratory study (Ioannidis et al. 2022). According to the experimental data (e.g. partition coefficient values, K_d), the sorption efficiency differed significantly between MP resin types. As it the case with other heavy metals, solution pH determined sorption efficiency to a large degree. Sorption efficiency in seawater declined due to the increased salinity and the presence of competing nuclide species. Even under seawater conditions, however, the sorption affinity of the MPs for the radionuclides is significantly higher than for seawater sediments (Ioannidis et al. 2022), thus suggesting that MPs are a significant radionuclide reservoir. This is a concern, given the mobility of plastics in marine and freshwater ecosystems. In contrast, the degree of ^{134}Cs and ^{85}Sr adsorption onto HDPE MPs was approximately 2–3 orders of magnitude lower than for sediment reference values (Johansen et al. 2018). The highest K_d was for Cs in freshwater (80 $mL \cdot g^{-1}$) and lowest for Sr in estuarine conditions (5 $mL \cdot g^{-1}$). However, adsorption occurred on all samples. Particle surfaces were covered by mineral elements such as Fe and Ti, indicating that adhered mineral/clay agglomerates may increase overall adsorption capacity.

SORPTION OF ANTIBIOTICS

Antibiotics are widely distributed in waterways such as rivers and seawater (Guo et al. 2019a). Urine and feces from persons using antibiotics contain significant residual active antibiotics which contaminate wastewater (Kiimmerer 2009a). Discharge of pharmaceutical industry waste without treatment may contain high levels of antibiotics (Larsson et al. 2007; Li et al. 2008; Sim et al. 2011), which pollute local waters (Fick et al. 2009). Direct disposal of medical waste from hospitals may add residual antibiotics and other medications in wastewater (Jarnheimer et al. 2004; Imran et al. 2019). Additionally, huge quantities of antibiotics are applied in animal production (i.e., livestock) to enhance their growth or support their survival in crowded and unsanitary environments. Livestock wastes contain antibiotics and, when used as soil amendments and fertilizers, ultimately reach water bodies as a consequence of surface runoff and erosion (Sarmah et al. 2006; Kiimmerer 2009a, 2009b; Pruden et al. 2013). Many antibiotics are known to occur in sewage sludges. In the United States, an estimated 4 million dry tons of biosolids are land-applied per year (Lu et al. 2012). The number is 3.6 million dry tons in the EU (Evans 2012) and 2.3 million dry tons in China (Wei et al. 2020). When applied to land, it is possible that substantial quantities of antibiotics may be lost via runoff of MPs.

Data on the sorption behavior of antibiotics, pharmaceuticals, and other PPCPs to MPs is emerging only recently. This family of compounds possesses a range of diverse properties; therefore, their sorption behavior may differ from that of POPs. The significant adsorption of several organic contaminants of water, including pesticides, personal care products, and pharmaceuticals, is attributed primarily to the hydrophobicity of the compound (Puckowski et al. 2021; Wu et al. 2016). However, other factors may be involved, including solution pH, salinity, and the presence of DOM (table 5.3).

The sorption capacity of several PPCPs (carbamazepine (CBZ), 4-methylbenzylidene camphor (4MBC), triclosan (TCS), and 17α-ethinyl estradiol (EE2), to PE debris (250 to 280 μm) was related to their hydrophobicity (Wu et al. 2016). Some compounds such as TCS experienced enhanced sorption with an increase in salinity. An increased concentration of humic acid reduced sorption of 4MBC, EE2, and TCS but not CBZ. When PE, PS, PP, PA, and PVC were reacted with five antibiotics [sulfadiazine (SDZ), amoxicillin (AMX), tetracycline (TC), ciprofloxacin (CIP), and trimethoprim (TMP)], polyamide MPs were found to effectively bind to AMX, TC, and CIP (Li et al. 2018). For the other four plastics, adsorption amounts decreased in the order: CIP > AMX > TMP > SDZ > TC. The K_f values correlated positively with octanol-water partition coefficients, K_{ow}. Tylosin, an antibiotic, was adsorbed efficiently on the surface of PE, PP, PS, and PVC MPs by up to more than 80% at 8 h (Guo et al. 2018). In contrast, sulfamethoxazole and tetracycline does not strongly interact with PS, PP, or PE MPs (Guo et al. 2019c; Xu et al. 2018). In these cases, solution pH and ionic strength of the solution played a key important role in sorption.

SORPTION OF MICROORGANISMS TO MPs

It has been suggested that biofilms occurring on MPs attract organisms to adhere, thus improving predatory efficiency and enhancing rates of energy, material, and information flow in the aquatic environment (Shen et al. 2019). A number of studies have identified microbial colonizers of MPs (table 5.4); a global survey showed that up to 50% of floating debris, including plastics and natural organic particles, are colonized by marine organisms (Barnes 2002; Ma et al. 2020).

One study (Harrison et al. 2014) detected bacterial colonization of LDPE MPs after exposure to marine sediment for seven days. Another study (Ogonowski et al. 2018) reported that plastics-associated communities were distinctly different from those growing on non-plastic substrates. For example, the abundance of Burkholderiales was twofold higher on plastic than on non-plastic media, while the latter had a significantly higher proportion of Actinobacteria and Cytophagia. The variation in community structure was linked to surface properties (specifically, hydrophobicity) of the substrate.

The microbial communities on marine plastics are known to differ consistently from surrounding seawater communities, and the term "plastisphere"

Table 5.3 Sorption of Antibiotics and Other Pharmaceuticals to Microplastics, Recent Data

MP Type	*Compound(s)*	*Concentration or Partition Coefficient*	*Comments*	*Reference*
PE and tire wear particles (TWP)	Chlortetracycline (CTC) and amoxicillin (AMX)	CTC: from -0.154 (fresh TWP) to 1.16 (aged TWP); AMX: from -0.16 (aged TWP) to 0.08 (fresh TWP) $L \cdot mg^{-1}$	Main adsorption modes of TWP and PE were surface adsorption and intraparticle diffusion, respectively. TWP might have stronger carrier effects on antibiotics compared to PE.	Fan et al. 2021
PS and hexabromocyclododecane composite (HBCD-PS) MPs	Tetracycline (TC)	PS: mean 0.005 HBCD-PS: mean 0.006 $L \cdot \mu mol^{-1}$	TC sorption was solution pH-dependent; effect of NaCl content on sorption was negligible. π-π and hydrophobic interactions regulated the sorption of TC onto MPs.	Lin et al. 2021
PE, PP	Enrofloxacin (ENR), ciprofloxacin (CIP), norfloxacin (NOR), 5-fluorouracil (5-FU), methotrexate (MET), flubendazole (FLU), fenbendazole (FEN),propranolol (PRO) and nadolol (NAD	Highest K_d value: 2.4 $L \cdot kg^{-1}$ for PRO on PP. Lowest K_d values: up to 0.2 $L \cdot kg^{-1}$ for NAD, 5-FU and MET.	Ionic strength of medium a significant factor influencing sorption potential.	Puckowski et al. 2021
PET, HDPE, PVC, LDPE, PP	Atenolol, sulfamethoxazole, and ibuprofen	PET, HDPE, PVC, LDPE, and PP showed very similar coefficients (~0.005 $L \cdot kg^{-1}$)	Surface area a major determinant of adsorption, regardless of material type.	Magadini et al. 2020

(*continued*)

Table 5.3 (Continued)

MP Type	*Compound(s)*	*Concentration or Partition Coefficient*	*Comments*	*Reference*
PE, ultra-high MW PE (UHMWPE), and PP	Non-steroidal anti-inflammatory drugs (NSAIDs): ibuprofen (IBU), Naproxen (NPX), and diclofenac (DCF)	Highest K_d value: 8.84 $L \cdot kg^{-1}$ for IBU on PE; lowest K_d value: 0.64 $L \cdot kg^{-1}$ for NPX on PE.	Sorption on the microplastic particles followed trend: DCF ≈ IBU > NPX, which coincides with the log K_{ow} values, DCF (4.51), IBU (3.97) and NPX (3.18).	Elizalde-Velázquez et al. 2020
PA, PE, PS, PET, PVC, and PP	Sulfamethazine (SMX)	K_d values: PA: 38.7 PE: 23.5 PS: 21.0 PET: 22.6 PVC: 18.6 PP: 15.1 $L \cdot kg^{-1}$	80% of SMX was sorbed to microplastics in 2 hours.	Guo et al. 2019b
PE, PS (naturally aged)	Sulfamethoxazole (SMX), Sulfamethazine (SMT), and Cephalosporin C (CEP-C)	K_d values ranged from 0.0236 $L \cdot g^{-1}$ to 0.0383 $L \cdot g^{-1}$	Main sorption mechanisms are hydrophobic, van der Waals, and electrostatic interactions.	Guo et al. 2019c
PS	Triclosan (TCS)	K_d values of 0.15–0.18 $L \cdot g^{-1}$ from 288–318 K.	TCS sorption on PS was higher within pH range 3.0–6.0; decrease occurred at pH > 6.0. No obvious variation in sorption at different NaCl concentrations.	Li et al. 2019
PE, PP, PS, PVC	Tylosin (TYL)	K_d ranged from 62.75–155.27 $L \cdot kg^{-1}$	Sorption capacity of TYL on MPs followed the order: PVC > PS > PP > PE	Guo et al. 2018
polyethylene (PE), polystyrene (PS), polypropylene (PP), polyamide (PA), and polyvinyl chloride (PVC)	sulfadiazine (SDZ), amoxicillin (AMX), tetracycline (TC), ciprofloxacin (CIP), and trimethoprim (TMP	(K_d) values ranged from 7.36 ±0.257 to 756 ±48.0 $L \cdot kg^{-1}$	results indicated that commonly observed polyamide particles can serve as a carrier of antibiotics in the aquatic environment.	Li et al. 2018

PE	Sulfamethoxazole (SMX), propranolol (PRP), and sertraline (SER)	SER: 28.61% PRP: 21.61% SMX: 15.31%	A total of 8% and 4% PRP and SER, respectively, were desorbed from the microplastics within 48 h; sorption of SMX was irreversible.	Razanajatovo et al. 2018
PE	Tetracycline (TC)	Total of 91.7 $\mu g \cdot g^{-1}$ TC adsorbed to aged MPs at pH 7.	Little impact of environmental factors (pH, ionic strength, temperature) on adsorptive capacity of MPs for TC. MPs aged in humic acid solution exhibited significant decreased adsorptive capacity for TC.	Shen et al. 2018
PE	Sulfamethoxazole (SMX)	K_d: 591.7 $L \cdot kg^{-1}$	Sorption of SMX to PE microplastics not significantly influenced by pH or salinity. Negligible effect of dissolved organic matter on sorption due to greater affinity to PE microplastics.	Xu et al. 2018
PS	Oxytetracycline	K_d: Virgin PS: 41.7±5.0 Beached PS: 428.4±15.2 $L \cdot kg^{-1}$	Maximum adsorption at pH 5. Effects of pH on adsorption to beached foams more pronounced to the virgin foams. Electrostatic interaction may have regulated adsorption.	Zhang et al. 2018
PE	carbamazepine (CBZ), 4-methylbenzylidene camphor (4MBC), triclosan (TCS), and 17α-ethinyl estradiol (EE2)	K_d: CBZ: 191.4 EE2: 311.5 TCS: 5140 4MBC: 53225 $L \cdot kg^{-1}$	Sorption capacity of PPCPs related to their hydrophobicity. Increase in dissolved organic matter content reduced sorption of 4MBC, EE2, and TCS.	Wu et al. 2016

Table 5.4 Microbial Species and Consortia Bound to Microplastics in Freshwater and Marine Environments, Recent Data

Location	*MP Type*	*Microbial Type(s)*	*Comments*	*Reference*
Laboratory study; seawater	PE microbeads, polyester microfibers	*Toxoplasma gondii, Cryptosporidium parvum, Giardia enterica*	More parasites adhered to microfibers compared to microbeads.	Zhang et al. 2022
Coastal regions, China (sediment)	PS	*Vibrio, Exiguobacterium*	Epiphytic bacteria on eelgrass (*Zostera marina*) leaves formed biofilms, ultimately increasing MP density and facilitating sinking.	Zhao et al. 2022
Río de la Plata estuary, Argentina		*Eschericia coli, Enterococci*	Fragments (69.6%), films (20.5%), pellets (9.4%). foam (0.4%). MPs > 1000 μm the most common size category.	Pazos et al. 2020
Mallorca, Spain	PE	*Roseobacter*-like (likely *Thalassococcus halodurans*), *Oleiphilus, Aestuariibacter*	Plastispheres vary with state of plastic weathering. Early colonizing communities enriched with taxa that can potentially degrade hydrocarbons.	Erni-Cassola et al. 2020
Haihe Estuary, China	PE, PP, PS, PET, PUR	*Halobacteriaceae, Pseudoalteromonadaceae, Pseudomonas* (potentially pathogenic), and *Bacillus* (potentially pathogenic)	Abundance of potentially pathogenic bacteria (e.g., *Pseudomonas* and *Bacillus*) on MPs significantly higher than in the ambient environment. MPs could be messengers, facilitating bacterial transport between water and sediment. Little effect of polymer type on bacterial community structure.	Wu et al. 2020
East Lothian, Scotland, UK. Five public beaches.	Pellets (nurdles)	*E. coli* and *Vibrio* spp.	Biofilms that colonize the plastisphere could be a reservoir for fecal indicator organisms.	Rodrigues et al. 2019

Singapore. Populated and pristine beaches.		*Erythrobacter* (21%), *Cohaesibacter* (12%), *Hyphomonas* (10%), *Arcobacter* (6%), *Albimonas* (5%), *Bacteroides* (4%), *Brachymonas* (5%), *Pseudomonas* (5%), *Sphingobium* (4%)	MPs primarily foam particles (55%) and fragments (35%). MPs significantly more abundant on heavily populated beaches compared to pristine beaches.	Curren and Leong 2019
Northern Corsica (Calvi Bay, Mediterranean Sea), (sediments and water columns)	FP, PE, PVC, PET, LDPE, PS	*Alcanivorax borkumensis, Microbulbifer* sp., *Rhodobacteraceae, Alcanivoraceae, Flavobacteraceae.*	Bacterial community composition is polymer-dependent.	Delacuvellerie et al. 2019
Coastal regions of Tamil Nadu, India	HDPE	*Pseudomonas, Bacillus*	Ten bacterial isolates considered to be potent HDPE degraders.	Sangeetha Devi et al. 2019
Baltic Sea	PE, PS	*Devosiaceae, Sphingomonadaceae, Pseudomonas, Alteromonadaceae, Rhodobacteraceae, Vibrio* spp.	Sample type was the most important factor structuring bacterial assemblages. Surface properties were less significant in differentiating attached biofilms on PE and PS. Abundances correlated with salinity.	Kesy et al. 2019
Northeast Germany (7 stations) (brackish ecosystems)	PE, PS	*Pfiesteri* (harmful dinoflagellate). More than 500 different taxa of eukaryotes identified on MPs, dominated by Alveolata, Metazoa, and Chloroplastida.	Eukaryotic community composition on MP was significantly distinct from the surrounding water.	Kettner et al. 2019

(*continued*)

Table 5.4 (Continued)

Location	*MP Type*	*Microbial Type(s)*	*Comments*	*Reference*
Helgoland, Germany	HDPE, LDPE, PP, PS, PET, PLA, SAN, PVC	*Aquibacter, Oceanococcus, Parvularcula, Nannocystaceae, Polycyclovorans, Ulvibacter, Phyllobacteriaceae, Labrenzia, Maricaulis, Planctomycetes* OM190*, Simiduia, Planctomycetes* BD7-11*, Winogradskyella, Dokdonia, Spongiibacter, Roseovarius, Rhizobiales* OCS116 *, Congregibacter, Saprospiraceae, Planctomycetes* SPG12- 401-411-B72*, Hirschia, Erythrobacter, Flexithris*.	The presence of plastic-specific microbes/assemblages identified, which could benefit from specific plastics properties.	Kirstein et al. 2019
King George Island (Antarctica)	PS piece	*Alteromonas hispanica, Pseudomonas balearica, Thalassospira lohafexi, Shewanella* sp., *Halomonas* sp., *Pseudoalteromonas* sp.	Results suggest that plastics can serve as vectors for spread of multiple resistances to antibiotics across Antarctic marine environments.	Lagana et al. 2019
Yangtze Estuary, China (intertidal zone)	PE, PP, PS	*Rhodobacterales, Sphingomonadales, Rhizobiales*	Metabolic pathway analysis suggested adaptations of bacterial assemblages to the plastic surface-colonization lifestyle, including greater xenobiotics biodegradation and metabolism.	Jiang et al. 2019
Tara-Mediterranean expedition	PE, PP, PS	*Cyanobacteria* (40.8%, mainly *Pleurocapsa*), *Alphaproteobacteria* (32%, mainly *Roseobacter*).	Taxonomic diversity higher in plastic marine debris communities compared with free-living communities.	Dussud et al. 2018

Bay of Brest, Brittany, France	PE, PP, PS	*Vibrio splendidus.* PE: *Pseudomonadales, Oceanospirillales, Propionispira* PP: *Alphaproteobacteria, Holophagae* PS: *Rhodospirillaceae, Roseovarius, Nitrosomonas*	*Vibrio splendidus* species harboring putative oyster pathogens detected on most MPs. No effect of MP size on community composition. Microbial assemblages on PS were distinct from those on PE and PP; suggest that additives are related to bacterial colonization.	Frere et al. 2018
Gulf of Oman	PET, PE	*Actinobacteria* *Microcyctis* (PET-specific) *Hydrogenophaga, Sphingomonas* and *Laceyella* (PE-specific).	Despite prevalence of the same bacterial groups on all substrates, genera-level differences were noted. SEM and FTIR analyses indicated PET and PE degradation.	Muthukrishnan et al. 2019
North Atlantic Sea	PET, PE, PS	*Planctomycetes, Alphaproteobacteria, Bacteroidetes, Cyanobacteria* and *Chloroplexi,* Gammaproteobacteria, *Rhodobacteraceae, Ralstonia*, unclassified *Gammaproteobacteria, Thiotrichiales, Muricauda, Pelomonas, Sphingomonas, Acinetobacter, Staphylococcus epidermis, Thalassospira*	Majority of MPs colonized by Betaproteobacteria. Bacteria inhabiting plastics harbored distinct metabolisms from those present in surrounding water.	Debroas et al. 2017
North Sea	PE	*Alpha-* and *Betaproteobacteria, Flavobacteria, Gammaproteobacteria, Ascomycota,* and *Basidiomycota.*	Identified and characterized fungal genera on plastic debris. Data suggest that biofilm formation is severely hampered in the natural environment.	De Tender et al. 2017

(*continued*)

Table 5.4 (Continued)

Location	*MP Type*	*Microbial Type(s)*	*Comments*	*Reference*
Baltic Sea	PE, PP, PS	*Vibrio* spp.	*V. parahaemolyticus* isolated from PE fibers and PE fragments.	Kirstein et al. 2016
Germany	PS	*Amphritea atlantica, Lentisphaera marina*	Study of whether bacterial assemblages on PS are selectively modified during their passage through the gut of the lugworm *Arenicola marina* and are able to develop pathogenic biofilms.	Kesy et al. 2016
Belgian North Sea	PE pellets	*Actinobacteria, Proteobacteria, Vibrionaceae, Pseudoalteromonadaceae, Alpha-* and *Gamma proteobacteria, Bacteroidetes, Mycobacteria*	Substantial differences in bacterial community composition compared to those in sediment and seawater. Intrinsic plastic properties such as pigment content may contribute to differences in bacterial colonization.	De Tender et al. 2015
Humber Estuary, UK	LDPE	*Arcobacter, Colwellia*	Bacteria in coastal marine sediments can rapidly colonize LDPE MPs. Evidence for successional formation of plastisphere-specific bacterial assemblages. Both *Arcobacter* and *Colwellia* spp. associated with degradation of hydrocarbon contaminants in certain marine environments.	Harrison et al. 2014

has been introduced for this unique habitat (Zettler et al. 2013). One study (Amaral-Zettler et al. 2015) has reported that plastisphere communities possess dominant taxa that are highly variable and diverse. In plastics sampled from surface waters in the North Sea (UK) diverse plastisphere communities were found, composed of bacteria belonging to Bacteroidetes, Proteobacteria, Cyanobacteria, and members of the eukaryotes Bacillariophyceae and Phaeophyceae (Oberbeckmann et al. 2014). Polymers collected and sampled from the sea surface were mainly PE, PS, and PP particles. Variation within plastisphere communities on the different polymer types was observed, but communities were primarily dominated by Cyanobacteria. Cyanobacteria have been found extensively on MPs collected in other marine locations as well (Dussud et al. 2018; Oberbeckmann et al. 2014).

The plastic chemical composition, combined with seasonal and geographic factors, are important for supporting microbial colonization of plastics in the ocean (Oberbeckmann et al. 2014). Fletcher and Loeb (1979) determined that substantial numbers of *Pseudomonas* sp. attached to hydrophobic plastics (PE, PS, PETE, Teflon). More recently, microscopic investigations revealed that microbes were most strongly attached to PP followed by LDPE, PS, HDPE, PUR prepolymer, PVC, styrene-acrylonitryle, PLA, and PET (Kirstein et al. 2018).

Sorption of Pathogenic Microorganisms

MPs can function as a vector for dispersal of pathogenic species. The dispersal of harmful microorganisms attached to plastic was first suggested by Masó et al. (2003), who identified potentially harmful microalgae such as *Ostreopsis* sp., *Coolia* sp., and *Alexandrium* on plastic debris collected from the northwestern Mediterranean.

In studies by Zettler et al. (2013), genetic sequences affiliated to *Vibrio* spp. were detected on marine plastic particles in the North Atlantic. The genus *Vibrio* harbors several species that are an integral part of the marine community and contribute to biofilm formation on various surfaces (Pruzzo et al. 2008; Romalde et al. 2014). Some *Vibrio* species are known to be animal pathogens, invading coral species and causing coral bleaching (Ben-Haim et al. 2003); others are classified as human pathogens (Morris and Acheson 2003). *Vibrio parahaemolyticus*, *V. vulnificus*, and *V. cholerae* are especially of concern as water-related human pathogens, as they cause wound infections associated with recreational bathing, and septicemia or diarrhea after ingestion of contaminated food or water (Thompson et al. 2004). *Vibrio* infections are common in tropical areas; however, over the last decade a significant increase in documented cases has occurred in more temperate northern waters (Eiler et al. 2006; Gras-Rouzet et al. 1996; Martinez-Urtaza et al. 2005).

De Tender et al. (2015) detected Vibrionaceae on marine plastics from the Belgian North Sea; however, identification to the species level was not conducted. Pathogenic species (e.g., *V. cholerae*, *V. vulnificus*, and *V. parahaemolyticus*) were detected on plastic debris in coastal and estuarine regions of the North Sea and Baltic Sea (Kirstein et al. 2016). In a later study (Rodrigues et al. 2019), nurdles were collected from five EU bathing beaches in eastern Scotland, UK, which occur downstream of an industrial town and the capital city, Edinburgh. Wastewater treatment plants, combined sewer overflows and emergency overflows, and rural catchments are present within the study area. Nurdles colonized by *E. coli* and *Vibrio* spp. were found at all five beaches, with high levels (>75%) of *Vibrio* colonization at four of them. Colonization by *E. coli* was consistently lower than colonization by *Vibrio* spp. It is likely that *E. coli* colonization occurred in fecal-contaminated seawater, arising from either agricultural runoff or sewage discharge (Rodrigues et al. 2019). The high colonization of nurdles in this study suggests that *Vibrio* spp. may make up part of the indigenous biofilm consortia on MPs found on beaches. In a recent laboratory study (Zhang et al. 2022), the zoonotic protozoan parasites *Toxoplasma gondii*, *Cryptosporidium parvum*, and *Giardia enterica* were capable of adsorbing to PE microbeads and polyester microfibers. The sticky biofilms present on the MPs were important in mediating the protozoan-MP association.

Microbial colonization of environmental plastics is influenced by the length of time the plastic has been exposed to water and soil; samples present in the environment for extended periods are more likely to have a well-developed biofilm and ultimately serve as host to a range of microbial types (Kirstein et al. 2018; Rodrigues et al. 2019). Older and more degraded plastic surfaces commonly are pitted and have grooves and scratches, which increase the surface area for microbial colonization (Fotopoulou and Karapanagioti 2012). *Vibrio* spp. are more likely to colonize plastics that already contain a biofilm compared with bare MPs (Yokota et al. 2017). Most beaches are subject to semidiurnal (i.e., half-day) high tides; consequently, beach cast MPs can be submerged twice per day. Submergence in contaminated water could support *E. coli* establishment on biofilms. It is not known whether *E. coli* prefers to bind to an existing biofilm, or whether they can efficiently bind to uncolonized plastic debris (Rodrigues et al. 2019).

It must be noted that not all studies have detected pathogenic species on plastic debris. In a study conducted at intertidal locations around the Yangtze Estuary in China, pathogenic *Vibrio* species were detected on only a few MPs samples (Jiang et al. 2018). No evidence was found of enrichment of potential pathogens on PE and PS incubated for two weeks in waters ranging from the Baltic Sea to wastewater treatment plants (Oberbeckmann et al. 2017; Ma et al. 2020).

Sorption of Drug-Resistant Bacteria

It is well documented that uncontrolled and haphazard use of antibiotics has led to the evolution of multiple drug resistant (MDR) bacteria (Kunkalekar et al. 2014). The occurrence of MDR bacterial pathogens has reduced the efficiency of antibiotics, thus posing a significant threat to human health. In recent years, environments impacted by sewage pollution, agriculture, aquaculture, industrial waste, and mining waste are considered important locations for development of antibiotic-resistant bacteria (Knapp et al. 2012; Su et al. 2015; Hu et al. 2016).

MP biofilms have been suggested to be potential hot spots of horizontal gene transfer (HGT), as they contain sites of increased nutrient availability and high densities of microbial cells, allowing for rigorous interactions (Aminov 2011; Sezonov et al. 2007; Shen et al. 2019). (See box below.)

An unusual recent finding by ecotoxicologists is that environments contaminated with both antibiotics and metals may cause selection toward antibiotic resistance. In many ecosystems, combined heavy metal and antibiotic contamination has driven the spread of antibiotic resistance and sometimes multiple antibiotic resistance among affected microbial consortia (Baker-Austin et al. 2006). For example, in agricultural soils amended with copper, the microbial selection of Cu resistance co-selects multi-antibiotic resistance (tetracycline, ampicillin, and chloramphenicol) (Berg et al. 2005). Likewise, Ni and Cd contamination are reported to enhance bacterial resistance toward ampicillin or chloramphenicol (Stepanauskas et al. 2006). Numerous reports now suggest that metal contamination is a chief selector of antibiotic resistance (Stepanauskas et al. 2006; Bednorz et al. 2013). The abundance of antibiotic resistance genes has, in many cases, been directly linked to metal concentration in the environment (Ji et al. 2012; Knapp et al. 2011). It has been reported that bacteria associated with MPs show higher frequency of HGT between genetically distinct bacteria, as compared to free-living bacteria; therefore, MPs may provide an ideal location for metal-driven co-selection of antibiotic resistance in human pathogenic bacteria.

IMPLICATIONS: THE VECTOR CAPABILITIES OF MPs

Based on evidence to date, MPs may impart a "vector capability" to microorganisms in several ways. First, MPs can accelerate the spread of microorganisms in the environment. Many MP types are buoyant and lightweight and therefore quite mobile (Khatmullina et al. 2017). These can support the drift and long-term existence of surface microorganisms in water. MPs could transfer a microbial community into a new habitat via ocean currents

(Keswani et al. 2016; Shen et al. 2019). Alien species, especially toxic and pathogenic bacteria, may invade new habitats and multiply quickly in a short time. This will alter the local community structure, which may affect water quality and threaten human health (Kirkpatrick et al. 2004; Kirstein et al. 2016; Zettler et al. 2013).

HORIZONTAL GENE TRANSFER

Horizontal gene transfer (HGT) is defined as the movement of genetic material between organisms other than by transmission of DNA from parent to offspring (i.e., reproduction). HGT is important in the evolution of many organisms. Conjugation is the main mechanism of HGT, a process in which two bacteria in close contact exchange genetic information via plasmid transfer from a donor to a recipient cell (Drudge and Warren 2012). This process can occur between distantly related organisms, thus influencing bacterial evolution and the spread of important traits such as antibiotic or heavy metal resistance genes (Carattoli 2013). It has been hypothesized (Arias-Andres et al. 2018) that pollution by MPs in aquatic ecosystems favors higher transfer frequencies of plasmids carrying antibiotic resistance genes.

A second aspect of the "vector function" concept is that MPs can increase gene exchange between attached biofilm communities, or between communities on biofilms and the local environment (Arias-Andres et al. 2018; Shen et al. 2019). New bacterial types may be created during gene exchange. Pathogenic and antibiotic-resistant bacteria contain a range of pathogenic and antibiotic resistance genes, which may be transferred by multiple pathways between communities on biofilms.

Resistant genes and bacteria have been detected in waters downstream of medical wastewater treatment plants (Karkman et al. 2018). Resistance genes have also been identified in locations where sewage sludge has been disposed to land (Burch et al. 2019; Shen et al. 2019). If a large number of resistance genes and bacteria enter the ocean, for example by attachment to MPs, horizontal transfer of resistance genes may be induced between communities or the surrounding environment (Rodrigues et al. 2019). A spread of pathogenic and antibiotic-resistant genes can lead to large-scale infection events (Shen et al. 2019).

CLOSING STATEMENTS

Assessing the hazards of sorbed contaminants on MPs in aquatic habitats requires knowledge of the types of polymers comprising the debris, location of the affected MPs, the length of time the MPs were present in the aquatic environment, and the possible transport of the debris.

REFERENCES

Aarkrog, A. (2003). Input of anthropogenic radionuclides into the World Ocean. *Deep Sea Research Part II: Topical Studies in Oceanography* 50(17–21):2597–2606.

Acosta-Coley, I., Mendez-Cuadro, D., Rodriguez-Cavallo, E., de la Rosa, J., and Olivero-Verbel, J. (2019). Trace elements in microplastics in Cartagena: A hotspot for plastic pollution at the Caribbean. *Marine Pollution Bulletin* 139:402–411.

Amaral-Zettler, L.A., Zettler, E.R., Slikas, B., Boyd, G.D., Melvin, D.W., Morrall, C.E., and Mincer, T.J. (2015). The biogeography of the Plastisphere: Implications for policy. *Frontiers in Ecology and the Environment* 13(10):541–546.

Aminov, R.I. (2011). Horizontal gene exchange in environmental microbiota. *Frontiers in Microbiology* 2:158.

Antunes, J.C., Frias, J.G.L., Micaelo, A.C., and Sobral, P. (2013). Resin pellets from beaches of the Portuguese coast and adsorbed persistent organic pollutants. *Estuarine, Coastal and Shelf Science* 130:62–69.

Arias-Andres, M., Klümper, U., Rojas-Jimenez, K., and Grossart, H.-P. (2018). Microplastic pollution increases gene exchange in aquatic ecosystems. *Environmental Pollution* 237:253–261.

Arms Control Association. (2019). *The Nuclear Testing Tally*. Washington (DC): Arms Control Association.

Ashton, K., Holmes, L., and Turner, A. (2010). Association of metals with plastic production pellets in the marine environment. *Marine Pollution Bulletin* 60(11):2050–2055.

Baker-Austin, C., Wright, M.S., Stepanauskas, R., and McArthur, J. (2006). Co-selection of antibiotic and metal resistance. *Trends in Microbiology* 14(4):176–182.

Bakir, A., Rowland, S.J., and Thompson, R.C. (2014). Transport of persistent organic pollutants by microplastics in estuarine conditions. *Estuarine, Coastal and Shelf Science* 140:14–21.

Barnes, D.K. (2002). Invasions by marine life on plastic debris. *Nature* 416(6883):808–809.

Bayo, J., Guillén, M., Olmos, S., Jiménez, P., Sánchez, E., and Roca, M.J. (2019). Microplastics as vector for persistent organic pollutants in urban effluents: The role of polychlorinated biphenyls. *International Journal of Sustainable Development and Planning* 134(4):671–682. DOI: 10.2495/SDP-V13-N4-671-682.

Ben-Haim, Y., Thompson, F., Thompson, C., Cnockaert, M., Hoste, B., Swings, J., and Rosenberg, E. (2003). *Vibrio coralliilyticus* sp. nov., a temperature-dependent pathogen of the coral *Pocillopora damicornis*. *International Journal of Systematic and Evolutionary Microbiology* 53(1):309–315.

Berg, J., Tom-Petersen, A., and Nybroe, O. (2005). Copper amendment of agricultural soil selects for bacterial antibiotic resistance in the field. *Letters in Applied Microbiology* 40(2):146–151.

Bradney, L., Wijesekara, H., Palansooriya, K.N., Obadamudalige, N., Bolan, N.S., Ok, Y.S., Rinklebe, J., Kim, K., and Kirkham, M. (2019). Particulate plastics as a vector for toxic trace-element uptake by aquatic and terrestrial organisms and human health risk. *Environment International* 131:104937.

Brandon, J., Goldstein, M., and Ohman, M.D. (2016). Long-term aging and degradation of microplastic particles: Comparing in situ oceanic and experimental weathering patterns. *Marine Pollution Bulletin* 110(1):299–308.

Brennecke, D., Duarte, B., Paiva, F., Caçador, I., and Canning-Clode, J. (2016). Microplastics as vector for heavy metal contamination from the marine environment. *Estuarine, Coastal and Shelf Science* 178:189–195.

Buesseler, K.O. (2014). Fukushima and ocean radioactivity. *Oceanography* 27(1):92–105.

Burch, K.D., Han, B., Pichtel, J., and Zubkov, T. (2019). Removal efficiency of commonly prescribed antibiotics via tertiary wastewater treatment. *Environmental Science and Pollution Research* 26(7):6301–6310.

Carattoli, A. (2013). Plasmids and the spread of resistance. *International Journal of Medical Microbiology* 303(6–7):298–304.

Curren, E., and Leong, S.C.Y. (2019). Profiles of bacterial assemblages from microplastics of tropical coastal environments. *Science of the Total Environment* 655:313–320. DOI: 10.1016/j.scitotenv.2018.11.250.

De Tender, C.A., Devriese, L.I., Haegeman, A., Maes, S., Ruttink, T., and Dawyndt, P. (2015). Bacterial community profiling of plastic litter in the Belgian part of the North Sea. *Environmental Science & Technology* 49(16):9629–9638.

De Tender, C.A., Devriese, L.I., Haegeman, A., Maes, S., Vangeyte, J., Cattrijsse, A., and Ruttink, T. (2017). Temporal dynamics of bacterial and fungal colonization on plastic debris in the North Sea. *Environmental Science & Technology* 51(13):7350–7360. DOI: 10.1021/acs.est.7b00697.

Debroas, D., Mone, A., and Ter Halle, A. (2017). Plastics in the North Atlantic garbage patch: A boat-microbe for hitchhikers and plastic degraders. *Science of the Total Environment* 599–600:1222–1232. DOI: 10.1016/j.scitotenv.2017.05.059.

Delacuvellerie, A., Cyriaque, V., Gobert, S., Benali, S., and Wattiez, R. (2019). The plastisphere in marine ecosystem hosts potential specific microbial degraders including *Alcanivorax borkumensis* as a key player for the low-density polyethylene degradation. *Journal of Hazardous Materials* 380:120899. DOI: 10.1016/j.jhazmat.2019.120899.

Di, M., and Wang, J. (2018). Microplastics in surface waters and sediments of the Three Gorges Reservoir, China. *Science of the Total Environment* 616:1620–1627.

Dobaradaran, S., Schmidt, T.C., Nabipour, I., Khajeahmadi, N., Tajbakhsh, S., Saeedi, R., Javad Mohammadi, M., Keshtkar, M., Khorsand, M., and Faraji Ghasemi, F. (2018). Characterization of plastic debris and association of metals with microplastics in coastline sediment along the Persian Gulf. *Waste Management* 78:649–658. DOI: 10.1016/j.wasman.2018.06.037.

Doong, R.A., and Lin, Y.T. (2004). Characterization and distribution of polycyclic aromatic hydrocarbon contaminations in surface sediment and water from Gao-ping River, Taiwan. *Water Research* 38(7):1733–1744.

Drudge, C.N., and Warren, L.A. (2012). Prokaryotic horizontal gene transfer in freshwater lakes: Implications of dynamic biogeochemical zonation. *Journal of Environmental Protection* 3(12):1634–1654.

Dussud, C., Meistertzheim, A., Conan, P., Pujo-Pay, M., George, M., Fabre, P., Coudane, J., Higgs, P., Elineau, A., Pedrottie, M., Gorski, G., and Ghiglione, J. (2018). Evidence of niche partitioning among bacteria living on plastics, organic particles and surrounding seawaters. *Environmental Pollution* 236:807–816.

Eiler, A., Johansson, M., and Bertilsson, S. (2006). Environmental influences on Vibrio populations in northern temperate and boreal coastal waters (Baltic and Skagerrak Seas). *Applied and Environmental Microbiology* 72(9):6004–6011.

Elizalde-Velazquez, A., Subbiah, S., Anderson, T.A., Green, M.J., Zhao, X., and Cañas-Carrell, J.E. (2020). Sorption of three common nonsteroidal anti-inflammatory drugs (NSAIDs) to microplastics. *Science of the Total Environment* 715:136974.

Endo, S., Takizawa, R., Okuda, K., Takada, H., Chiba, K., Kanehiro, H., Ogi, H., Yamashita, R., and Date, T. (2005). Concentration of polychlorinated biphenyls (PCBs) in beached resin pellets: Variability among individual particles and regional differences. *Marine Pollution Bulletin* 50(10):1103–1114.

Erni-Cassola, G., Wright, R.J., Gibson, M.I., and Christie-Oleza, J.A. (2020). Early colonization of weathered polyethylene by distinct bacteria in marine coastal seawater. *Microbial Ecology* 79(3):517–526. DOI: 10.1007/s00248-019-01424-5.

Evans, T. (2012). Biosolids in Europe. 26th Water Environment Federation. Annual Residuals and Biosolids Conference. Raleigh, NC.

Fan, X., Gan, R., Liu, J., Xie, Y., Xu, D., Xiang, Y., and Hou, J. (2021). Adsorption and desorption behaviors of antibiotics by tire wear particles and polyethylene microplastics with or without aging processes. *Science of the Total Environment* 771:145451.

Faure, F., Demars, C., Wieser, O., Kunz, M., and De Alencastro, L.F. (2015). Plastic pollution in Swiss surface waters: Nature and concentrations, interaction with pollutants. *Environmental Chemistry* 12(5):582–591.

Fick, J., Söderström, H., Lindberg, R.H., Phan, C., Tysklind, M., and Larsson, D.J. (2009). Contamination of surface, ground, and drinking water from pharmaceutical production. *Environmental Toxicology and Chemistry* 28(12):2522–2527.

Fisner, M., Majer, A., Taniguchi, S., Bícego, M., Turra, A., and Gorman, D. (2017). Colour spectrum and resin-type determine the concentration and composition of

Polycyclic Aromatic Hydrocarbons (PAHs) in plastic pellets. *Marine Pollution Bulletin* 122(1–2):323–330.

Fletcher, M., and Loeb, G.I. (1979). Influence of substratum characteristics on the attachment of a marine Pseudomonad to solid surfaces. *Applied and Environmental Microbiology* 37(1):67–72.

Fotopoulou, K.N., and Karapanagioti, H.K. (2012). Surface properties of beached plastic pellets. *Marine Environmental Research* 81:70–77.

Frere, L., Maignien, L., Chalopin, M., Huvet, A., Rinnert, E., Morrison, H., Kerninon, S., Cassone, A.L., Lambert, C., Reveillaud, J., and Paul-Pont, I. (2018). Microplastic bacterial communities in the Bay of Brest: Influence of polymer type and size. *Environmental Pollution* 242:614–625. DOI: 10.1016/j.envpol.2018.07.023.

Friedman, C.L., Burgess, R.M., Perron, M.M., Cantwell, M.G., Ho, K.T., and Lohmann, K. (2009). Comparing polychaete and polyethylene uptake to assess sediment resuspension effects on PCB bioavailability. *Environmental Science & Technology* 43(8):2865–2870.

Fries, E., and Zarfl, C. (2012). Sorption of polycyclic aromatic hydrocarbons (PAHs) to low and high density polyethylene (PE). *Environmental Science and Pollution Research* 19(4):1296–1304.

Gao, F., Li, J., Sun, C., Zhang, L., Jiang, F., Cao, W., and Zheng, L. (2019). Study on the capability and characteristics of heavy metals enriched on microplastics in marine environment. *Marine Pollution Bulletin* 144:61–67.

Ghosal, S., Chen, M., Wagner, J., Wang, Z.-M., and Wall, S. (2018). Molecular identification of polymers and anthropogenic particles extracted from oceanic water and fish stomach – A Raman micro-spectroscopy study. *Environmental Pollution* 223:1113–1124.

Godoy, V., Blazquez, G., Calero, M., Quesada, L., and Martin-Lara, M.A. (2019). The potential of microplastics as carriers of metals. *Environmental Pollution* 255(Pt 3):113363. DOI: 10.1016/j.envpol.2019.113363.

Goldstein, M.C., Carson, H.S., and Eriksen, M. (2014). Relationship of diversity and habitat area in North Pacific plastic-associated rafting communities. *Marine Biology* 161(6):1441–1453.

Gorman, D., Moreira, F.T., Turra, A., Fontenelle, F.R., Combi, T., Bícego, M.C., and Martins, C.d.C. (2019). Organic contamination of beached plastic pellets in the South Atlantic: Risk assessments can benefit by considering spatial gradients. *Chemosphere* 223:608–615.

Gras-Rouzet, S., Donnio, P., Juguet, F., Plessis, P., Minet, J., and Avril, J. (1996). First European case of gastroenteritis and bacteremia due to *Vibrio hollisae*. *European Journal of Clinical Microbiology and Infectious Diseases* 15(11):864–866.

Guo, X., Pang, J., Chen, S., and Jia, H. (2018). Sorption properties of tylosin on four different microplastics. *Chemosphere* 209:240–245.

Guo, X., and Wang, J. (2019a). The chemical behaviors of microplastics in marine environment: A review. *Marine Pollution Bulletin* 142:1–14.

Guo, X., Liu, Y., and Wang, J. (2019b). Sorption of sulfamethazine onto different types of microplastics: A combined experimental and molecular dynamics simulation study. *Marine Pollution Bulletin* 145:547–554.

Guo, X., and Wang, J. (2019c). Sorption of antibiotics onto aged microplastics in freshwater and seawater. *Marine Pollution Bulletin* 149:110511.

Guo, X., Wang, X., Zhou, X., Kong, X., Tao, S., and Xing, B. (2012). Sorption of four hydrophobic organic compounds by three chemically distinct polymers: Role of chemical and physical composition. *Environmental Science & Technology* 46(13):7252–7259.

Hamblin, J.D. (2008). *Poison in the Well: Radioactive Waste in the Oceans at the Dawn of the Nuclear Age*: New Brunswick (NJ): Rutgers University Press.

Harrison, J.P., Schratzberger, M., Sapp, M., and Osborn, A.M. (2014). Rapid bacterial colonization of low-density polyethylene microplastics in coastal sediment microcosms. *BMC Microbiology* 14(1):1–15.

He, Y., Jin, L., Sun, F., Hu, Q., and Chen, L. (2016). Antibiotic and heavy-metal resistance of *Vibrio parahaemolyticus* isolated from fresh shrimps in Shanghai fish markets, China. *Environmental Science and Pollution Research* 23(15):15033–15040.

Herzke, D., Anker-Nilssen, T., Nøst, T.H., Götsch, A., Christensen-Dalsgaard, S., Langset, M., Fangel, F., and Koelmans, A.A. (2016). Negligible impact of ingested microplastics on tissue concentrations of persistent organic pollutants in northern fulmars off coastal Norway. *Environmental Science & Technology* 50(4):1924–1933.

Hodson, M.E., Duffus-Hodson, C.A., Clark, A., Prendergast-Miller, M.T., and Thorpe, K.L. (2017). Plastic bag derived-microplastics as a vector for metal exposure in terrestrial invertebrates. *Environmental Science & Technology* 51(8):4714–4721.

Holmes, L.A., Turner, A., and Thompson, R.C. (2012). Adsorption of trace metals to plastic resin pellets in the marine environment. *Environmental Pollution* 160:42–48.

Holmes, L.A., Turner, A., and Thompson, R.C. (2014). Interactions between trace metals and plastic production pellets under estuarine conditions. *Marine Chemistry* 167:25–32.

Hu, H.W., Wang, J.T., Li, J., Li, J.J., Ma, Y.B., Chen, D., and He, J.Z. (2016). Field-based evidence for copper contamination induced changes of antibiotic resistance in agricultural soils. *Environmental Microbiology* 18(11):3896–3909.

Hu, J.-Q., Yang, S.-Z., Guo, L., Xu, X., Yao, T., and Xie, F. (2017). Microscopic investigation on the adsorption of lubrication oil on microplastics. *Journal of Molecular Liquids* 227:351–355.

Hua, X., Dong, D., Liu, L., Gao, M., and Liang, D. (2012). Comparison of trace metal adsorption onto different solid materials and their chemical components in a natural aquatic environment. *Applied Geochemistry* 27(5):1005–1012.

Imran, M., Das, K.R., and Naik, M.M. (2019). Co-selection of multi-antibiotic resistance in bacterial pathogens in metal and microplastic contaminated environments: An emerging health threat. *Chemosphere* 215:846–857.

Ioannidis, I., Anastopoulos, I., and Pashalidis, I. (2022). Microplastics as radionuclide (U-232) carriers. *Journal of Molecular Liquids* 351:118641.

Iwata, H., Tanabe, S., Sakai, N., and Tatsukawa, R. (1993). Distribution of persistent organochlorines in the oceanic air and surface seawater and the role of ocean on their global transport and fate. *Environmental Science & Technology* 27(6):1080–1098.

Jarnheimer, P.-Å., Ottoson, J., Lindberg, R., Stenström, T.-A., Johansson, M., Tysklind, M., and Olsen, B. (2004). Fluoroquinolone antibiotics in a hospital sewage line; Occurrence, distribution and impact on bacterial resistance. *Scandinavian Journal of Infectious Diseases* 36(10):752–755.

Ji, X., Shen, Q., Liu, F., Ma, J., Xu, G., Wang, Y., and Wu, M. (2012). Antibiotic resistance gene abundances associated with antibiotics and heavy metals in animal manures and agricultural soils adjacent to feedlots in Shanghai; China. *Journal of Hazardous Materials* 235:178–185.

Jiang, P., Zhao, S., Zhu, L., and Li, D. (2018). Microplastic-associated bacterial assemblages in the intertidal zone of the Yangtze Estuary. *Science of the Total Environment* 624:48–54.

Johansen, M.P., Prentice, E., Cresswell, T., and Howell, N. (2018). Initial data on adsorption of Cs and Sr to the surfaces of microplastics with biofilm. *Journal of Environmental Radioactivity* 190:130–133.

Karapanagioti, H., Endo, S., Ogata, Y., and Takada, H. (2011). Diffuse pollution by persistent organic pollutants as measured in plastic pellets sampled from various beaches in Greece. *Marine Pollution Bulletin* 62(2):312–317.

Karkman, A., Do, T.T., Walsh, F., and Virta, M.P. (2018). Antibiotic-resistance genes in waste water. *Trends in Microbiology* 26(3):220–228.

Kedzierski, M., d'Almeida, M., Magueresse, A., Le Grand, A., Duval, H., César, G., Sire, O., Bruzaud, S., and Le Tilly, V. (2018). Threat of plastic aging in marine environment: Adsorption/desorption of micropollutants. *Marine Pollution Bulletin* 127:684–694.

Keswani, A., Oliver, D.M., Gutierrez, T., and Quilliam, R.S. (2016). Microbial hitchhikers on marine plastic debris: Human exposure risks at bathing waters and beach environments. *Marine Environmental Research* 118:10–19.

Kesy, K., Oberbeckmann, S., Kreikemeyer, B., and Labrenz, M. (2019). Spatial environmental heterogeneity determines young biofilm assemblages on microplastics in Baltic Sea Mesocosms. *Frontiers in Microbiology* 10:1665. DOI: 10.3389/fmicb.2019.01665.

Kesy, K., Oberbeckmann, S., Muller, F., and Labrenz, M. (2016). Polystyrene influences bacterial assemblages in *Arenicola marina*–populated aquatic environments in vitro. *Environmental Pollution* 219:219–227. DOI: 10.1016/j.envpol.2016.10.032.

Kettner, M.T., Oberbeckmann, S., Labrenz, M., and Grossart, H.P. (2019). The eukaryotic life on microplastics in brackish ecosystems. *Frontiers in Microbiology* 10:538. DOI: 10.3389/fmicb.2019.00538.

Khatmullina, L., Bagaev, A., and Chubarenko, I. (2017). Microplastics in the Baltic Sea water: Fibers everywhere. Paper presented at the EGU General Assembly Conference Abstracts. EGU2017, Vienna, Austria. https://ui.adsabs.harvard.edu/abs/2017EGUGA..19.1050K/abstract.

Kirkpatrick, B., Fleming, L.E., Squicciarini, D., Backer, L.C., Clark, R., Abraham, W., Benson, J., Cheng, Y., Johnson, D., Pierce, R., Zaias, J., Bossart, G., and Baden, D.R. (2004). Literature review of Florida red tide: Implications for human health effects. *Harmful Algae* 3(2):99–115.

Kirstein, I.V., Kirmizi, S., Wichels, A., Garin-Fernandez, A., Erler, R., Löder, M., and Gerdts, G. (2016). Dangerous hitchhikers? Evidence for potentially pathogenic *Vibrio* spp. on microplastic particles. *Marine Environmental Research* 120:1–8.

Kirstein, I.V., Wichels, A., Gullans, E., Krohne, G., and Gerdts, G. (2019). The Plastisphere – Uncovering tightly attached plastic "specific" microorganisms. *PLoS ONE* 14(4):e0215859. DOI: 10.1371/journal.pone.0215859.

Kirstein, I.V., Wichels, A., Krohne, G., and Gerdts, G. (2018). Mature biofilm communities on synthetic polymers in seawater – Specific or general? *Marine Environmental Research* 142:147–154.

Knapp, C.W., McCluskey, S.M., Singh, B.K., Campbell, C.D., Hudson, G., and Graham, D.W. (2011). Antibiotic resistance gene abundances correlate with metal and geochemical conditions in archived Scottish soils. *PLoS ONE* 6(11):e27300.

Kümmerer, K. (2009a). Antibiotics in the aquatic environment–A review–Part I. *Chemosphere* 75(4):417–434.

Kümmerer, K. (2009b). Antibiotics in the aquatic environment–A review–Part II. *Chemosphere* 75(4):435–441.

Kunkalekar, R., Prabhu, M., Naik, M., and Salker, A. (2014). Silver-doped manganese dioxide and trioxide nanoparticles inhibit both gram positive and gram negative pathogenic bacteria. *Colloids and Surfaces B: Biointerfaces* 113:429–434.

Lagana, P., Caruso, G., Corsi, I., Bergami, E., Venuti, V., Majolino, D., La Ferla, R., Azzaro, M., and Cappello, S. (2019). Do plastics serve as a possible vector for the spread of antibiotic resistance? First insights from bacteria associated to a polystyrene piece from King George Island (Antarctica). *International Journal of Hygiene and Environmental Health* 222(1):89–100. DOI: 10.1016/j.ijheh.2018.08.009.

Larsson, D.J., de Pedro, C., and Paxeus, N. (2007). Effluent from drug manufactures contains extremely high levels of pharmaceuticals. *Journal of Hazardous Materials* 148(3):751–755.

Li, J., Zhang, K., and Zhang, H. (2018). Adsorption of antibiotics on microplastics. *Environmental Pollution* 237:460–467.

Li, Y., Li, M., Li, Z., Yang, L., and Liu, X. (2019). Effects of particle size and solution chemistry on Triclosan sorption on polystyrene microplastic. *Chemosphere* 231:308–314.

Lin, L., Tang, S., Wang, X., Sun, X., and Liu, Y. (2021). Sorption of tetracycline onto hexabromocyclododecane/polystyrene composite and polystyrene microplastics: Statistical physics models, influencing factors, and interaction mechanisms. *Environmental Pollution* 284:117164.

Livingston, H.D., and Povinec, P.P. (2000). Anthropogenic marine radioactivity. *Ocean & Coastal Management* 43(8–9):689–712.

Lu, Q., He, Z.L., and Stofella, P.J. (2012). Biosolids soil application: Agronomic and environmental implications. *Applied and Environmental Soil Science*. Article ID 201462. DOI: 10.1155/2012/201462.

Ma, Y., Wang, L., Wang, T., Chen, Q., and Ji, R. (2020). Microplastics as vectors of chemicals and microorganisms in the environment. In: *Particulate Plastics in Terrestrial and Aquatic Environments*. Boca Raton (FL): CRC Press, pp. 209–230.

Magadini, D.L., Goes, J.I., Ortiz, S., Lipscomb, J., Pitiranggon, M., and Yan, B. (2020). Assessing the sorption of pharmaceuticals to microplastics through in-situ experiments in New York City waterways. *Science of the Total Environment* 729:138766.

Maršić-Lučić, J., Lušić, J., Tutman, P., Varezić, D.B., Šiljić, J., and Pribudić, J. (2018). Levels of trace metals on microplastic particles in beach sediments of the island of Vis, Adriatic Sea, Croatia. *Marine Pollution Bulletin* 137:231–236.

Martinez-Urtaza, J., Simental, L., Velasco, D., DePaola, A., Ishibashi, M., Nakaguchi, Y., Nishibuchi, M., Carrera-Flores, D., Rey-Alvarez, C., and Pousa, A. (2005). Pandemic *Vibrio parahaemolyticus* O3: K6, Europe. *Emerging Infectious Diseases* 11(8):1319.

Masó, M., Garcés, E., Pagès, F., and Camp, J. (2003). Drifting plastic debris as a potential vector for dispersing Harmful Algal Bloom (HAB) species. *Scientia Marina* 67(1):107–111.

Mato, Y., Isobe, T., Takada, H., Kanehiro, H., Ohtake, C., and Kaminuma, T. (2001). Plastic resin pellets as a transport medium for toxic chemicals in the marine environment. *Environmental Science & Technology* 35(2):318–324.

Mizukawa, K., Takada, H., Ito, M., Geok, Y.B., Hosoda, J., Yamashita, R., Saha, M., Suzuki, S., Miguez, C., Frias, J., Antunes, J.C., Sobral, P., Santos, I., Micaelo, C., and Ferreira, A.M. (2013). Monitoring of a wide range of organic micropollutants on the Portuguese coast using plastic resin pellets. *Marine Pollution Bulletin* 70(1–2):296–302.

Morris, J.G., Jr., and Acheson, D. (2003). Cholera and other types of vibriosis: A story of human pandemics and oysters on the half shell. *Clinical Infectious Diseases* 37(2):272–280.

Müller, A., Becker, R., Dorgerloh, U., Simon, F.-G., and Braun, U. (2018). The effect of polymer aging on the uptake of fuel aromatics and ethers by microplastics. *Environmental Pollution* 240:639–646.

Munier, B., and Bendell, L. (2018). Macro and micro plastics sorb and desorb metals and act as a point source of trace metals to coastal ecosystems. *PLoS ONE* 13(2):e0191759.

Muthukrishnan, T., Al Khaburi, M., and Abed, R.M.M. (2019). Fouling microbial communities on plastics compared with wood and steel: Are they substrate- or location-specific? *Microbial Ecology* 78(2):361–374. DOI: 10.1007/s00248-018-1303-0.

Noik, V.J., Tuah, P.M., Seng, L., and Sakari, M. (2015). Fingerprinting and quantification of selected heavy metals in meso-and microplastics sampled from

Santubong and Trombol beach. Kuching, Sarawak, Malaysia. Paper presented at the 2nd International Conference on Agriculture, Environment and Biological Sciences.

Oberbeckmann, S., Kreikemeyer, B., and Labrenz, M. (2018). Environmental factors support the formation of specific bacterial assemblages on microplastics. *Frontiers in Microbiology* 8:2709.

Oberbeckmann, S., Loeder, M.G., Gerdts, G., and Osborn, A.M. (2014). Spatial and seasonal variation in diversity and structure of microbial biofilms on marine plastics in Northern European waters. *FEMS Microbiology Ecology* 90(2):478–492.

Ogata, Y., Takada, H., Mizukawa, K., Hirai, H., Iwasa, S., Endo, S., and Nakashima, A. (2009). International Pellet Watch: Global monitoring of persistent organic pollutants (POPs) in coastal waters. 1. Initial phase data on PCBs, DDTs, and HCHs. *Marine Pollution Bulletin* 58(10):1437–1446.

Ogonowski, M., Motiei, A., Ininbergs, K., Hell, E., Gerdes, Z., Udekwu, K.I., Bacsik., Z., and Gorokhova, E. (2018). Evidence for selective bacterial community structuring on microplastics. *Environmental Microbiology* 20(8):2796–2808.

Patterson, J., Jeyasanta, K.I., Sathish, N., Edward, J.K.P., and Booth, A.M. (2020). Microplastic and heavy metal distributions in an Indian coral reef ecosystem. *Science of the Total Environment* 744:140706. DOI: 10.1016/j.scitotenv.2020.140706.

Pazos, R.S., Suárez, J.C., and Gómez, N. (2020). Study of the plastisphere: Biofilm development and presence of faecal indicator bacteria on microplastics from the Río de la Plata estuary. *Ecosistemas* 29(3). DOI: 10.7818/ecos.2069.

Pichtel, J. (2017). Biofilms for remediation of xenobiotic hydrocarbons–A technical review. In: Ahmad, I., and F.M. Husain, editors. *Biofilms in Plant and Soil Health*. Berlin/Heidelberg (Germany): Springer.

Pozo, K., Urbina, W., Gómez, V., Torres, M., Nuñez, D., Přibylová, P., Audy, O., Clarke, B., Arias, A., Tombesi, N., Guida, Y., and Klánová, J. (2020). Persistent organic pollutants sorbed in plastic resin pellet – "Nurdles" from coastal areas of Central Chile. *Marine Pollution Bulletin* 151:110786.

Pruden, A., Larsson, D.J., Amézquita, A., Collignon, P., Brandt, K.K., Graham, D.W., Lazorchak, J., Suzuki, S., Silly, P., Snape, J., Topp, E., Zhang, T., and Zhu, Y. (2013). Management options for reducing the release of antibiotics and antibiotic resistance genes to the environment. *Environmental Health Perspectives* 121(8):878–885.

Prunier, J., Maurice, L., Perez, E., Gigault, J., Pierson Wickmann, A.C., Davranche, M., and Halle, A.T. (2019). Trace metals in polyethylene debris from the North Atlantic subtropical gyre. *Environmental Pollution* 245:371–379. DOI: 10.1016/j.envpol.2018.10.043.

Pruzzo, C., Vezzulli, L., and Colwell, R.R. (2008). Global impact of *Vibrio cholerae* interactions with chitin. *Environmental Microbiology* 10(6):1400–1410.

Puckowski, A., Cwięk, W., Mioduszewska, K., Stepnowski, P., and Białk-Bielińska, A. (2021). Sorption of pharmaceuticals on the surface of microplastics. *Chemosphere* 263:127976.

Qiu, Y., Zheng, M., Wang, L., Zhao, Q., Lou, Y., Shi, L., and Qu, L. (2019). Sorption of polyhalogenated carbazoles (PHCs) to microplastics. *Marine Pollution Bulletin* 146:718–728.

Razanajatovo, R.M., Ding, J., Zhang, S., Jiang, H., and Zou, H. (2018). Sorption and desorption of selected pharmaceuticals by polyethylene microplastics. *Marine Pollution Bulletin* 136:516–523.

Rios, L.M., and Jones, P.R. (2015). Characterization of microplastics and toxic chemicals extracted from microplastic samples from the North Pacific Gyre. *Environmental Chemistry* 12(5):611–617.

Rios, L.M., Moore, C., and Jones, P.R. (2007). Persistent organic pollutants carried by synthetic polymers in the ocean environment. *Marine Pollution Bulletin* 54(8):1230–1237.

Robards, M.D., Gould, P.J., and Piatt, J.F. (1997). The highest global concentrations and increased abundance of oceanic plastic debris in the North Pacific: Evidence from seabirds. In: *Marine Debris.* Berlin/Heidelberg (Germany): Springer, pp. 71–80.

Rochman, C.M., Hentschel, B.T., and Teh, S.J. (2014). Long-term sorption of metals is similar among plastic types: Implications for plastic debris in aquatic environments. *PLoS ONE* 9(1):e85433.

Rodrigues, A., Oliver, D.M., McCarron, A., and Quilliam, R.S. (2019). Colonisation of plastic pellets (nurdles) by *E. coli* at public bathing beaches. *Marine Pollution Bulletin* 139:376–380.

Romalde, J.L., Diéguez, A.L., Lasa, A., and Balboa, S. (2014). New *Vibrio* species associated to molluscan microbiota: A review. *Frontiers in Microbiology* 4:413.

Romeo, T., D'Alessandro, M., Esposito, V., Scotti, G., Berto, D., Formalewicz, M., Noventa, S., and Renzi, M. (2015). Environmental quality assessment of Grand Harbour (Valletta, Maltese Islands): A case study of a busy harbour in the Central Mediterranean Sea. *Environmental Monitoring and Assessment* 187(12):1–21.

Sangeetha Devi, R., Ramya, R., Kannan, K., Robert Antony, A., and Rajesh Kannan, V. (2019). Investigation of biodegradation potentials of high density polyethylene degrading marine bacteria isolated from the coastal regions of Tamil Nadu, India. *Marine Pollution Bulletin* 138:549–560. DOI: 10.1016/j.marpolbul.2018.12.001.

Sarkar, D.J., Das Sarkar, S., Das, B.K., Sahoo, B.K., Das, A., Nag, S.K., Manna, R.K., Behera, B.K., and Samanta, S. (2021). Occurrence, fate and removal of microplastics as heavy metal vector in natural wastewater treatment wetland system. *Water Research* 192:116853. DOI: 10.1016/j.watres.2021.116853.

Sarmah, A.K., Meyer, M.T., and Boxall, A.B. (2006). A global perspective on the use, sales, exposure pathways, occurrence, fate and effects of veterinary antibiotics (VAs) in the environment. *Chemosphere* 65(5):725–759.

Schwantes, J.M., Orton, C.R., and Clark, R.A. (2012). Analysis of a nuclear accident: Fission and activation product releases from the Fukushima Daiichi nuclear facility as remote indicators of source identification, extent of release, and state of damaged spent nuclear fuel. *Environmental Science & Technology* 46(16):8621–8627.

Sezonov, G., Joseleau-Petit, D., and d'Ari, R. (2007). *Escherichia coli* physiology in Luria-Bertani broth. *Journal of Bacteriology* 189(23):8746–8749.

Shen, M., Zhu, Y., Zhang, Y., Zeng, G., Wen, X., Yi, H., Ye, S., Ren, X., and Song, B. (2019). Micro (nano) plastics: Unignorable vectors for organisms. *Marine Pollution Bulletin* 139:328–331.

Shen, X.-C., Li, D.-C., Sima, X.-F., Cheng, H.-Y., and Jiang, H. (2018). The effects of environmental conditions on the enrichment of antibiotics on microplastics in simulated natural water column. *Environmental Research* 166:377–383.

Sim, W.-J., Lee, J.-W., Lee, E.-S., Shin, S.-K., Hwang, S.-R., and Oh, J.-E. (2011). Occurrence and distribution of pharmaceuticals in wastewater from households, livestock farms, hospitals and pharmaceutical manufactures. *Chemosphere* 82(2):179–186.

Stepanauskas, R., Glenn, T.C., Jagoe, C.H., Tuckfield, R.C., Lindell, A.H., King, C.J., and McArthur, J. (2006). Coselection for microbial resistance to metals and antibiotics in freshwater microcosms. *Environmental Microbiology* 8(9):1510–1514.

Su, L., Xue, Y., Li, L., Yang, D., Kolandhasamy, P., Li, D., and Shi, H. (2016). Microplastics in taihu lake, China. *Environmental Pollution* 216:711–719.

Ta, A.T., and Babel, S. (2020). Microplastic contamination on the lower Chao Phraya: Abundance, characteristic and interaction with heavy metals. *Chemosphere* 257:127234. DOI: 10.1016/j.chemosphere.2020.127234.

Taniguchi, S., Colabuono, F.I., Dias, P.S., Oliveira, R., Fisner, M., Turra, A., Izar, G.M., Abessa, D.M., Saha, M., Hosoda, J., Yamashita, R., Takada, H., Lourenço, R.A., Magalhães, C.A., Bícego, M.C., and Montone, R.C. (2016). Spatial variability in persistent organic pollutants and polycyclic aromatic hydrocarbons found in beach-stranded pellets along the coast of the state of São Paulo, southeastern Brazil. *Marine Pollution Bulletin* 106(1–2):87–94.

Tazaki, K., Nakano, M., Takehara, T., Ishigaki, Y., and Nakagawa, H. (2015). Experimental analysis of plastic materials containing radionuclides for decontamination viability. *Chikyu Kagaku (Earth Science)* 69(2):99–108.

Teuten, E.L., Rowland, S.J., Galloway, T.S., and Thompson, R.C. (2007). Potential for plastics to transport hydrophobic contaminants. *Environmental Science & Technology* 41(22):7759–7764.

Thompson, F.L., Iida, T., and Swings, J. (2004). Biodiversity of vibrios. *Microbiology and Molecular Biology Reviews* 68(3):403–431.

Tourinho, P.S., do Sul, J.A.I., and Fillmann, G. (2010). Is marine debris ingestion still a problem for the coastal marine biota of southern Brazil? *Marine Pollution Bulletin* 60(3):396–401.

Turner, A., Holmes, L., Thompson, R.C., and Fisher, A.S. (2020). Metals and marine microplastics: Adsorption from the environment versus addition during manufacture, exemplified with lead. *Water Research* 173:115577. DOI: 10.1016/j.watres.2020.115577.

Turner, A., and Holmes, L.A. (2015). Adsorption of trace metals by microplastic pellets in fresh water. *Environmental Chemistry* 12(5):600–610.

Vedolin, M., Teophilo, C., Turra, A., and Figueira, R. (2018). Spatial variability in the concentrations of metals in beached microplastics. *Marine Pollution Bulletin* 129(2):487–493.

Wang, F., Shih, K.M., and Li, X.Y. (2015). The partition behavior of perfluorooctanesulfonate (PFOS) and perfluorooctanesulfonamide (FOSA) on microplastics. *Chemosphere* 119:841–847.

Wang, F., Wong, C.S., Chen, D., Lu, X., Wang, F., and Zeng, E.Y. (2018). Interaction of toxic chemicals with microplastics: A critical review. *Water Research* 139:208–219.

Wang, J., Peng, J., Tan, Z., Gao, Y., Zhan, Z., Chen, Q., and Cai, L. (2017). Microplastics in the surface sediments from the Beijiang River littoral zone: Composition, abundance, surface textures and interaction with heavy metals. *Chemosphere* 171:248–258.

Wang, L.C., Chun-Te Lin, J., Dong, C.D., Chen, C.W., and Liu, T.K. (2021). The sorption of persistent organic pollutants in microplastics from the coastal environment. *Journal of Hazardous Materials* 420:126658.

Wang, P., Shang, H., Li, H., Wang, Y., Li, Y., Zhang, H., Zhang, Q., Liang, Y., and Jiang, G. (2016). PBDEs, PCBs and PCDD/Fs in the sediments from seven major river basins in China: Occurrence, congener profile and spatial tendency. *Chemosphere* 144:13–20.

Wang, W., and Wang, J. (2018a). Comparative evaluation of sorption kinetics and isotherms of pyrene onto microplastics. *Chemosphere* 193:567–573.

Wang, W., and Wang, J. (2018b). Different partition of polycyclic aromatic hydrocarbon on environmental particulates in freshwater: Microplastics in comparison to natural sediment. *Ecotoxicology and Environmental Safety* 147:648–655.

Wei, L., Zhu, F., Li, Q., Xue, C., Xia, X., Yu, H., Zhao, Q., Jiang, J., and Bai, S. (2020). Development, current state and future trends of sludge management in China: Based on exploratory data and CO_2-equivaient emissions analysis. *Environment International* 144:106093.

Wijesekara, H., Bolan, N.S., Bradney, L., Obadamudalige, N., Seshadri, B., Kunhikrishnan, A., Dharmarajan, R., Ok, Y., Rinkle, J., Kirkham, M., and Vithanage, M. (2018). Trace element dynamics of biosolids-derived microbeads. *Chemosphere* 199:331– 339.

Wu, C., Zhang, K., Huang, X., and Liu, J. (2016). Sorption of pharmaceuticals and personal care products to polyethylene debris. *Environmental Science and Pollution Research* 23(9):8819–8826.

Wu, N., Zhang, Y., Zhao, Z., He, J., Li, W., Li, J., Ma, Y., and Niu, Z. (2020). Colonization characteristics of bacterial communities on microplastics compared with ambient environments (water and sediment) in Haihe Estuary. *Science of the Total Environment* 708:134876. DOI: 10.1016/j.scitotenv.2019.134876.

Wurl, O., and Obbard, J.P. (2004). A review of pollutants in the sea-surface microlayer (SML): A unique habitat for marine organisms. *Marine Pollution Bulletin* 48(11–12):1016–1030.

Xie, Q., Li, H.X., Lin, L., Li, Z.L., Huang, J.S., and Xu, X.R. (2021). Characteristics of expanded polystyrene microplastics on island beaches in the Pearl River Estuary: Abundance, size, surface texture and their metals-carrying capacity. *Ecotoxicology* 30(8):1632–1643. DOI: 10.1007/s10646-020-02329-7.

Xu, B., Liu, F., Brookes, P.C., and Xu, J. (2018). The sorption kinetics and isotherms of sulfamethoxazole with polyethylene microplastics. *Marine Pollution Bulletin* 131:191–196.

Xu, P., Ge, W., Chai, C., Zhang, Y., Jiang, T., and Xia, B. (2019). Sorption of polybrominated diphenyl ethers by microplastics. *Marine Pollution Bulletin* 145:260–269.

Yeo, B.G., Takada, H., Yamashita, R., Okazaki, Y., Uchida, K., Tokai, T., Tanaka, K., and Trenholm, N. (2020). PCBs and PBDEs in microplastic particles and zooplankton in open water in the Pacific Ocean and around the coast of Japan. *Marine Pollution Bulletin* 151:110806.

Yokota, K., Waterfield, H., Hastings, C., Davidson, E., Kwietniewski, E., and Wells, B. (2017). Finding the missing piece of the aquatic plastic pollution puzzle: Interaction between primary producers and microplastics. *Limnology and Oceanography Letters* 2(4):91–104.

Yoshida, N., and Kanda, J. (2012). Tracking the Fukushima radionuclides. *Science* 336(6085):1115–1116.

Yu, W., He, J., Lin, W., Li, Y., Men, W., Wang, F., and Huang, J. (2015). Distribution and risk assessment of radionuclides released by Fukushima nuclear accident at the northwest Pacific. *Journal of Environmental Radioactivity* 142:54–61.

Zettler, E.R., Mincer, T.J., and Amaral-Zettler, L.A. (2013). Life in the "plastisphere": Microbial communities on plastic marine debris. *Environmental Science & Technology* 47(13):7137–7146.

Zhang, E., Kim, M., Rueda, L., Rochman, C., VanWormer, E., Moore, J., and Shapiro, K. (2022). Association of Zoonotic protozoan parasites with microplastics in seawater: Implications for human and wildlife health. *Scientific Reports* 12:1–11. DOI: 10.1038/s41598-022-10485-5.

Zhang, H., Wang, J., Zhou, B., Zhou, Y., Dai, Z., Zhou, Q., and Luo, Y. (2018). Enhanced adsorption of oxytetracycline to weathered microplastic polystyrene: Kinetics, isotherms and influencing factors. *Environmental Pollution* 243:1550–1557.

Zhang, W., Ma, X., Zhang, Z., Wang, Y., Wang, J., Wang, J., and Ma, D. (2015). Persistent organic pollutants carried on plastic resin pellets from two beaches in China. *Marine Pollution Bulletin* 99(1–2):28–34.

Zhang, X., Zheng, M., Wang, L., Lou, Y., Shi, L., and Jiang, S. (2018). Sorption of three synthetic musks by microplastics. *Marine Pollution Bulletin* 126:606–609.

Zhao, L., Ru, S., He, J., Zhang, Z., Song, X., Wang, D., Li, X., and Wang, J. (2022). Eelgrass (*Zostera marina*) and its epiphytic bacteria facilitate the sinking of microplastics in the seawater. *Environmental Pollution* 292:118337. DOI: 10.1016/j.envpol.2021.118337.

Zou, J., Liu, X., Zhang, D., and Yuan, X. (2020). Adsorption of three bivalent metals by four chemical distinct microplastics. *Chemosphere* 248:126064.

Chapter 6

Collection, Identification, and Quantification of Microplastics

INTRODUCTION

As discussed in previous chapters, numerous papers have been published concerning microplastic (MP) pollution in marine, freshwater, and terrestrial ecosystems; unfortunately, the methods used in these studies lack standardization, so it is often difficult to compare results. This chapter will address analytical methods including sampling, identification, and quantitation, commonly used in MP research.

As stated earlier, MPs are typically defined as plastic particles measuring less than 5 mm; however, no recognized standards exist to date. The debate over the appropriate size for defining MPs appears in chapter 1.

Tracking particles and accurately measuring MP concentrations in water, sediment, and soil continue to be a challenge. To appreciate the distribution of MPs, and to collect MPs that are representative of an affected location, it is essential that they be correctly identified and quantified. A number of methods for MP identification and quantification are available (and continue to evolve) (figure 6.1); however, in the absence of standard protocols for sampling and analysis, unreliable or non-comparable data on MP concentrations and polymer compositions continue to appear across studies (Hidalgo-Ruz et al. 2012).

FIELD SAMPLING

Sampling comprises the initial stage of field research; appropriate sampling protocols are essential for obtaining representative MP samples at the site of interest (table 6.1). Various methods, however, can produce results that

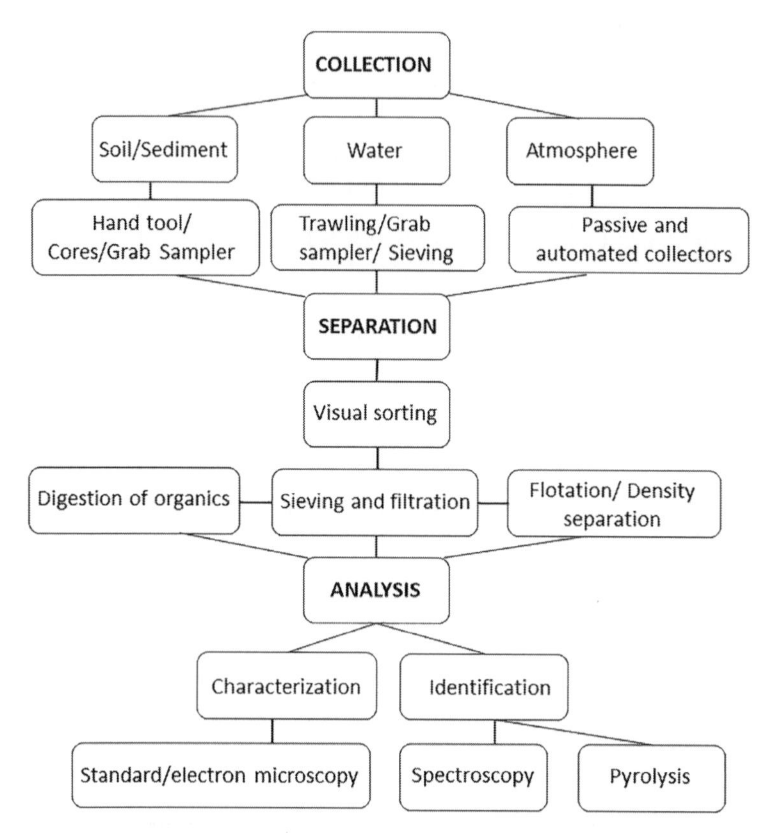

Figure 6.1 Common Methods for MPs Collection in Water, Sediment, and the Atmosphere

may not be representative of the MP distribution in the environment; it is, therefore, important to consider the limitations of each method.

Sampling in Water

MPs occur along the length of the water column as a function of particle properties such as density, shape, size, presence of biofilms, and adsorption of chemicals; and on environmental conditions such as water density, currents, and wave action. Therefore, the types and quantities of MPs recovered are highly dependent on sampling location and depth (Prata et al. 2019). Most MPs tend to be buoyant; therefore, water surface layers are commonly sampled in order to determine the occurrence of MPs (Eriksen et al. 2013; Yu et al. 2016; Zhao et al. 2014). Results are commonly expressed as total items per unit of sample (e.g., particles per liter); some authors provide detailed classifications

Table 6.1 Common Methods for MPs Collection in Water, Sediment, and the Atmosphere

Sample	*Method*	*Advantages*	*Disadvantages*
Water	Neuston/Paired Neuston/Manta/Plankton/Bongo	Simple to use; can sample range of water volumes (dependent on opening); comparative sampling (paired/bongo); customizable; mesh size 10 μm–500 μm.	Cost of larger nets; not representative of media at higher mesh sizes; can clog frequently at lower mesh sizes; quality control difficult.
	Grab sample/bulk sample	Strong quality controls if collecting below surface; captures samples < 1 um; will capture microfibers.	Small volume of water; variability in samples; small sample area.
	Pumps	Multiple sampling sizes and large volumes of water allow for representative samples; less time-consuming; point sampling possible by changing sieve sizes; good quality control.	Small sample area; requires energy source; clogging with smaller sieve sizes.
Sediment/Soil	Water grab sample (Ponar, Peterson, Orange Peel, Van Veen)	Simple to use; provides access to subsamples; small hand-held samplers available.	Small sample area; poor quality control; lack of depth control; lacks deposition history.
	Soil grab sample	Quick; simple to operate; inexpensive.	Five-point sampling required; mixed sample.
	Core sampler (terrestrial/freshwater/marine)	Manual hand-held and motorized options; increased depth of collection; reconstruct deposition history; discrete sampling.	Small sample areas; motorized samplers are costly; larger samplers require rig and vessel.
	Box corer (marine)	Reconstructs deposition history; less variability on depth of sample compared to grab sampler.	Deep water sediments only; requires vessel.
	Spoon, shovel, trowel	Rapid, large sample size, inexpensive.	Does not provide deposition history; sampling bias.
Atmosphere	Passive collector (Petri dish, buckets)	Simple setup, low cost; total deposition.	Total deposition; variability due to weather.
	Automated atmospheric deposition sampler	Control of wet/dry deposition.	Requires energy source; mechanical failures possible in remote study areas.

of size classes, color, and shape (fiber, particle, fragment) (Prata et al. 2019). In MPs research no standardized depth for water sampling has been established; therefore, different sampling depths have been used by researchers.

One of the most common methods for collecting MPs in marine and freshwater environments is trawling with nylon nets (Fok et al. 2020; Ivleva et al. 2017; Wang and Wang 2018); these include manta, paired plankton nets (bongo nets) and neuston nets (figure 6.2). Nets are often towed or suspended for a specified period at a few centimeters below the water surface. Weights can be added to sample certain desired depths of the water column. The net construction consists of either a rectangular opening (60 × 25 cm) (manta/neuston net) or a circular opening (30–100 cm in diameter) (plankton net). Mesh sizes range from 50–500 μm.

Neuston nets are common tools for collecting water occurring in the surface microlayer (see box below). These devices consist of a large, rectangular

Figure 6.2 Common Net Types Used for MPs Collection in Water: (a) Bongo, (b) Manta, and (c) and (d) Neuston

Source: Figures 6.2a, 6.2b, and 6.2d: NOAA.gov. Fig. 6.2c adapted from Campanale et al. (2020). Creative Commons Attribution (CC BY) license (http://creativecommons.org/licenses/by/4.0/).

net frame and a long trailing net for sampling large water volumes. The most widely accepted mesh size for MP sampling is 300 μm (Mai et al. 2018). Some neuston systems are designed with paired nets, each having different mesh sizes; these allow for sampling different-sized particles. When sampling the vast surfaces of open seas and lakes and given the advantages of small volumes of final samples, neuston/manta trawls or nets are commonly employed for depths of 0–0.5 m (Anderson et al. 2017; Eriksen et al. 2013; Ivar Do Sul et al. 2014).

A technical challenge for trawling is that MPs smaller than the net aperture size pass through trawls, resulting in underestimation of MP abundance. On the other hand, nets having fine mesh sizes may become clogged with other particulates (e.g., plankton). Some researchers suggest installing a lightweight coarse mesh aluminum cover over the net to strengthen the trawl and prevent damage during lengthy sampling periods (Mai et al. 2018). Song et al. (2014) used metal sieves (2 mm mesh) to collect seawater at microlayer depths (150–400 μm). They found the mean abundances of recovered MPs by method to be in the order: surface microlayer sampler > hand net > bulk water > manta trawl net.

Nylon nets have been recommended for large-scale MP surface water sampling in marine and freshwater environments due to their ability to filter large volumes of water. It is recommended that mesh aperture size, trawling speed, and sampling time be standardized in order to acquire comparable data from different locations. A drawback of nylon nets is their limited ability to capture microfibers (Prata et al. 2019a). The diameter of a typical fiber is 5–10 μm which can readily pass through the most commonly used mesh size (300 μm) (Mai et al. 2018). As mesh size is reduced, the ability to capture fibers increases significantly (Dris et al. 2015). While the smaller mesh size produces a more representative sample, it will clog at a far higher rate, which will significantly reduce the volume that can be sampled. Studies that use nets of a single mesh size are likely to misrepresent MP concentrations; this is primarily due to their inability to capture particles smaller than the given mesh size.

THE SEA SURFACE MICROLAYER

The sea surface microlayer is the top 1000 micrometers of the ocean surface. This is the critical boundary layer where the exchange of matter occurs between the atmosphere and the ocean. The chemical, physical, and biological properties of the sea surface microlayer differ substantially from the sub-surface water just a few cm below.

A submersible pump, equipped with stainless-steel filters, can recover MPs having sizes below the smallest mesh size of nylon nets. The pump can be used at various depths but is typically used to sample 10 cm below the surface. It is generally fitted with a high-capacity motor that allows for large volumes of water to be sampled; however, manual operation is also possible. Pumps are fitted with digital or electromagnetic flowmeters to measure water volume passing through the filter. Multi-step filtering with up to three different mesh sizes is possible.

While pumps increase sampling efficiency, they are limited to small sampling areas compared to trawling, and encounter problems of clogging at small sieve sizes. Setälä et al. (2016) compared two water sampling methods: trawling of surface water (333 μm mesh size) and collecting subsurface water by pumping followed by passage through a 300- or 100 μm mesh sieve. A higher MP concentration was recovered using a pump with a 100 μm filter than with a pump with a 300 μm filter, or a trawl.

Sampling surface water alone may provide incomplete data for MP concentrations, morphologies, and sizes in a selected location (Lenaker et al. 2019). A review of studies conducted in the Mediterranean Sea found that MP concentrations were generally higher near the surface than any other layer, and that fibers tend to be the dominant shape (Cincinelli et al. 2019). Although most MPs are buoyant, high-density plastic particles or those with additives or sorbed materials often sink to significant depths. MP numbers were demonstrated to decrease from the surface down through the water column; regardless, they were still detected at substantial depth (Reisser et al. 2015). Lenaker et al. (2019) sampled three Milwaukee-area streams, an estuary, and nearshore Lake Michigan at depths of 13.7 m. They demonstrated that microfibers, fragments, and pellets (nurdles and microbeads) occur in the surface, sub-surface, and sediment to varying degrees. Sampling of the surface microlayer of a lake or ocean may, therefore, result in a markedly different composition of polymer types, particle sizes, and concentrations as compared with vertical sampling of the water column. It is important to understand the variability not only in sampling techniques but also with sampling location and depth in aquatic systems.

Sieving is another common method for separating MPs and will be discussed in greater detail later in this chapter, as the method is applied in both aquatic and terrestrial settings. Sieving is rapid and simple; the method involves using multiple sieves, often stacked, with mesh sizes that can range from 0.2 μm to 5 mm (Tirkey et al. 2021). The materials recovered on the sieve are removed for subsequent analysis. Using various mesh sizes allows for size selection which is beneficial when studying MPs of a specific size range. Song et al. (2015b) studied the distribution of MPs in the seawater microlayer in South Korea using metal sieves. The MP concentration detected was markedly higher than that using other sampling methods such as trawling and pumping.

Although not as common as nets and sieves, grab samples can also be used to sample MPs from marine and freshwater environments. Grab samples using 1–2.2 L containers increase the likelihood of capturing the smallest plastic particles. Green et al. (2018) demonstrated that grab samples captured up to three times as many microfibers as the more commonly used nets. Grab samples are accomplished by placing the container just below the water surface. The sample is sealed while submerged to avoid atmospheric contamination. The small volume sampled and variability between samples are two primary considerations with this collection method (Barrows et al. 2017).

To improve the variability of deep-sea sampling, certain sophisticated technologies (e.g., remote sensors) have been developed. Edson and Patterson (2015) developed an autonomous sensor for deep-sea MP monitoring. This instrument collects seawater continuously and can determine MP concentrations for 28 discrete samples in a single deployment. Each sample provides GPS coordinates, temperature, and salinity of the location. The deployment of multiple instruments offers great potential for development of MP dispersion models of the deep oceans.

When collecting water samples, a combination of techniques will increase the validity of the sampling process. One of the greatest hurdles in MP research is collecting representative samples from the field. Improving this component of MP research is crucial for establishing accurate risk assessments and understanding how particles migrate within aquatic environments. The development of new technologies and standardization of sampling methods may possibly overcome certain difficulties when working in the field.

Sampling Sediment and Soil

Sediment is sampled for MPs either from the surficial layer of coasts and shorelines or from the seafloor/lake bottom. MP sampling on beaches is simple and convenient; therefore, shoreline sampling has been used in studies worldwide to monitor MP contamination. Examples include the Tamar Estuary in the United Kingdom (Browne et al. 2010), the Belgian coast (Claessens et al. 2011), the North Sea (Dekiff et al. 2014), and the coastlines of Singapore (Ng and Obbard 2006), Hong Kong (Fok and Cheung 2015), the Bohai Sea (Yu et al. 2016), and a lake in China (Zhang et al. 2016).

To date, there is no standardized procedure for beach sampling—sampling depth, sampling volume, and tide lines have varied across different studies (Mai et al. 2018). Some researchers have selected sampling sites randomly along a beach (Nuelle et al. 2014), while others collected samples by the tide (Mathalon and Hill 2014). Depth of sampling has varied among studies, with the top 5 cm being the most common (Hidalgo-Ruz et al. 2012).

A quadrat (i.e., a small frame of varying dimensions which marks out a sampling area) is commonly used for sediment sampling. Several quadrats may be evenly placed along the coastline for sampling (Van et al. 2012), or a number of representative quadrats may be selectively positioned based on the study location (Liebezeit and Dubaish 2012).

For sampling sediment deep below a body of water, the most frequently used tool is the box corer (Vianello et al. 2013). Grab samplers have also been used in rivers (Castañeda et al. 2014). With the continual deposition of particulate matter and fluctuation of pore pressure, MPs accumulate in sediment to varying depths. Thus, stratified sediment samples can be collected with cores to reconstruct the depositional history of MPs (Hidalgo-Ruz et al. 2012). Carson et al. (2011) reported that the top 15 cm of beach sediment (Kamilo Beach, Hawai'i islands) accounted for 95% of total detected MPs, with more than half in the top 5 cm. Mai et al. (2018) recommend that the geochronology of sediment be taken into account when considering the depth of sediment sampling, because MPs in sediment can be transported by bioturbation (e.g., activities of worms, crabs, and other benthic organisms).

Soil sampling methods include collection with spoons, trowels, shovels, split spoon samplers, and corers (figure 6.3). Surface application of black plastic mulch film (most commonly PE) is a common agricultural practice to eliminate weeds and warm the soil. This practice, however, could result in the

Figure 6.3 Split Spoon Samplers Are Effective for Collecting Sediment to Certain Depths

presence of substantial quantities of MPs. Isari et al. (2021) used stainless-steel shovels to investigate possible MP contamination of tomato and watermelon fields. The study found a significant disparity in MP content between the watermelon field (301 ± 140 items·kg^{-1}) and tomato fields (69 ± 38 items·kg^{-1}), which was attributed to a second planting of watermelon during the sampling season, and not to variability in the sampling method.

The lack of standardization in sampling sediments and soil could be simplified by implementing current EPA and USDA-NRCS protocols (EPA 2020; Schoeneberger et al. 2012; LSAD 2020) (or those of EU or other governing bodies) for soil and amending as needed in terms of quality control.

Collection of Atmospheric Samples

Several recent studies (Allen et al. 2019; Liu et al. 2019; Wright et al. 2020; Zhang et al. 2019; Zhang et al. 2020) have addressed collection and identification of atmospheric MPs; however, the number is small in comparison to aquatic MP studies. MPs have been recovered from both indoor and outdoor atmospheres (Allen et al. 2019; Huang et al. 2021; Lao et al. 2021; Uddin et al. 2022; Yao et al. 2022; Zhang et al. 2020). The USGS has identified the occurrence of MPs in wet deposition collected in the Rocky Mountains above 3000 m a.s.l. (Wetherbee et al. 2019).

Collection of atmospheric MPs using passive collection is a relatively simple and low-cost process. A passive collector relies upon gravitational forces and is the most common method for collecting suspended atmospheric MPs. This method employs buckets, glass bottles, Petri dishes, or other containers capable of capturing atmospheric deposition indoors or outdoors. Results represent total deposition (wet and dry) within an established timeframe. The possibility of wind lifting particles from collectors must be considered when using most passive collectors. An automated alternative is an atmospheric precipitation sampler (figure 6.4) or some variation of this technology. This device allows for controlled wet and dry deposition sampling via the use of an infrared sensor. In the case of collecting wet deposition, the sensor detects precipitation and opens the receptacle to begin collection; once precipitation has ended, the cover returns and thus limits exposure to dry deposition. The sensors are sufficiently sensitive to detect light rain, snow, and heavy fog. Samplers can be fitted with solar panels, thus allowing for sampling in remote areas (Porter et al. 2007).

Allen et al. (2019) estimated that MPs recovered in a remote catchment in the Pyrenees had travelled a minimum 95 km prior to deposition. Their approach to estimating MP transfer and deposition calculated both wind speed and wind direction of individual storms and individual wind events; it was the first study to reveal a possible correlation between wind speed, wind

Figure 6.4 Automated Atmospheric Wet Deposition Sampler
Source: US National Park Service.

direction, and precipitation on atmospheric MP deposition rates. Allen et al. (2021) used a total suspended particle sampler to test for high-altitude MPs measuring < 50 μm. Examination of atmospheric samples collected at the Pic du Midi Observatory had counts of 0.09–0.66 MPs particles·m^{-3}. The observatory is located at 2877 m above sea level.

Measuring deposition in remote regions, where plastic production is presumably non-existent, offers insight into the extent that MPs can travel from their original source. Knowledge gaps exist concerning the sources that contribute to atmospheric contamination of MPs as well as degree of MP distribution and deposition (Chen et al. 2020). Atmospheric studies could be well served to rely upon established particulate sampling plans designed by the US EPA, EU Environmental Commission, or other governing bodies, in order to compile consistent and reproducible data.

SEPARATION OF MPs FROM ENVIRONMENTAL MEDIA

Once MPs samples are collected, they must be prepared for subsequent identification and analysis. Several methods are in common use for separation of MPs from water, soil, and sediment. Prospective methods should consider the size and chemical composition of the particles being investigated. The most common techniques for separation of MPs from environmental media are floatation/density separation, digestion, and filtration (Ngyuen et al. 2019). A combination of these methods can separate a wide range of MP sizes in a solution.

Flotation/Density Separation

Separation by density is a popular method for isolating low-density particles from higher-density sand, mud, sediment, and other matrices (Dillon 1964). Many MPs, such as PP and PE, have densities lower than seawater (~ 1.10 $g \cdot cm^{-3}$). High-density plastics, for example, PVC, have densities up to 1.40 $g \cdot cm^3$ or greater depending on the presence of additives. Various high-density solutions have been used to separate MPs from environmental media (table 6.2). The most commonly used is saturated NaCl, which is inexpensive and nonhazardous, and has a density of approximately 1.20 $g \cdot cm^{-3}$ (Fries et al. 2013). Separation by NaCl is applied to MPs greater than > 1 mm in size and becomes significantly less effective for MPs under 1 mm (Ivleva et al. 2017).

A disadvantage of NaCl for density separation is that plastic particles having densities greater than 1.20 $g \cdot cm^{-3}$ may not be completely recovered. Qi et al. (2020) used a neuston tow net (333 microns) and bulk sampling for collection of water samples in Haiku Bay. Sediment samples (top 5 cm of the surface) were also collected. H_2O_2 and NaCl were used for digestion and separation, respectively. The predominant (46%) size range of MPs was 1–1.9 mm for all samples collected. Polymer type in order of amount identified was PE (90.3%), PS (1.3%), PP (2.6%), PET (2.6%), and nylon (1.3%). The sediment sample had a broader range of polymers which included polyolefin (25.50%), rayon (23.53%), PE (28.41%), (PP-PE) (6.85%), polyester (4.91%), PS (2.92%), PMMA (1.97%), PVC (1.97%), and PP (1.97%). Measurement of MP concentration was limited by the density separation method used (Qi et al. 2020).

Table 6.2 Examples of Flotation/Density Separation Methods Used to Extract MPs

Separation Solution	*Density ($g \cdot cm^{-3}$)*	*Suitable Polymer Types*	*Reference*
Sodium chloride (NaCl)	1.2	PE, PP, PS, PA, PMMA	Nava et al. (2021)
Sodium iodide (NaI)	1.6	Most polymers	Nava et al. (2021)
Zinc chloride ($ZnCl_2$)	1.5–1.8	PE, PS, PMMA, PP, PET, PA	Rodrigues et al. (2020)
Potassium carbonate (K_2CO_3)	1.8	Most polymers	Ghola et al. (2020)
Calcium chloride ($CaCl_2$)	1.4	PE, PS, PMMA, PP, PET, PA	Duong et al. (2022)
Castor oil	1.0	PE, PS, PMMA, PP	Mani et al. (2020)
Pine oil	1.0	PE, PS, PMMA, PP	Lechthaler et al. (2020)

To overcome this drawback, Nuelle et al. (2014) developed a two-step method where an air-induced overflow with NaCl was used, followed by NaI mixture (1.80 g·cm^3) for additional flotation. Recovery rates ranged from 67% (expanded polystyrene, EPS) to 99% (PE) for different polymer types (Nuelle et al. 2014). Another commonly used flotation solution is $ZnCl_2$ (density 1.50–1.70 g·cm^3), which is capable of extracting almost all MPs (Liebezeit and Dubaish 2012).

Procedures for density separation are generally similar. The sample material is mixed with the preferred solution and shaken for a desired period to homogenize the slurry. The slurry is left for several hours so that denser materials (e.g., sand) settle to the bottom of the reaction vessel. The floating fraction is collected for further analysis. To enhance extraction efficiency, the solution above the sediment can undergo filtration (Stolte et al. 2015; Zobkov and Esiukova 2017a).

Shen et al. (2002) state that because MPs tend to be hydrophobic, surfactants can be applied to separate MPs from water. This same approach can be used for MP extraction from sediment. Using density separation with canola oil, Crichton et al. (2017) observed 96% recovery for total MPs, with recovery rates of 92% for fibers and 99% for particles. Mani et al. (2019) used castor oil to separate PP, PS, PMMA, and PET from spiked and non-spiked Rhine River samples. Recovery rate was 99 ± 4% for spiked samples and 74 ± 13% for non-spiked. Strongly biofouled samples had a lower recovery rate; it was therefore suggested to chemically digest the organic debris prior to the oil separation process.

A combination of flotation and other extraction methods is suggested for optimizing density separation of MPs. Some sophisticated devices have been designed for extraction by density separation (see chapter 8 for discussion of the Munich Plastic Sediment Separator).

Sieving and Filtration

Stainless-steel sieves or glass fiber filters can be used for MP separation by size. A wet sieving process was reported by Masura et al. (2015), who poured samples through a 5.6 mm stainless-steel mesh sieve followed by a 0.3 mm mesh sieve. The desired MP sizes were separated for further sorting and identification. To minimize clogging of sieve openings, a medium coarse mesh sieve of 1 mm is suggested (Nuelle et al. 2014; van Cauwenberghe et al. 2013). A 500 µm sieve is often used to segregate size fractions of > 500 and < 500 µm. This size fractionation cutoff is considered reasonable because MPs in the > 500 µm fraction can be identified visually (Mai et al. 2018; Hidalgo-Ruz et al. 2012).

Separation by sieving has its share of disadvantages, including blocking of sieve apertures, difficulty obtaining a representative sample, and lengthy sample processing time (Mai et al. 2018). Stacks of sieves having different aperture sizes may be used to fractionate MPs. Yang et al. (2021) found that the minimum size of MPs recovered across 17 studies ranged from 0.1 to 2 mm based on mesh sizes used. It would be beneficial, for comparing MP size fractions, to establish a standardized series of aperture sizes.

Membrane filtration is popular due to its combined low cost and ease of implementation. Creating a multi-step filtration process allows for separation of plastics ranging from 25 to 1000 μm and can avoid clogging with heavy sediment samples (Schwaferts et al. 2019). The recovered filtrate can be analyzed for submicron- and nanometer-sized MPs. Applying a second multi-step process across the size range of 0.1 to 25 μm has isolated plastic particles as small as 24–52 nm (Schwaferts et al. 2019). Centrifugation can be applied to the filtrate of a larger sample.

Once separation is completed, the sample will need further preparation as the separation process segregates all organic material in addition to the plastic. Digestion of the remaining matrix is feasible; however, due to the fine size of the remaining plastic particles, the possibility of aggregation and/or damage to particles increases (Li et al. 2018).

SAMPLE PROCESSING: REMOVAL OF RESIDUAL ORGANIC MATTER

Purification is essential for accurate identification of MPs. Given the sorption of organic molecules and formation of biofilms on MPs, it is important to remove the organic matter attached to surfaces of recovered particles to allow for accurate identification of polymer type. Without adequate pretreatment, recovery of MPs may be lower than expected.

Various digestion reagents including oxidizing agents (30% H_2O_2/35% H_2O_2/5% NaClO), acids (65% HNO_3/37% HCl), alkalis (10 *M* NaOH/10% KOH) and enzymes (trypsin/proteinase K) have been used for particle purification (Lusher et al. 2017; Liebezeit and Dubaish 2012; Mathalon and Hill 2014; Stolte et al. 2015; Zhao et al. 2014). Nuelle et al. (2014) immersed PP, PVC, PET, and PU pellets for one week in 35% H_2O_2 and found that it performed better than other solutions (30% H_2O_2, 20% HCl, and 20–50% NaOH) in dissolving organic matter; however, particle colors faded. Cole et al. (2014) compared HCl, NaOH, and enzymatic (proteinase-K) digestion treatments on a range of MP samples from plankton-rich seawater. The enzymatic protocol digested > 97% of the material present in the samples without destroying any MPs. Fenton's reagent is also effective for oxidizing organic compounds (see box below) (Anderson et al. 2017; Masura et al. 2015).

According to Van et al. (2012), Fenton-based treatment destroys persistent organic pollutants (POPs) sorbed to MPs.

Certain drawbacks are associated with all the digestion methods outlined above. MPs having resin types PA, PET, and PS can experience significant degradation during acid digestion. When acid treatment is necessary due to high carbonate content on particle surfaces, it is recommended to use only a dilute concentration of HCl (e.g., 10%) at 20°C; this level was demonstrated to preserve the polymer. Degradation of PET occurred in solutions of NaOH and KOH (Pfeiffer and Fischer 2020). NaOH was also demonstrated to decompose esters prior to Fourier-transformed infrared spectroscopy (FTIR) analysis; therefore, plastics that contain ester groups would likely be found at lower concentrations or at non-detectable sizes when digesting samples using NaOH.

It is currently unknown whether any of the digestion solutions cause chemical changes to the organic additives included in polymers. Multi-step pretreatments are suggested to ensure complete removal of biofilms and other organic debris attached to MP surfaces (Mai et al. 2018). Sequential reaction using proteinase K and H_2O_2 is a useful pretreatment for dissolving biofilms and organic material.

FENTON REACTION

Fenton's reagent is a solution of hydrogen peroxide (H_2O_2) with ferrous iron (typically iron sulfate, $FeSO_4$) as a catalyst. The reagent is used to oxidize a wide range of organic substances. The Fenton process is described in detail in Chapter 9. Care must be taken when implementing various digestion techniques, as they can cause post-collection aging of the polymers.

VISUAL SORTING OF MPs

After collection and purification of particles, they must be sorted from the remaining solids. Large plastics can be sorted directly; however, smaller MPs require observation under a dissecting microscope (typically a stereomicroscope) after samples are oven-dried, commonly at 60°C (de Carvalho and Baptista Neto 2016; Fok and Cheung 2015; Mathalon and Hill 2014). Norén (2007) suggested criteria for visual sorting of MPs (large particles); these include: (1) no cellular or organic structures are visible on the plastic particle/fiber; (2) if the particle is a synthetic fiber, it should be of uniform thickness, that is, not taper toward the ends, and have three-dimensional bending (fibers that are entirely straight imply biological origin); and (3) synthetic particles will be clear or homogeneously colored (blue, red, black, etc.).

According to published literature, the MPs extracted from sediment, water, biota, WWTPs, and air (dust) are typically grouped into two categories after visual identification: (1) large MPs (5 mm–500 μm) and (2) small MPs (<500 μm) (Veerasingam et al. 2020).

Visual sorting of fine particles may not always be applicable: this technique may under- or overestimate the abundance of MPs. Studies have demonstrated that visual analysis can result in a 40–70% false positive rate; furthermore, as MP particle size decreases, the rate of error increases (Eriksen et al. 2013; Woo et al. 2020). It may be necessary to apply instrumental techniques such as FTIR or Raman spectroscopy for more thorough identification (Cheung et al. 2016) (see below).

The number of plastic particles identified will vary with different analytical methods. Lenz et al. (2015) found that less than 69% of visually identified particles were MPs following analysis by Raman spectroscopy; furthermore, rates of successful identification of different particles or fibers were dependent upon an array of variables. The percentage of MPs identified varied with type, color, and size of the particle. Fibers had a higher identification rate (75%) than did particles (64%). Song et al. (2015) found that more MP particles were identified under FTIR than with an optical microscope.

ANALYTICAL TECHNIQUES FOR POLYMER IDENTIFICATION

Several analytical methods are available for accurate characterization of the polymer composition of recovered MPs (table 6.3). Certain methods are time-consuming and involve considerable sample preparation; some are destructive of the sample as well (Veerasingam et al. 2020).

FTIR Spectroscopy

FTIR is a powerful and commonly used technique for MP identification; it has the capability to identify all molecular and functional groups occurring in plastic polymers (Bhargava et al. 2003; Mecozzi et al. 2016). FTIR spectroscopy measures the infrared (IR) radiation absorbed by a sample which leads to determination of molecular composition. An infrared spectrum (figure 6.5) represents the unique "fingerprint" of a sample; the IR absorption "peaks" correspond to types of chemical bonds between the atoms which comprise the polymer (table 6.4) (Veerasingam et al. 2020). Each polymer is a unique combination of atoms; therefore, no two compounds produce exactly the

Table 6.3 Common Analytical Instrumentation for MPs Analysis

Instrument	*Methodology*	*Limit of Detection*	*Advantages*	*Limitations*
Light Microscope	Samples placed on slide and identified based on physical characteristics.	500 nm	Limited sample prep; inexpensive; rapid for manual separation of plastics >1 mm.	Must be paired with chemical identification. Physical test of samples < 1 mm difficult.
ATR-FTIR	Samples placed directly onto a crystal. Light is sent into crystal at or below the critical angle. The IR spectrum is produced by collecting absorption information (vibrations) that results as light interacts with the first few microns on sample surface layer.	200 μm	Rapid analysis of solid MPs and powders; limited to no sample prep; non-destructive; simple to use; FTIR the most common instrument used in polymer analysis in both commercial and academic settings.	Expertise for manual analysis; samples can be easily contaminated; requires LCD screen to increase LOD; coatings on sample can completely absorb IR; poor crystal contact results in poor spectra; library analysis is expensive; water can interfere with signal.
FTIR microspectroscopy	Scans conducted in transmitted or reflected light mode. In transmission mode the IR beam and resulting transmitted IR signal are recorded by detector. In reflectance micro-FTIR, IR radiation reflects from a polished and smooth solid sample surface, and resulting signal is processed to achieve similar spectra as transmission micro-FTIR spectra.	5 μm	Target specific areas; Can be coupled with focal plane array detectors for whole sample mapping, reducing sampling time to minutes; non-destructive.	Imaging can be time-consuming; sample preparation required; for transmission mode samples must be sufficiently thin to transmit light which requires a level of expertise; reflectance mode can result in light scattering with fiber samples.

Raman microspectroscopy	Samples placed on slide and irradiated with monochromatic light (laser source). Raman scattering occurs as light interacts with sample. Raman shift occurs when molecule experiences a change in its polarizability that creates molecular fingerprint of sample.	1 µm	Limited sample preparation; Non-destructive; water does not interfere with signal; as of this publication, has lowest limit of detection of individual particles.	Fluorescence can completely mask spectrum; sample must be purified; manual analysis requires expertise; library analysis is expensive; weak signal can cause lengthy analysis time; limited studies compared to FTIR.
Pyrolysis-GC/MS	Sample thermally decomposed at a specific temperature. Molecules released by pyrolysis are analyzed by gas chromatography coupled with mass spectrometry.	ng - µg	Identifies polymers and additives; only method for semi-quantitative analysis of sub-micron MPs; best option for mass balance studies.	Polymer additives can limit detection; preconcentrating may be necessary; not useful with larger MPs; interference from environmental matrices; destructive.
SEM	Samples exposed to a beam of electrons to form an image. Secondary electrons are emitted by atoms that have absorbed the energy of the beam and are then detected to produce image.	Sub-micron	High-resolution images that allow visual investigation of physical weathering (cracks and abrasions).	Must be coupled with EDS for elemental analysis; expensive.
Fluorescence microscope	Dyed samples are irradiated with filtered UV light at a predetermined wavelength. Sample produces fluorescence that can be recorded.	Down to µm range	Rapid identification if in solution lacking NOM; could potentially spot unidentified plastics; simple and quick step that can be added to separation process; inexpensive.	Matrices with high organic content will require aggressive digestion such as Fenton reaction; Organic content that is not removed can sorb dye and result in overestimation of MPs; not all polymers will sorb dye; no method available for complete removal of organic material from environmental sample.

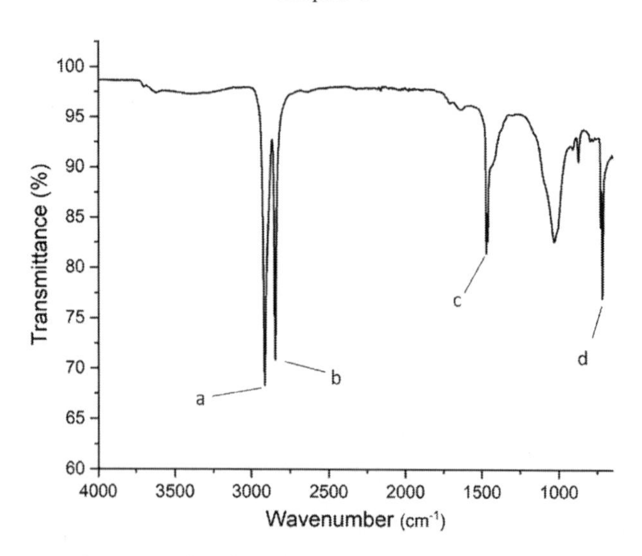

Figure 6.5 FTIR Spectrum Showing Characteristic Absorption Bands in PE Collected from the Texas Gulf Coast

same infrared spectrum. The FTIR spectrum of a sample particle is compared with that of known plastic polymers in a spectral library for identification. MPs measuring > 200 μm are often identified by ATR-FTIR (attenuated total reflection-FTIR), whereas smaller MPs (particles down to 10 μm) are characterized by micro-FTIR.

FTIR provides accurate spectral data for individual particles; however, agglomerates or films of smaller particles may be analyzed as well (Hernandez et al. 2017). FTIR and its optimized technologies (e.g., ATR-FTIR, micro-FTIR, focal plane array detector-based micro-FTIR imaging) have been successfully used in MP studies (Veerasingam et al. 2020). Harrison et al. (2012) compared reflectance micro-FTIR and ATR-FTIR for analysis of PE MPs. Both were effective for accurate identification of polymer composition (Harrison et al. 2012). Micro-FTIR increases the limit of detection (LOD) to single particles measuring ~10 μm and greater. The use of a focal plane array

Table 6.4 Diagnostic Peaks for Commonly Found MPs Using FTIR

Polymer Type	*Diagnostic Peaks cm^{-1}*
PET	1721, 1245, 1100
HDPE	2915, 2848, 730, 720
LDPE	2915, 2848, 1377, 720
Polyamide/Nylon 6-6 (PA)	3081, 3059, 3025, 2923, 2850, 1600, 1492, 756, 698
Poly(methyl methacrylate)	1730, 1242, 1191, 1149
PP	2956, 2921, 2875, 2840, 720

detector–based micro-FTIR has the benefit of increased LOD and can also complete detection of all polymers on a 10 × 10 mm filter in one hour.

In a study of six beaches around Hong Kong (Li et al. 2020), PE, PP, and PS were the dominant polymers identified by comparing ATR-FTIR spectra against an FTIR library with a 70% sensitivity threshold. The lower the threshold, the higher likelihood that the IR library will identify both mixed polymers, and polymers that produce a poor spectrum. It is recommended that users become familiar with the diagnostic peaks of common polymers and conduct visual identification on any spectrum that does not meet the threshold requirement.

A study on the gastrointestinal contents of eight species of birds of prey used micro-FTIR to identify polymers (Carlin et al. 2020). Cellulose was the most common polymer identified (37%), followed by PET (16%) and a polymer blend (4:1) of polyamide-6 and poly(ethylene-*co*-polypropylene) (11%). Fibers constituted 86% of the identified items, demonstrating that micro-FTIR is capable of readily identifying both particles and fibers. Bingxu Nan et al. (2020) collected *Paratya australiensis* (shrimp) and raw water samples at ten sites within the Goulburn River catchment (Australia). A total of 199 possible polymer particles were identified in the shrimp; using micro-FTIR, a total of 72 items had a match in the polymer library. Eleven different polymers were found, with the highest concentrations for rayon (22.6%, ten sites) and polyester (7.5%, nine sites). The dominant polymers in the river samples were polyester (30.6%) at seven sites, polyamide (12.9%) at four sites, and rayon (8.1%) at three sites. Shrimp contained mean values of MPs at 0.52 ± 0.55 MPs·sample or 2.4 ± 3.1 MPs·g^{-1}. The mean value for MPs in water was 0.40 ± 0.27 items·l^{-1}.

Raman Spectroscopy

Raman spectroscopy is another effective analytical technique for MP detection (van Cauwenberghe et al. 2013; Zhang et al. 2016); this surface analytical technique allows for the study of large and visually sorted particles. Raman is a "scattering" method: laser light of a single wavelength is applied to excite the molecules on a surface, and the radiation interaction with the sample is detected. A requirement for Raman spectroscopy is change in the polarizability of a chemical bond; therefore, this technique is best suited for compounds having aromatic bonds, C–H and C=C double bonds (figure 6.6) (Li et al. 2018).

The primary benefits of Raman spectroscopy are that: (1) particles below ~20 μm can be examined, (2) the method has a better response to non-polar plastic functional groups than do other analytical methods (Lenz et al. 2015; Ribeiro-Claro et al. 2017), and (3) particle shape and thickness do not

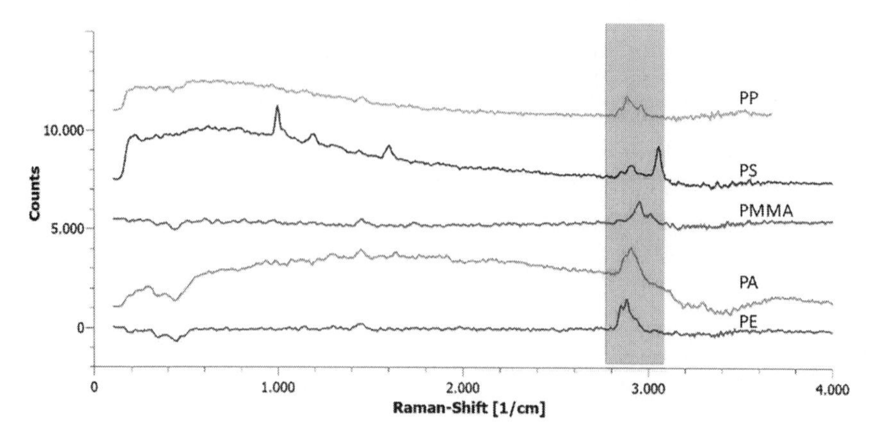

Figure 6.6 Raman Spectrum of Multiple Polymers with Highlighted Spectral Range Characteristic for Each Polymer

Source: Kniggendorf et al. (2019). Creative Commons License 4.0 (http://creativecommons.org/licenses/by/4.0/).

influence measurements. These advantages make Raman microscopy a potentially more sensitive tool to identify MPs compared with FTIR.

The Raman signal is strongly affected by the presence of dyes and microbiological, organic, and inorganic substances (Elert et al. 2017; Nguyen et al. 2019; Li et al. 2018). To minimize false signals with Raman micro-spectrometry, effective sample purification is recommended.

Asensio-Montesinos et al. (2020) collected plastic from 40 beaches along the Atlantic coast of Cádiz Province, Spain, which were characterized using micro-Raman. A total of 14,275 plastic items were counted, and 57 plastic litter groups were identified. Cellulose acetate represented 79% of the total material identified followed by PP (9%) and PE (8%). Unidentifiable items (0.29% to 12.86% of total) were attributed to aging and the proprietary blends of individual polymers. As stated in chapter 3, many polymer types experience significant changes in functional groups due to weathering. This is a common issue when identifying polymers with FTIR and Raman. FTIR is known to be less affected by the presence of additives.

Micro-Raman was used to investigate the presence of MPs in 57 beverages purchased from commercial suppliers: beer (26), cold tea (4), soft drinks (19), and energy drinks (8) (Shruti et al. 2020). Identification and characterization were accomplished with an epifluorescence microscope, scanning electron microscopy–energy dispersive x-ray analysis (SEM-EDS), and micro-Raman. Results were as follows:

- All four tea samples contained MPs (total abundance of 11 ± 5.26 per bottle)

- 16 out of 19 soft drinks contained MPs (total abundance 40 ± 24.53)
- Five energy drinks contained MPs (total abundance 14 ± 5.79)
- 23 of 26 beers contained MPs (total abundance 152 ± 50.97)

Beer samples had the most significant variance between samples. PA, PEA, PET, and ABS were identified by micro-Raman.

Dong et al. (2020) collected MP samples along the Zhenzhu River (China) and examined surface characteristics of 155 items from sediment. Micro-Raman and ATR-FTIR were employed to characterize individual polymers. A significant issue when analyzing weathered plastics with Raman spectroscopy is a weakening of key bands used for identification. Bands representing methyl functional groups that are essential for identifying PE can become almost non-existent in Raman spectra. The loss of peak intensity could result in misidentification and cause false negatives if searches are conducted with a standard Raman library (Asensio-Montesinos et al. 2020; Dong et al. 2020). The authors suggest developing a "weathered plastic database" to increase the accuracy of identification. This issue emphasizes the need to pair Raman with FTIR. The latter technology is unaffected by aging of the polymer and will routinely produce strong spectra under most conditions.

Pyrolysis-GC/MS

In contrast to spectroscopic methods like FTIR and Raman, thermoanalytical methods are considered destructive techniques. A sample is thermally decomposed under controlled conditions using specialized devices such as pyrolyzers or thermogravimetric systems. The products formed from the pyrolysis reactions are subsequently analyzed via gas chromatography coupled with mass spectrometry (GC/MS). The specific polymer type is identified by its characteristic decomposition products. Thermoanalytical methods can be performed qualitatively for single MP particles (Fries et al. 2013); such methods are also applied to complex environmental samples to identify and quantify MP polymer types. Pyr-GC/MS may be used in concert with FTIR or Raman spectroscopy for comprehensive analysis of MPs (Fischer and Scholz-Böttcher 2017).

A major benefit of Pyr-GC/MS analysis is that it can analyze both polymer type and organic additives of MPs simultaneously (Jansson et al. 2007). A disadvantage of the technique is that it allows only one particle to proceed through the pyrolysis tube at a time, which is time-consuming.

Wahl et al. (2021) studied the possible presence of nanoplastics in soil. Particles from soil suspensions were separated using asymmetric flow-field flow fractionation coupled to UV spectroscopy and static light scattering (AF4), and chemical composition was determined using Pyr-GC/MS. PE, PP, and PVC nanoplastics were identified. The work of Wahl et al. (2021) is preliminary and must be replicated and refined before any firm conclusions can be drawn regarding environmental contamination by nanoplastics. Pyr-GC/MS was applied to identify the composition of sub-micron plastic particles in human blood (Leslie et al. 2022); particles of PET, PE, styrene, and PMMA were identified.

Sediment and surface water samples were collected from Lake Superior beaches along the US-Canada border (Minor et al. 2020). Using pyr-GC/MS, all particles with identifiable mass spectra were PE. An issue with pyr-GC/MS is that the accompanying organic matrix can produce interferences; therefore, having experience with GC-MS interpretation is valuable in such investigations.

Electron Microscopy

In SEM, electrons are generated by an electron source at the top of a column under high vacuum pressure. The electron beam scans the sample, and electrons reflected from the near-surface region are used to create an image. The shorter wavelength of electrons produced by the SEM creates an image far more enhanced than that of a light microscope (figure 6.7) (Thermofisher 2020).

Figure 6.7 SEM Image of Polystyrene Particles
Source: Schmieg et al. (2020) http://creativecommons.org/licenses/by/4.0/.

The use of SEM for identification of MPs provides clear and high-magnification images, thus allowing for discrimination of MPs from organic particles. SEM is often coupled with energy-dispersive x-ray analysis (EDS), which can obtain the elemental composition of the sample. This also serves to identify and separate carbon-dominant plastics from inorganic particles (Silva et al. 2018). SEM-EDS is an expensive technology. Sample preparation is laborious and time-consuming for proper examination of samples, hence limiting the number of particles that may be analyzed in a given timeframe. Additionally, particle color cannot be used as an identifier in SEM-EDS analysis; therefore, SEM-EDS is recommended for surface characterization of either previously identified polymers and/or elemental analysis (Silva et al. 2018).

Dehghani et al. (2017) state that SEM is appropriate for accurate identification of MP particles of different sizes and shapes (e.g., fiber, spherule, hexagonal, irregular polyhedron); furthermore, trace amounts of Al, Na, Ca, Mg, and Si can be detected by EDS. This can help determine the presence of polymer additives and/or debris adsorbed to MP surfaces. One drawback associated with EDS spectra is that elemental signatures between additives of plastic polymers and adsorbed debris on MP surfaces may not be differentiated.

Fluorescence

Nile red (NR) dye is considered a promising approach for rapid identification of MPs in environmental samples (Maes et al. 2017). NR binds to various plastics and fluoresces in a range of colors that are a function of the hydrophobicity of the polymer. Samples are typically stained with a solution of NR (1–2 $\mu g \cdot mL^{-1}$ in methanol) for 30 to 60 minutes. Fluorescence intensity increases with increased incubation time. Maes et al. (2017) found that incubation with 10 $\mu g \cdot ml^{-1}$ NR for 30 minutes produced the best combination of speed and visibility of the polymer.

Konde et al. (2020) tested the degree that fluorescence from NR allows for rapid identification of PP, PE, PET, and PVC. They included materials known to interfere with field studies including fine-grained quartz sand, sea snail eggs and shell, cellulose tissue, coralline red algae, and others. NR was found to be excellent for use in lab controlled studies but questionable when applied to field studies.

Muller et al. (2020) carried out a comparative study of popular analytical methods for MP analysis. Results from 17 laboratories in eight countries were compared. Reference particles were of five different MP types ranging in diameter from 8 μm to 140 μm. Optical microscopy, micro-FTIR, micro-Raman, Pyr-GC/MS, and SEM were evaluated and compared for determination of total particle number, polymer type, number of particles and/or particle mass for each polymer type. Micro-Raman and Pyr-GC/MS performed best for identification of polymer type. Quantification of polymer mass for identified polymer types was questionable with most methods. Quantification of total particle numbers was best for optical microscopy and to a lesser extent for micro-FTIR. As mentioned earlier, care must be taken when using optical microscopy due to the rate of false positives that can occur during counting. Substantial variance of results was noted between methods and also within methods. The authors emphasized the need for standardization and validation of analytical methods in MP research, both on a global scale as well as within individual laboratories.

QUALITY CONTROL

Control of external contamination of samples is an essential step in the analysis of MPs. Depending on collection, separation, and identification methods applied, the degree of contamination can vary significantly. The size of particles being analyzed also impacts the extent of contamination.

Fine MP particles can be airborne; therefore, conducting laboratory work under a fume hood is recommended to avoid atmospheric contamination of samples. Lang et al. (1988) found that dust (10 μm–50 μm) can easily result in contamination of lab samples. Their investigation of the dust demonstrated that a range of synthetic materials was present.

The amount of time that a sample is exposed should be noted. All personnel must wash any exposed areas of skin with deionized (DI) water before working with samples. Clothing should be white cotton, and lab coats are to be worn at all times to prevent sample contamination. Plastic gloves are to be thoroughly rinsed under DI water prior to touching samples.

Sample preparation must be conducted in conjunction with lab blanks; this will allow researchers to establish possible contamination within the laboratory environment. Blanks should be placed alongside samples that are being examined. All instrument contact points and containers must be rinsed in DI water both in the field and in the laboratory. Containers should be metal or glass and heated at 120°C for 4 hours to remove possible contamination (Yan et al. 2019). When possible, all methods should be lab-tested for accuracy with spiked samples before use in the field. While the quality controls listed herein are numerous, they in no way exhaust all controls that can comprise a MP study. Without clearly defined experimental controls, the validity of an entire study can be called into question (Pinto da Costa et al. 2019).

CLOSING STATEMENTS

A lack of consistent standards has likely led to over- or underrepresentation of the quantity of MPs occurring in various sampling locations. Researchers must establish standards for sampling, separation, identification, and quantification of MPs as well as acknowledge the limitations of these tasks which are cited in certain published works. MPs analysis is an intricate process that must be tailored to the particle size and geographic location being studied. Beyond the specific method for MP collection and analysis, care must be implemented to avoid external contamination of samples. Where contamination is inevitable, strict protocols must be applied at every step of the process.

REFERENCES

Allen, S., Allen, D., and Baladima, F. (2021). Evidence of free tropospheric and long-range transport of microplastic at Pic du Midi observatory. *Nature Communications* 12:7242.

Allen, S., Allen, D., Phoenix, V.R., Le Roux, G., Durántez Jiménez, P., Simonneau, A., Binet, S., and Galop, D. (2019). Atmospheric transport and deposition of microplastics in a remote mountain catchment. *Nature Geoscience* 12(5):339–344.

Anderson, P.J., Warrack, S., Langen, V., Challis, J.K., Hanson, M.L., and Rennie, M.D. (2017). Microplastic contamination in lake Winnipeg, Canada. *Environmental Pollution* 225:223–231.

Andrady, A.L. (2011). Microplastics in the marine environment. *Marine Pollution Bulletin* 62(8):1596–1605.

Ariza-Tarazona, M.C., Villarreal-Chiu, J.F., Barbieri, V., Siligardi, C., and Cedillo-González, E.I. (2019). New strategy for microplastic degradation: Green photocatalysis using a protein-based porous N-TiO_2 semiconductor. *Ceramics International* 45(7):9618–9624.

Asensio-Montesinos, F., Oliva Ramirez, M., Gonzalez-Leal, M., Carrizo, J.M., and Anfuso, D.G. (2020). Characterization of plastic beach litter by Raman spectroscopy in South-western Spain. *Science of the Total Environment* 744:140890.

Barrows, P., Neumann, W., Bergera, C.A., and Shawad, M.L. (2017). Grab vs. neuston tow net: A microplastic sampling performance comparison and possible advances in the field. *Analytical Methods* 9:1446–1453.

Bhargava, R., Wang, S-Q., and Koenig, J.L. (2003). FTIR microspectroscopy of polymeric systems. In: *Liquid Chromatography/FTIR Microspectroscopy/Microwave Assisted Synthesis*. Berlin/Heidelberg (Germany): Springer, pp. 137–191.

Biver, T., Bianchi, S., Carosi, M.R., Ceccarini, A., Corti, A., Manco, E., and Castelvetro, V. (2018). Selective determination of poly (styrene) and polyolefin microplastics in sandy beach sediments by gel permeation chromatography coupled with fluorescence detection. *Marine Pollution Bulletin* 136:269–275.

Browne, M.A., Galloway, T.S., and Thompson, R.C., (2010). Spatial patterns of plastic debris along estuarine shorelines. *Environmental Science & Technology* 44(9):3404–3409.

Cabernard, L., Roscher, L., Lorenz, C., Gerdts, G., and Primpke, S. (2018). Comparison of Raman and Fourier transform infrared spectroscopy for the quantification of microplastics in the aquatic environment. *Environmental Science & Technology* 52(22):13279–13288.

Carlin, J., Craig, C., Little, S., Donnelly, M., Fox, D., Zhai, L., and Walters, L. (2020). Microplastic accumulation in the gastrointestinal tracts in birds of prey in central Florida, USA. *Environmental Pollution* 264:114633.

Carson, H.S., Colbert, S.L., Kaylor, M.J., and McDermid, K.J. (2011). Small plastic debris changes water movement and heat transfer through beach sediments. *Marine Pollution Bulletin* 62(8):1708–1713.

Castañeda, R.A., Avlijas, S., Simard, M.A., and Ricciardi, A. (2014). Microplastic pollution in St. Lawrence river sediments. *Canadian Journal of Fisheries and Aquatic Sciences* 71(12):1767–1771.

Catarino, A.I., Thompson, R., Sanderson, W., and Henry, T.B. (2017). Development and optimization of a standard method for extraction of microplastics in mussels by enzyme digestion of soft tissues. *Environmental Toxicology and Chemistry* 36(4):947–951.

Chalmers, J.M. (2006). Infrared spectroscopy in analysis of polymers and rubbers. In: *Encyclopedia of Analytical Chemistry: Applications, Theory and Instrumentation*. New York: Wiley.

Chen, G., Feng, Q., and Wang, J. (2020). Mini-review of microplastics in the atmosphere and their risks to humans. *Science of the Total Environment* 703:135504.

Cheung, P.K., Cheung, L.T.O., and Fok, L. (2016). Seasonal variation in the abundance of marine plastic debris in the estuary of a subtropical macro-scale drainage basin in South China. *Science of the Total Environment* 562:658–665.

Cincinelli, A., Martellni, T., Guerranti, C., Scopetani, C., Chelazzi, D., and Giarrizzo, T. (2019). A potpourri of microplastics in the sea surface and water column of the Mediterranean Sea. *Trends in Analytical Chemistry* 110:321–326.

Claessens, M., De Meester, S., Van Landuyt, L., De Clerck, K., and Janssen, C.R. (2011). Occurrence and distribution of microplastics in marine sediments along the Belgian coast. *Marine Pollution Bulletin* 62(10):2199–2204.

Claessens, M., Van Cauwenberghe, L., Vandegehuchte, M.B., and Janssen, C.R. (2013). New techniques for the detection of microplastics in sediments and field collected organisms. *Marine Pollution Bulletin* 70(1–2):227–233.

Cole, M., Webb, H., Lindeque, P.K., Fileman, E.S., Halsband, C., and Galloway, T.S. (2014). Isolation of microplastics in biota-rich seawater samples and marine organisms. *Scientific Reports* 4:4528.

Collard, F., Gilbert, B., Eppe, G., Parmentier, E., and Das, K. (2015). Detection of anthropogenic particles in fish stomachs: An isolation method adapted to identification by Raman spectroscopy. *Archives of Environmental Contamination and Toxicology* 69(3):331–339.

Coppock, R.L., Cole, M., Lindeque, P.K., Queirós, A.M., and Galloway, T.S. (2017). A small-scale, portable method for extracting microplastics from marine sediments. *Environmental Pollution* 230:829–837.

Crichton, E.M., Noël, M., Gies, E.A., and Ross, P.S. (2017). A novel, density-independent and FTIR-compatible approach for the rapid extraction of microplastics from aquatic sediments. *Analytical Methods* 9(9):1419–1428.

de Carvalho, D.G., and Neto, J.A.B. (2016). Microplastic pollution of the beaches of Guanabara Bay, Southeast Brazil. *Ocean & Coastal Management* 128:10–17.

Dehghani, S., Moore, F., and Akhbarizadeh, R. (2017). Microplastic pollution in deposited urban dust, Tehran metropolis, Iran. *Environmental Science and Pollution Research International* 24:20360–20371.

Dekiff, J.H., Remy, D., Klasmeier, J., and Fries, E. (2014). Occurrence and spatial distribution of microplastics in sediments from Norderney. *Environmental Pollution* 186:248–256.

Desforges, J.-P.W., Galbraith, M., Dangerfield, N., and Ross, P.S. (2014). Widespread distribution of microplastics in subsurface seawater in the NE Pacific Ocean. *Marine Pollution Bulletin* 79(1–2):94–99.

Dillon, W.P. (1964). Flotation technique for separating fecal pellets and small marine organisms from sand. *Limnology and Oceanography* 9(4):601–602.

Do Sul, J.A.I., Costa, M.F., and Fillmann, G. (2014). Microplastics in the pelagic environment around oceanic islands of the Western Tropical Atlantic Ocean. *Water, Air, & Soil Pollution* 225(7):2004.

Dong, M., Zhang, Q., Xing, X., Chen, W., She, Z., and Luo, Z. (2020). Raman spectra and surface changes of microplastics weathered under natural environments. *Science of the Total Environment* 739:139990.

Doyle, M.J., Watson, W., Bowlin, N.M., and Sheavly, S.B. (2011). Plastic particles in coastal pelagic ecosystems of the Northeast Pacific ocean. *Marine Environmental Research* 71(1):41–52.

Dris, R., Gasperi, J., Saad, M., Mirande, C., and Tassin, B. (2016). Synthetic fibers in atmospheric fallout: A source of microplastics in the environment? *Marine Pollution Bulletin* 104(1–2):290–293.

Dümichen, E., Barthel, A.-K., Braun, U., Bannick, C.J., Brand, K., Jekel, M., and Senz, R. (2015). Analysis of polyethylene microplastics in environmental samples, using a thermal decomposition method. *Water Research* 85:451–457.

Edson, E.C., and Patterson, M.R. (2015). MantaRay: A novel autonomous sampling instrument for in situ measurements of environmental microplastic particle concentrations. In: *OCEANS 2015-MTS/IEEE*. Washington: IEEE.

Elert, A.M., Becker, R., Duemichen, E., Eisentraut, P., Falkenhagen, J., Sturm, H., and Braun, U. (2017). Comparison of different methods for mp detection: What can we learn from them, and why asking the right question before measurements matters? *Environmental Pollution* 231:1256–1264.

Enders, K., Lenz, R., Stedmon, C.A., and Nielsen, T.G. (2015). Abundance, size and polymer composition of marine microplastics ≥10 μm in the Atlantic Ocean and their modelled vertical distribution. *Marine Pollution Bulletin* 100(1):70–81.

Eriksen, M., Mason, S., Wilson, S., Box, C., Zellers, A., Edwards, W., Farley, H., and Amato, S. (2013). Microplastic pollution in the surface waters of the Laurentian Great Lakes. *Marine Pollution Bulletin* 77(1–2):177–182.

Fischer, M., and Scholz-Böttcher, B.M. (2017). Simultaneous trace identification and quantification of common types of microplastics in environmental samples by pyrolysis-gas chromatography–mass spectrometry. *Environmental Science & Technology* 51(9):5052–5060.

Foekema, E.M., De Gruijter, C., Mergia, M.T., van Franeker, J.A., Murk, A.J., and Koelmans, A.A. (2013). Plastic in North Sea fish. *Environmental Science & Technology* 47(15):8818–8824.

Fok, L., and Cheung, P.K. (2015). Hong Kong at the Pearl River Estuary: A hotspot of microplastic pollution. *Marine Pollution Bulletin* 99(1–2):112–118.

Fries, E., Dekiff, J.H., Willmeyer, J., Nuelle, M-T., Ebert, M., and Remy, D. (2013). Identification of polymer types and additives in marine microplastic particles using pyrolysis-GC/MS and scanning electron microscopy. *Environmental Science: Processes & Impacts* 15(10):1949–1956.

Galgani, F., Hanke, G., Werner, S., Oosterbaan, L., Nilsson, P., Fleet, D., Kinsey, S., Thompson, R.C., Van Franeker, J., and Vlachogianni, T. (2013). *Guidance on Monitoring of Marine Litter in European Seas*. Luxembourg: Publications Office of the European Union.

Green, D.S., Kregting, L., Boots, B., Blockley, D.J., Brickle, P., da Costa, M., and Crowle, Q. (2018). A comparison of sampling methods for seawater microplastics and a first report of the microplastic litter in coastal waters of Ascension and Falkland Islands. *Marine Pollution Bulletin* 137:695–701.

Harrison, J.P., Ojeda, J.J., and Romero-González, M.E. (2012). The applicability of reflectance micro-Fourier-transform infrared spectroscopy for the detection of synthetic microplastics in marine sediments. *Science of the Total Environment* 416:455–463.

Hidalgo-Ruz, V., Gutow, L., Thompson, R.C., and Thiel, M. (2012). Microplastics in the marine environment: A review of the methods used for identification and quantification. *Environmental Science & Technology* 46(6):3060–3075.

Huang Y, He, T., Yan, M., Yang, L., Gong, H., Wang, W., Qing, X., and Wang, J. (2021). Atmospheric transport and deposition of microplastics in a subtropical urban environment. *Journal of Hazardous Material* 416:126168.

Imhof, H.K., Schmid, J., Niessner, R., Ivleva, N.P., and Laforsch, C. (2012). A novel, highly efficient method for the separation and quantification of plastic particles in sediments of aquatic environments. *Limnology and Oceanography: Methods* 10(7):524–537.

Isari, E., Papaioannou, A.D., Kalavrouziotis, I.K., and Karapanagioti, H.K. (2021). Microplastics in agricultural soils: A case study in cultivation of watermelons and canning tomatoes. *Water* 13(16):2168.

Ivleva, N.P., Wiesheu, A.C., and Niessner, R. (2017). Microplastic in aquatic ecosystems. *Angewandte International Edition Chemie* 56(7):1720–1739.

Jansson, K.D., Zawodny, C.P., and Wampler, T.P. (2007). Determination of polymer additives using analytical pyrolysis. *Journal of Analytical and Applied Pyrolysis* 79(1):353–361.

Jung, M.R., Horgen, F.D., Orski, S.V., Rodriguez, V., Beers, K.L., Balazs, G.H., Jones, T.T., Work, T.M., Brignac, K.C., and Royer, S.J. (2018). Validation of ATR FT-IR to identify polymers of plastic marine debris, including those ingested by marine organisms. *Marine Pollution Bulletin* 127:704–716.

Käppler, A., Fischer, D., Oberbeckmann, S., Schernewski, G., Labrenz, M., Eichhorn, K-J., and Voit, B. (2016). Analysis of environmental microplastics by vibrational microspectroscopy: FTIR, Raman or both? *Analytical and Bioanalytical Chemistry* 408(29):8377–8391.

Kniggendorf, A.-K., Wetzel, C., and Roth, B. (2019). Microplastics detection in streaming tap water with Raman spectroscopy. *Sensors* 19(8):1839.

Koelmans, A.A., Besseling, E., and Shim, W.J. (2015). Nanoplastics in the aquatic environment. Critical review. In: Bergmann, M., Gutow, L., and Klages, M., editors. *Marine Anthropogenic Litter*. Berlin/Heidelberg (Germany): Springer, pp. 325–340.

Konde, S., Ornik, J., Prume, J. A., Taiber, J., and Koch, M. (2020). Exploring the potential of photoluminescence spectroscopy in combination with Nile Red staining for microplastic detection. *Marine pollution bulletin* 159: 111475.

Lambert, S., and Wagner, M. (2016). Characterisation of nanoplastics during the degradation of polystyrene. *Chemosphere* 145:265–268.

Lang P.L., Katon J.E., and Bonanno A.S. (1988). Identification of Dust Particles by Molecular Microspectroscopy. *Applied Spectroscopy* 42(2):313–317.

Lee, P.-K., Lim, J., Jeong, Y.-J., Hwang, S., Lee, J.-Y., and Choi B-Y. (2021). Recent pollution and source identification of metal(loid)s in a sediment core from Gunsan Reservoir, South Korea. *Journal of Hazardous Materials* 416:126204.

Lenaker, P. L., Baldwin, A. K., Corsi, S. R., Mason, S. A., Reneau, P. C., and Scott, J. W. (2019). Vertical Distribution of Microplastics in the Water Column and Surficial Sediment from the Milwaukee River Basin to Lake Michigan. *Environment Science Technology* 53(21):12227–12237.

Lenz, R., Enders, K., Stedmon, C.A., Mackenzie, D.M., and Nielsen, T.G. (2015). A critical assessment of visual identification of marine microplastic using Raman spectroscopy for analysis improvement. *Marine Pollution Bulletin* 100(1):82–91.

Leslie, H.A., van Velzen, M.J.M., Brandsma, S.H., Vethaak, A.D., Garcia-Vallejo, J.J., and Lamoree, M.H. (2022). Discovery and quantification of plastic particle pollution in human blood. *Environment International* 163:107199.

Li, J., Liu, H., and Chen, J.P. (2018). Microplastics in freshwater systems: A review on occurrence, environmental effects, and methods for microplastics detection. *Water Research* 137:362–374.

Li, J., Qu, X., Su, L., Zhang, W., Yang, D., Kolandhasamy, P., Li, D., and Shi, H. (2016). Microplastics in mussels along the coastal waters of China. *Environmental Pollution* 214:177–184.

Li, W., Lo, H.S., Wong, H.M., Zhou, M., Wong, C.Y., Tam, N.F., and Cheung, S.G. (2020). Heavy metals contamination of sedimentary microplastics in Hong Kong. *Marine Pollution Bulletin* 153:110977.

Liao, Z., Ji, X., Ma, Y., Lv, B., Huang, W., Zhu, X., Fang, M., Wang, Q., Wang, X., Dahlgren, R., and Shang, X. (2021). Airborne microplastics in indoor and outdoor environments of a coastal city in Eastern China. *Journal of Hazardous Materials* 417:126007.

Liebezeit, G., and Dubaish, F. (2012). Microplastics in beaches of the East Frisian islands Spiekeroog and Kachelotplate. *Bulletin of Environmental Contamination and Toxicology* 89(1):213–217.

Liu, K., Wang, X., Wei, N., Song, Z., and Li, D. (2019). Accurate quantification and transport estimation of suspended atmospheric microplastics in megacities: Implications for human health. *Environment International* 132:105127.

Löder, M.G., and Gerdts, G. (2015). Methodology used for the detection and identification of microplastics—A critical appraisal. In: *Marine Anthropogenic Litter*. Berlin/Heidelberg (Germany): Springer, pp. 201–227.

Löder, M.G.J., Kuczera, M., Mintenig, S., Lorenz, C., and Gerdts, G. (2015). Focal plane array detector-based micro-Fourier-transform infrared imaging for the analysis of microplastics in environmental samples. *Environmental Chemistry* 12(5):563–581.

LSAD. (2020). Operating procedure for soil sampling. LSASDPROC-300-R4.

Maes, T., Jessop, R., Wellner, N., Haupt, N.K., and Mayes, A.G. (2017). A rapid-screening approach to detect and quantify microplastics based on fluorescent tagging with Nile Red. *Scientific Report* 7:44501.

Mai, L., Bao, L-J., Shi, L., Wong, C.S., and Zeng, E.Y. (2018). A review of methods for measuring microplastics in aquatic environments. *Environmental Science and Pollution Research* 25(12):11319–11332.

Majewsky, M., Bitter, H., Eiche, E., and Horn, H. (2016). Determination of microplastic polyethylene (PE) and polypropylene (PP) in environmental samples using thermal analysis (TGA-DSC). *Science of the Total Environment* 568:507–511.

Mallikarjunachari, G., and Ghosh, P. (2016). Analysis of strength and response of polymer nano thin film interfaces applying nanoindentation and nanoscratch techniques. *Polymer* 90:53–66.

Mani, T., Frehland, S., Kalberer, A., and Burkhardt-Holm, P. (2019). Using castor oil to separate microplastics from four different environmental matrices. *Analytical Methods* 11:1788–1794.

Masura, J., Baker, J.E., Foster, G.D., Arthur, C., and Herring, C. (2015). Laboratory methods for the analysis of microplastics in the marine environment: Recommendations for quantifying synthetic particles in waters and sediments. NOAA Marine Debris Program. Technical Memorandum NOS-OR&R-48. Silver Spring, MD.

Mathalon, A., and Hill, P. (2014). Microplastic fibers in the intertidal ecosystem surrounding Halifax Harbor, Nova Scotia. *Marine Pollution Bulletin* 81(1):69–79.

Mecozzi, M., Pietroletti, M., and Monakhova, Y.B. (2016). FTIR spectroscopy supported by statistical techniques for the structural characterization of plastic debris in the marine environment: Application to monitoring studies. *Marine Pollution Bulletin* 106(1–2):155–161.

Minor, E.C., Lin, R., Burrows, A., Cooney, E.M., Grosshuesch, S., and Lafrancois, B. (2020). An analysis of microlitter and microplastics from Lake Superior beach sand and surface-water. *Science of the Total Environment* 744:140824.

Mohamed Nor, N.H., and Obbard, J.P. (2014). Microplastics in Singapore's coastal mangrove ecosystems. *Marine Pollution Bulletin* 79(1–2):278–283.

Müller, Y.K., Wernicke, T., Pittroff, M., Witzig, C.S., Storck, F.R., Klinger, J., and Zumbülte, N. (2020). Microplastic analysis – Are we measuring the same?

Results on the first global comparative study for microplastic analysis in a water sample. *Analytical and Bioanalytical Chemistry* 412(3):555–560.

Nan, B., Su, L., Kellar, C., Craig, N.J., Keough, M.J., and Pettigrove, V. (2020). Identification of microplastics in surface water and Australian freshwater shrimp *Paratya australiensis* in Victoria, Australia. *Environmental Pollution* 259:113865.

Nel, H., and Froneman, P. (2015). A quantitative analysis of microplastic pollution along the south-eastern coastline of South Africa. *Marine Pollution Bulletin* 101(1):274–279.

Ng, K., and Obbard, J. (2006). Prevalence of microplastics in Singapore's coastal marine environment. *Marine Pollution Bulletin* 52(7):761–767.

Nguyen, B., Claveau-Mallet, D., Hernandez, L.M., Xu, E.G., Farner, J.M., and Tufenkji, N. (2019). Separation and analysis of microplastics and nanoplastics in complex environmental samples. *Accounts of Chemical Research* 52(4):858–866.

Norén, F. (2007). *Small Plastic Particles in Coastal Swedish Waters*. KIMO Sweden.

Nuelle, M.-T., Dekiff, J.H., Remy, D., and Fries, E. (2014). A new analytical approach for monitoring microplastics in marine sediments. *Environmental Pollution* 184:161–169.

Peez, N., Janiska, M-C., and Imhof, W. (2019). The first application of quantitative ^{1}H NMR spectroscopy as a simple and fast method of identification and quantification of microplastic particles (PE, PET, and PS). *Analytical and Bioanalytical Chemistry* 411(4):823–833.

Pfeiffer, F., and Fischer, E.K. (2020). Various digestion protocols within microplastic sample processing—Evaluating the resistance of different synthetic polymers and the efficiency of biogenic organic matter destruction. *Frontiers in Environmental Science* 8:572424.

Pinto da Costa, J., Reis, V., Paço, A., Costa, M., Duarte, A.C., and Rocha-Santos, T. (2019). Micro(nano)plastics–Analytical challenges towards risk evaluation. *Trends in Analytical Chemistry* 111:173–184.

Porter, E., and Morris, K. (2007). *Wet Deposition Monitoring Protocol: Monitoring Atmospheric Pollutants in Wet Deposition*. National Park Service, U.S. Department of the Interior L30002-2.

Prata, P.C., da Costa, J.P., Duarte, A.C., and Rocha-Santos, T. (2019). Methods for sampling and detection of microplastics in water and sediment: A critical review. *Trends in Analytical Chemistry* 110:150–159.

Qi, H., Fu, D., Wang, Z., Gao, M., and Peng, L. (2020). Microplastics occurrence and spatial distribution in seawater and sediment of Haikou Bay in the northern South China Sea. *Estuarine: Coastal and Shelf Science* 239:106757.

Reisser, J.W., Slat, B., Noble, K.D., Plessis, K.D., Epp, M., Proietti, M.C., Sonneville, J.D, Becker, T., and Pattiaratchi, C. (2015). The vertical distribution of buoyant plastics at sea: An observational study in the North Atlantic Gyre. *Biogeosciences Discussions* 11:16207–16226.

Ribeiro, D.R.G., Faccin, H., Dal Molin, T.R., de Carvalho, L.M., and Amado, L.L. (2017b). Metal and metalloid distribution in different environmental compartments of the middle Xingu River in the Amazon, Brazil. *Science of the Total Environment* 605:66–74.

Ribeiro-Claro, P., Nolasco, M.M., and Araújo, C. (2017a). Characterization of microplastics by Raman spectroscopy. *Comprehensive Analytical Chemistry* 75:119–151.

Roch, S., and Brinker, A. (2017). Rapid and efficient method for the detection of microplastic in the gastrointestinal tract of fishes. *Environmental Science & Technology* 51(8):4522–4530.

Schmieg, H., Huppertsberg, S., Knepper, T.P., Krais, S., Reitter, K., Rezbach, F., Ruhl, A.S. Köhler, H-R., and Triebskorn, R. (2020). Polystyrene microplastics do not affect juvenile brown trout (*Salmo trutta* f. fario) or modulate effects of the pesticide methiocarb. *Environmental Sciences Europe* 32:49.

Schoeneberger, P.J., Wysocki, D.A., Benham, E.C., and Soil Survey Staff. (2012). Field book for describing and sampling soils. Version 3.0. In: *Natural Resources Conservation Service*. Lincoln (NE): National Soil Survey Center.

Setälä, O., Magnusson, K., Lehtiniemi, M., and Norén, F. (2016). Distribution and abundance of surface water microlitter in the Baltic Sea: A comparison of two sampling methods. *Marine Pollution Bulletin* 110(1):177–183.

Shen, H., Pugh, R., and Forssberg, E. (2002). Floatability, selectivity and flotation separation of plastics by using a surfactant. *Colloids and Surfaces A: Physicochemical and Engineering Aspects* 196(1):63–70.

Shruti, V.C., Perez-Guevara, F., Elizalde-Martinez, I., and Kutralam-Muniasamy, G. (2020). First study of its kind on the microplastic contamination of soft drinks, cold tea and energy drinks: Future research and environmental considerations. *Science of the Total Environment* 726:138580.

Silva, A.B., Bastos, A.S., Justino, C.I.L., da Costa, J.P., Duarte, A.C., and Rocha-Santos, T.A. (2018). Microplastics in the environment: Challenges in analytical chemistry - A review. *Analytical Chimica Acta* 1017:1–19.

Song, Y.K., Hong, S.H., Jang, M., Han, G.M., Rani, M., Lee, J., and Shim, W.J. (2015a). A comparison of microscopic and spectroscopic identification methods for analysis of microplastics in environmental samples. *Marine Pollution Bulletin* 93(1–2):202–209.

Song, Y.K., Hong, S.H., Jang, M., Han, G.M., and Shim, W.J. (2015b). Occurrence and distribution of microplastics in the sea surface microlayer in Jinhae Bay, South Korea. *Archives of Environmental Contamination and Toxicology* 69(3):279–287.

Song, Y.K., Hong, S.H., Jang, M., Kang, J-H., Kwon, O.Y., Han, G.M., and Shim, W.J. (2014). Large accumulation of micro-sized synthetic polymer particles in the sea surface microlayer. *Environmental Science & Technology* 48(16):9014–9021.

Stolte, A., Forster, S., Gerdts, G., and Schubert, H. (2015). Microplastic concentrations in beach sediments along the German Baltic coast. *Marine Pollution Bulletin* 99(1–2):216–229.

Sweden, K. (2007). Small plastic particles in Coastal Swedish waters. N-Research Report. Commissioned by KIMO Sweden.

Thermofisher. (2020). What is SEM? Scanning electron microscopy explained. https://www.thermofisher.com/blog/microscopy/what-is-sem-scanning-electron-microscopy-explained.

Tirkey, A., and Upadhyay, L.S.B. (2021). Microplastics: An overview on separation, identification and characterization of microplastics. *Marine Pollution Bulletin* 170:112604.

Turner, A., and Holmes, L.A. (2015). Adsorption of trace metals by microplastic pellets in fresh water. *Environmental Chemistry* 12(5):600–610.

Uddin, S., Fowler, S.W., Habibi, N., Sajid, S., Dupont, S., and Behbehani, M. (2022). A preliminary assessment of size-fractionated microplastics in indoor aerosol-Kuwait's baseline. *Toxics* 10(2):71.

US Environmental Protection Agency. (2020). *Soil Sampling (LSASDPROC-300-R4)*. Athens (GA): US EPA.

Van, A., Rochman, C.M., Flores, E.M., Hill, K.L., Vargas, E., Vargas, S.A., and Hoh, E. (2012). Persistent organic pollutants in plastic marine debris found on beaches in San Diego, California. *Chemosphere* 86(3):258–263.

Van Cauwenberghe, L., Vanreusel, A., Mees, J., and Janssen, C.R. (2013). Microplastic pollution in deep-sea sediments. *Environmental Pollution* 182:495–499.

Veerasingam, S., Ranjani, M., Venkatachalapathy, R., Bagaev, A., Mukhanov, V., Litvinyuk, Mugilarasan, D., Gurumoorthi, K., Guganathan, L., and Aboobacker, V. (2020). Contributions of Fourier transform infrared spectroscopy in microplastic pollution research: A review. *Critical Reviews in Environmental Science and Technology* 51(22):2681–2743.

Vianello, A., Boldrin, A., Guerriero, P., Moschino, V., Rella, R., Sturaro, A., and Da Ros, L. (2013). Microplastic particles in sediments of Lagoon of Venice, Italy: First observations on occurrence, spatial patterns and identification. *Estuarine, Coastal and Shelf Science* 130:54–61.

Wahl, A., Le Juge, C., Davranche, M., El Hadri, H., Grassl, B., Reynaud, S., and Gigault, J. (2021). Nanoplastic occurrence in a soil amended with plastic debris. *Chemosphere* 262:127784.

Wessel, C.C., Lockridge, G.R., Battiste, D., and Cebrian, J. (2016). Abundance and characteristics of microplastics in beach sediments: Insights into microplastic accumulation in northern Gulf of Mexico estuaries. *Marine Pollution Bulletin* 109(1):178–183.

Wetherbee, G.A., Baldwin, A.K., and Ranville, J.F. (2019). It is raining plastic. U.S. Geological Survey Open-File Report 2019–1048.

Woo, H.; Seo, K., Choi, Y., Kim, J., Tanaka, M., Lee, K., and Choi, J. (2021). Methods of analyzing microsized plastics in the environment. *Applied Sciences* 11:10640.

Woodall, L.C., Sanchez-Vidal, A., Canals, M., Paterson, G.L., Coppock, R., Sleight, V., Calafat, A., Rogers, A.D., Narayanaswamy, B.E., and Thompson, R.C. (2014). The deep sea is a major sink for microplastic debris. *Royal Society Open Science* 1(4):140317.

Wright, S.L, Ulke, J., Font, A., Chan, K.L.A., and Kelly, F.J. (2020). Atmospheric microplastic deposition in an urban environment and an evaluation of transport. *Environment International* 136:105411.

Wu, C.-C., Bao, L-J., Tao, S., and Zeng, E.Y. (2016). Significance of antifouling paint flakes to the distribution of dichlorodiphenyltrichloroethanes (DDTs) in estuarine sediment. *Environmental Pollution* 210:253–260.

Yan, M., Nie, H., Xu, K., He, Y., Hu, Y., Huang, Y., and Wang, J. (2019). Microplastic abundance, distribution and composition in the Pearl River along Guangzhou city and Pearl River estuary, China. *Chemosphere* 217:879–886.

Yang, L., Zhang, Y., Kang, S., Wang, Z., and Wu, C. (2021). Microplastics in soil: A review on methods, occurrence, sources, and potential risk. *Science of the Total Environment* 780:146546.

Yao, Y., Glamoclija, M., Murphy, A., and Gao, Y. (2022). Characterization of microplastics in indoor and ambient air in northern New Jersey. *Environmental Research* 207:112142.

Yu, X., Peng, J., Wang, J., Wang, K., and Bao, S. (2016). Occurrence of microplastics in the beach sand of the Chinese inner sea: The Bohai Sea. *Environmental Pollution* 214:722–730.

Zhang, K., Su, J., Xiong, X., Wu, X., Wu, C., and Liu, J. (2016). Microplastic pollution of lakeshore sediments from remote lakes in Tibet plateau, China. *Environmental Pollution* 219:450–455.

Zhang, K., Xiong, X., Hu, H., Wu, C., Bi, Y., Wu, Y., Zhou, B., Lam, P.K., and Liu, J. (2017a). Occurrence and characteristics of microplastic pollution in Xiangxi Bay of Three Gorges Reservoir, China. *Environmental Science & Technology* 51(7):3794–3801.

Zhang, Y., Gao, T., Kang, S., and Sillanpaa, M. (2019). Importance of atmospheric transport for microplastics deposited in remote areas. *Environmental Pollution* 254:112953.

Zhang, J., Wang, L., and Kannan, K. (2020). Microplastics in house dust from 12 countries and associated human exposure. *Environmental International* 134:105314.

Zhang, Q., Zhao, Y., Du, F., Cai, H., Wang, G., and Shi, H. (2020). Microplastic fallout in different indoor environments. *Environmental Science & Technology* 54(11):6530–6539.

Zhao, S., Zhu, L., and Li, D. (2015). Characterization of small plastic debris on tourism beaches around the South China Sea. *Regional Studies in Marine Science* 1:55–62.

Zhao, S., Zhu, L., Wang, T., and Li, D. (2014). Suspended microplastics in the surface water of the Yangtze Estuary System, China: First observations on occurrence, distribution. *Marine Pollution Bulletin* 86(1–2):562–568.

Zobkov, M., and Esiukova, E. (2017a). Microplastics in Baltic bottom sediments: Quantification procedures and first results. *Marine Pollution Bulletin* 114(2):724–732.

Zobkov, M., and Esiukova, E. (2017b). Evaluation of the Munich Plastic Sediment Separator efficiency in extraction of microplastics from natural marine bottom sediments. *Limnology and Oceanography: Methods* 15(11):967–978.

Part II

SEPARATION OF MICROPLASTICS FROM WATER AND SEDIMENT

The pervasive presence of microplastics (MPs) in water, sediment, and soil has been revealed only recently. Scientists have only begun to understand how MPs behave in these media.

Part II of this book focuses on technologies for the processing and ultimate separation of MPs from water and sediment. Some systems are already in use; for example, wastewater treatment plants (WWTPs) are documented to recover substantial quantities of MPs. Other technologies, which are applied for processing of industrial wastewater, may be feasible as well. Some researchers have estimated that the vast majority of MPs in the global environment, perhaps more than 90%, have accumulated in sediments. Certain technologies are already established for separation of plastics from municipal and other wastes, and may be applicable for MP removal from sediments. Additionally, technologies currently in use for processing of minerals or solid waste fractions and may be applied for separation of MPs.

Chapter 7 presents a detailed discussion of MP removal from water, with emphasis on the capabilities of unit operations of WWTPs. In chapter 8, technologies for recovery of MPs from sediment are introduced. Some are already in use in various industries, while other, experimental techniques, are at the pilot scale.

Chapter 7

Removal of Microplastics from Wastewater

INTRODUCTION

For the past decade the volumes, chemical types, and morphologies of microplastics (MPs) entering wastewater and drinking water treatment plants (WWTPs and DWTPs, respectively) have been studied extensively. Both primary and secondary MPs comprise inputs to these facilities. Primary MP beads originate in cleaning and cosmetics/toothpaste/PCCPs and are composed primarily of PE, PP, and PS. Secondary MPs, predominantly breakdown products from washing of synthetic textiles, occur as fibers and are composed mainly of polyester, acrylic, and polyamide (Browne et al. 2011). Other secondary MPs are fragments and fibers from breakdown of macroplastics, tire and road wear particles, and others. Quantities and types of MPs entering WWTPs and DWTPs vary widely. One review (Gatidou et al. 2019) concluded that WWTP influents had estimated MP concentrations varying from 1 to 3,160 particles per liter.

As will be shown, MP removal rate is substantial at many WWTPs and DWTPs; but even so, WWTPs are recognized as a significant contributor of MPs to the environment due to the large volumes discharged (Talvitie et al. 2017a). For example, Murphy et al. (2016) estimated a daily discharge of 6.5×10^7 MPs into receiving water by a Scottish WWTP serving 650,000 inhabitants (treating 260,954 $m^3{\cdot}d^{-1}$). Carr et al. (2016) calculated a discharge of 0.93×10^6 MPs per day from a WWTP treating 1.06×10^6 m^3 of wastewater per day. Talvitie et al. (2017a) estimated high volumes of MPs in effluents of a Finnish WWTP treating 270,000 m^3 wastewater per day: quantities ranged from 1.7×10^6 to 1.4×10^8 particles$\cdot d^{-1}$, depending on day of the week. Ziajahromi et al. (2017) estimated MP discharges as high as 1×10^7 per day from an Australian WWTP that serves 150,000 inhabitants.

Disagreement persists regarding the ability of WWTPs to remove MPs from influent wastewater. Some researchers claim that WWTPs remove only limited quantities of MPs, in part due to their small size and buoyancy (Leslie et al. 2017; Browne et al. 2011; McCormick et al. 2014; Burns and Boxall 2018; Wang et al. 2018). Others, however, have documented near-complete removal of MPs from WWTPs (Hidayaturrahman and Lee 2019; Lares et al. 2018; Yang et al. 2019; Zhang and Chen 2020).

In order to determine the fate of MPs during wastewater treatment and the possible contribution to aquatic ecosystems by WWTPs, the mass of particles entering these facilities per unit volume of water must be determined accurately, as well as that released with the effluent. Studies in the United States, Europe, China, Russia, and elsewhere have measured MP flows in WWTPs (Talvitie and Heinonen 2014; Magnusson and Norén 2014; Schneiderman 2015) (table 7.1). It has been difficult, however, to compare MP removal efficiencies from different facilities operating in various countries. The divergence in researchers' findings is a consequence of several variables, including composition of influent water (e.g., natural organic matter [NOM], pH, electrolyte concentration), daily processing volume, types of wastewater treatment unit operations applied, particle size of MPs, physical forms of MPs in wastewater (e.g., particle versus fiber), and analytical techniques used for particle identification (Padervand et al. 2019). The necessary steps and technical challenges associated with MPs identification and quantification were discussed in chapter 6.

WWTPs are not designed for the dedicated removal of MPs from wastewater influent; regardless, it has been observed that MP removal efficiency varies at rates ranging from 25% to as high as 99.9% (table 7.1) (Bayo et al. 2020b; Park et al. 2020; Yang et al. 2019; Leslie et al. 2017; Carr et al. 2016). Bayo et al. (2020b) found that MPs comprised 46.6% of all micro litter entering a WWTP, with a removal rate of 90.3%. At a WWTP in Sweden the combined mechanical, chemical, and biological treatment processes removed about 99% of MPs—influent contained 15,000 particles$\cdot l^{-1}$ which was reduced to 8 particles$\cdot l^{-1}$ in effluent (Magnusson and Norén 2014). At a Russian WWTP the influent concentration of MPs as textile fibers was 467 particles$\cdot l^{-1}$, which declined to 16 particles$\cdot l^{-1}$ in effluent, a 96% reduction (Talvitie and Heinonen 2014). At the Viikinmäki WWTP in Finland, influent contained 180 and 430 MP fibers and particles, respectively, per liter. In tertiary-treated effluent average concentrations of 4.9 and 8.6 microfibers and particles per liter, respectively, were detected (~98% removal after tertiary treatment) (Talvitie et al. 2015).

Table 7.1 Removal of MPs from WWTPs and DWTPs, Recent Data

Location	*Type of Facility*	*Treatment Processes*	*Influent* $MP \cdot l^{-1}$	*Effluent* $MP \cdot l^{-1}$	*Microplastic Removal (%)*	*Reference*
Australia	WWTP	Screening and grit removal precedes primary treatment	–	–	69–79%	Ziajahromi et al. 2021
China	WWTP	A/O and A/A/O process	0.70–8.72	0.07–0.78	89.2–93.6%	Zhang et al. 2021
India	DWTP	Pulse clarification and sand filtration	17.88	2.75 ± 0.92	63% and 85% for pulse clarification and sand filtration, respectively.	Sarkar et al. 2021
Indonesia	DWTP	Aeration, pre-sedimentation, coagulation, flocculation-sedimentation, filtration, disinfection	26.8–35	8.5–12.3	54–76%.	Radityaningrum et al. 2021
Canada	DWTP	Alum-based coagulation-flocculation-sedimentation (CFS)	–	–	With an alum dose of 30 mg L^{-1}, 75.6 and 85.2% of 6 μm PS microspheres were removed from Grand River and Lake Erie water, respectively.	Xue et al. 2021
Mersin Bay, Turkey	WWTP	Karaduvar WWTP (tertiary treatment), Silifke and Tarsus WWTPs (secondary treatment)	2.8 at Karaduvar; 1.5 at Silifke; 3.1 at Tarsus.	1.6, 0.7, and 0.6 in Karaduvar, Tarsus and Silifke, respectively.	55–97%	Akarsu et al. 2020

(*continued*)

Table 7.1 (Continued)

Location	*Type of Facility*	*Treatment Processes*	*Influent* $MP \cdot l^{-1}$	*Effluent* $MP \cdot l^{-1}$	*Microplastic Removal (%)*	*Reference*
Northwest Turkey	WWTP	Flotation and settling units consisting of coarse screens, fine screens, aerated grit chambers (AGCs), and activated sludge.	–	–	AGCs removed 59% of influent concentration of large-sized films (1–5 mm); settling tanks removed approx. 90% of influent concentration of fragments. Observed MP removal rates: fragments > films > fibers (90% > 85% > 70%, respectively).	Bilgin et al.2020
China	ADWTP	Coagulation combined with sedimentation and GAC filtration	–	–	40.5–54.5%, mainly for fiber removal. Presence of GAC filtration reduced MP abundance by 56.8–60.9%, mainly for small MPs.	Wang et al. 2020b
China	WWTP	–	18–890	6–26	35–98%	Wang et al. 2020b
Spain	WWTP	MBR	4.4 ± 1.01	0.92 ± 0.21	79.0%	Bayo et al. 2020a
Spain	WWTP	RSF	4.4 ± 1.01	1.08 ± 0.28	75.5%	Bayo et al. 2020a
Spain	WWTP	Primary clarifier followed by A^2O biotreatment	–	12.8 ± 6.3	93.7%	Edo et al. 2020
Korea	WWTP	Primary Secondary (50 plants) AS, A^2O, MLE	10–470	0.004–0.51	98.7–99.9%	Park et al. 2020

Czech Republic	DWTP	Coagulation/flocculation-sedimentation, deep-bed filtration through clay-based material, GAC filtration	23 ± 2 at one DWTP, 1296 ± 35at second DWTP.	14 ± 1 at one DWTP; 151 ± 4 at second DWTP.	Higher total removal at DWTP with higher initial MP load and more complicated treatment (removal 88% versus 40%); coagulation-flocculation-sedimentation, deep-bed filtration through clay-based material, and granular activated carbon filtration contributed to MP elimination by 62%, 20%, and 6%, respectively	Pivokonský et al. 2020
Australia	WWTP	AS	11.80 + 1.10	2.76 + 0.11	76.6%	Raju et al. 2020
UK	WWTP	Primary Secondary Tertiary	–	–	96%	Blair et al. 2019
United States	WWTP	AS	49,600–83,500	1–30	74.8–98.1%	Conley et al. 2019
China	WWTP	–	1.57–13.69	0.20–1.73	79.3–97.8%	Long et al. 2019
China	WWTP	OD and MBR	–	–	99.5% in MBR system versus 97% in OD system on basis of plastic mass; 82.1% versus 53.6% based on MP number.	Lv et al. 2019
China	WWTP	–	196.00 ± 11.89	9.04 ± 1.12	over 90%	Xu et al. 2019

(*continued*)

Table 7.1 (Continued)

Location	*Type of Facility*	*Treatment Processes*	*Influent MP·l⁻¹*	*Effluent MP·l⁻¹*	*Microplastic Removal (%)*	*Reference*
China	WWTP	AS	79.9	28.4	64.4%	Liu et al. 2019
Italy	WWTP	Sedimentation, sand filter treatment, and disinfection	2.5	0.4	84%	Magni et al. 2019
China	WWTP	Aerated grit chamber, A^2O, tertiary treatments (ozonation, UV, other)	12.03	0.59	95%	Yang et al. 2019
Vancouver, Canada	WWTP	Secondary treatment	–	–	97–99%	Gies et al. 2018
Turkey	WWTP	Aeration	12–36	3.3–6.9	73–79%	Gündoğdu et al. (2018)
Denmark	WWTP	Ten WWTPs: Primary, Secondary, Tertiary (RSF)	8149	19	98.3–99.3%	Simon et al. 2018
Poland	WWTP	–	19.4–552.2	0.03–1.0	95–99%	Wisniowska et al. 2018
Finland	WWTP	AS Advanced MBR	0.1–57.6	1.0 ± 0.4	98.3%	Lares et al. 2018
Netherlands	WWTP	7 WWTPs: Primary MBR	65–238	9–91	25.0%	Leslie et al. 2017
Finland	WWTP	RSF DAF MBR	0.7 (RSF) 2.0 (DAF) 6.9 (MBR)	0.02 (RSF) 0.1 (DAF) 0.005 (MBR)	97% (RSF) 95% (DAF) 99.9% (MBR)	Talvitie et al. 2017a
Finland	WWTP	Activated sludge process, biologically active filter	597.9–675.5	0.4–1.6	99.9%	Talvitie et al. 2017b
Australia	WWTP	Primary Secondary Reverse osmosis	12	0.28	98.3%	Ziajahromi et al. 2017

Germany	WWTP	Primary secondary	–	–	97%	Mintenig et al. 2017
United States	WWTP	1 secondary, 7 tertiary WWTPs Aeration, activated sludge	1	8.8×10^{-4}	99.9%	Carr et al. 2016
United States	WWTP	Screening, grit removal, activated sludge	133	5.9	95.6%	Michielssen et al. 2016
Scotland	WWTP	Coarse screening, grit removal, sedimentation, aeration	15.7	0.25	98.4%	Murphy et al. 2016
France	WWTP	Primary secondary (biofilter)	260–320	14–50	84–94%	Dris et al. 2015
Sweden	WWTP	Primary secondary	15	–	99.9%	Magnusson and Norén 2014

Treatment processes: Secondary treatment: conventional activated sludge process expected for where specified.

AnMBR—anaerobic membrane bioreactor; A^2O—anaerobic–anoxic–aerobic treatment process; AS—activated sludge treatment process; BAF—biological aerated filter; DAF—dissolved air flotation; DF—disc filters; GF—granular filter; MBR—membrane bioreactor; MLE—modified Ludzack–Ettinger process; MP—maturation pond; OD—oxidation ditch; RO—reverse osmosis; RSF—rapid sand filter.

UNIT OPERATIONS AT WWTPS

Worldwide, WWTPs share a basic common design where several distinct unit operations and processes are incorporated. Depending on design and scale of the facility, plants are equipped with primary, secondary, and tertiary treatment unit operations to treat wastewater via physical, biological, and chemical processes, respectively. Conventional processes and their functions are shown in figures 7.1 and table 7.2.

Each unit operation serves a specific function in the conversion of wastewater to water suitable for release to an aquatic ecosystem. Individual facilities, however, often differ in design configuration based on variables such as population size, industrial inputs, and presence of non-point sources such as feedlots and cropland (Habib et al. 2020). Numerous published sources reveal the capture of MPs within primary, secondary, and/or tertiary phases at WWTPs, each with varying degrees of success (Xu et al. 2021). No unit operation at a WWTP specifically targets removal of MPs, but it has been reported that 35–59% of MPs are removed during preliminary treatment (Michielssen et al. 2016; Xu et al. 2021; Sun et al. 2019). Several studies (Xu et al. 2021; Sun et al. 2019; Murphy et al. 2016; Talvitie et al. 2017a) have reported MP removal efficiency with primary treatment ranging from 50–98%. Secondary treatment is associated with a modest decrease in MPs, ranging from 2–55% (Lv et al. 2019; Yang et al. 2019; Murphy et al. 2016; Alvim et al. 2020) and tertiary treatment appears to have only a minor effect on reducing the concentration of MPs (typically less than 1–5%) (Hou et al. 2021; Talvitie 2017b; Habib 2020; Cristaldi et al. 2020).

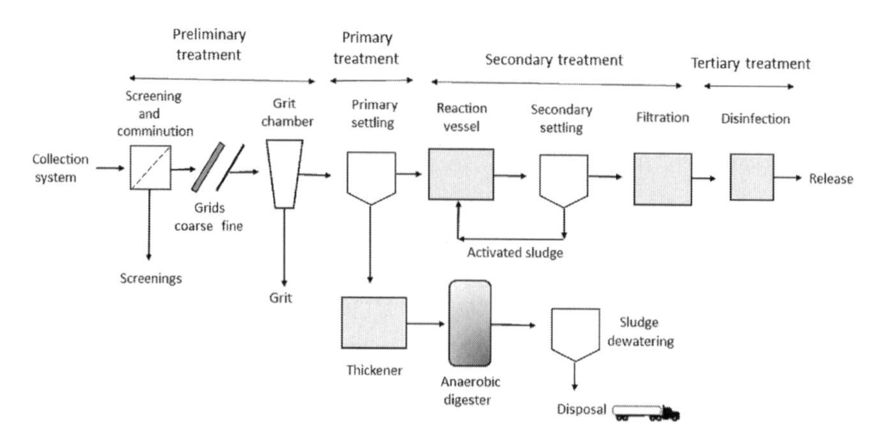

Figure 7.1 Schematic of Typical Municipal WWTP

Table 7.2 Key Functions of WWTP Unit Operations

Stage	*General Description*	*Example Unit Operations*
Preliminary	Physical processes that remove large debris at the start of the treatment process. Solids are removed based on particle density and size.	Screening Grit removal Sedimentation Aeration Flotation and skimming Degasification
Primary	Physical processes that remove large solids; homogenize remaining effluent. Removes large proportion of settleable, suspended, and floating materials.	Adsorption Sedimentation Flotation
Secondary	Biological processes for removal of biochemical demand for oxygen.	Activated sludge Trickling filters Rotating biological contactors Lagoons Oxidation ponds Aerobic digestion Anaerobic digestion
Tertiary	Physical, chemical, and biological processes that remove excess nutrients, SS and COD. Odor and color removal; further oxidation.	Rapid sand filtration Membrane filtration Adsorption (e.g., activated carbon) DAF Coagulation/flocculation Ion exchange Nitrification/denitrification Ozone oxidation Disinfection Reverse osmosis

MP Removal during Pretreatment

Raw influent entering the treatment plant contains diverse trash and debris. At many WWTPs, influent first passes through preliminary treatment units before primary and secondary treatment begin (US EPA 2004). Pretreatment of wastewater typically includes screening, grit removal, sedimentation, and flotation. Such preliminary processing serves to protect facility equipment by removing the large and coarse materials that may clog pumps, small pipes, and downstream processes, as well as cause excessive abrasion and wear to facility equipment. Removal of large debris early in the treatment process additionally saves valuable space inside the treatment plant (Spellman and Drinan 2003).

Screening

The purpose of the screening step is to remove large floating debris. Screens are constructed with parallel steel bars and are divided into different mesh sizes. Typical sizes are coarse (mesh size 50–100 mm), middle (10–40 mm), and fine (2.5–10 mm). Given that MPs are defined as particles of size less than 5 mm, they will pass through the coarse and middle screens; however, particle sizes in excess of 2.5 mm should be recovered in fine screens (Zhang and Chen 2020; US EPA 2004).

Some WWTPs use a device known as a comminutor which functions as both a screen and a grinder. This device traps and then cuts or shreds large floating material. The pulverized matter remains in the wastewater flow to be removed downgradient in a primary settling tank (US EPA 2004). It is also possible, however, that this unit can actually produce additional MPs due to simple mechanical breakage of macroplastic articles.

Grit Chamber

The grit chamber removes high-density contaminants such as sand and small stones; these are mineral particles having a density of approximately 2.65 $g \cdot ml^{-1}$. As stated in chapter 2, the density of most common MPs ranges between 0.8 and 1.6 $g \cdot ml^{-1}$; therefore, many MPs will float and pass through the grit chamber to subsequent stages of treatment. A small proportion of polymers have greater densities, however, and will be captured with the grit.

Two types of grit chambers are applied in wastewater treatment: horizontal and aerated. In the horizontal flow grit chamber, grit is removed by maintaining a constant upstream velocity of about 0.3 $m \cdot s^{-1}$ (1 $ft \cdot s^{-1}$). Velocity is controlled by weirs. Removal of particles is a function of the velocity of water flow (V) and traveling distance of the water (D). When particle settling time is less than D/V, MPs are removed; otherwise, they will remain in the flow (Zhang et al. 2020). The heavy grit particles settle to the base of the channel while the lighter, predominantly organic particles remain suspended or are resuspended for transport out of the channel. Some MPs may become attached to sand particles and settle along with the grit for removal. Grit is removed via conveyor with scrapers, buckets, or plows. Screw conveyors or bucket elevators elevate the grit for disposal (US EPA 2003) (figure 7.2).

In aerated grit chambers, the grit is removed by causing wastewater to flow in a spiral pattern (figure 7.3). Air is introduced into the chamber along one side, thus forcing wastewater to flow through the tank in a spiral pattern. Heavier particles are accelerated, depart from the forward flow, and fall to the bottom of the tank, while lighter particles remain suspended and are eventually transported out of the tank (US EPA 2003).

Some researchers have reported that MP removal in the grit chamber is low; Blair et al. (2019) measured less than 6% of MPs in the grit chamber at a WWTP in Scotland. In contrast, however, Yang et al. (2019) found that approximately 60% of MPs in wastewater was removed in aerated grit.

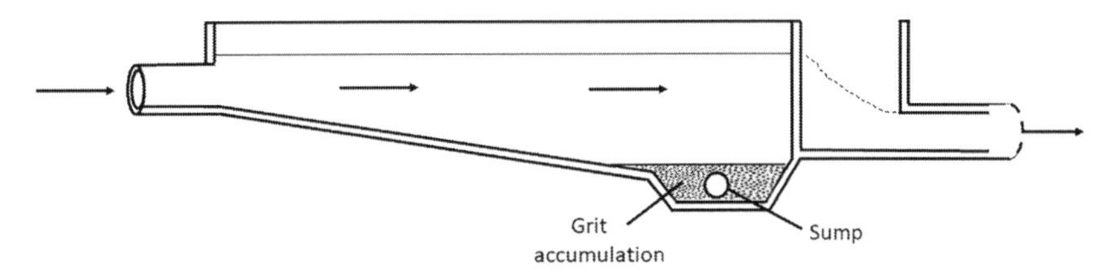

Figure 7.2 Horizontal Flow Grit Chamber
Source: Adapted from Chukwulozie et al. (2018). Design of sewage treatment plant for CBN Housing Estate Trans-Ekulu Enugu Nigeria. *Advances in Research,* 13(3), 1-18. Creative Commons Attribution License (http://creativecommons.org/licenses/by/4.0).

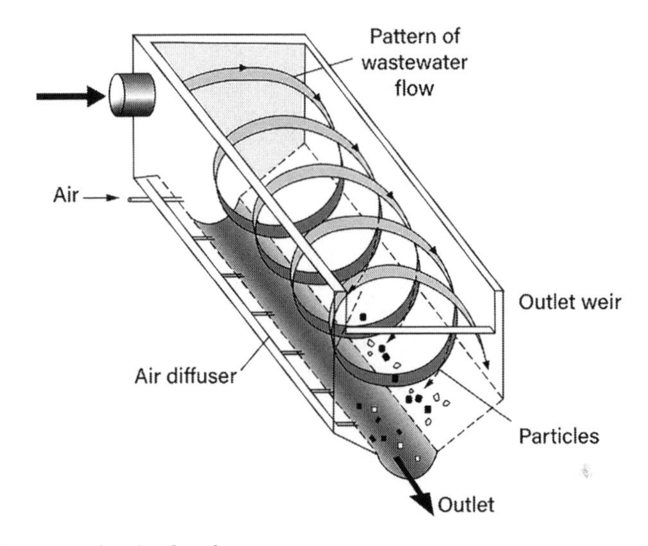

Figure 7.3 Aerated Grit Chamber
Source: Reproduced with kind permission of Ron Crites and George Tchobanoglous. (Adapted from Crites, R., and Tchobanoglous, G. 1998. *Small and Decentralized Wastewater Management Systems,* WCB McGraw-Hill. Inc. Boston, MA.)

Various data suggest that the aerated grit chamber is more effective in removing MPs as compared to the horizontal grit chamber (Zhang et al. 2020).

MP Removal in Primary Treatment

After screening and removal of grit, wastewater still contains large quantities of dissolved organic and inorganic constituents along with suspended solids (SS). The SSs consist of fine particles (including MPs) that can be removed by applying additional treatments such as sedimentation, chemical coagulation, or filtration (US EPA 2004).

Primary treatment is a physical process that reduces the organic loading on downstream processes by removing a large proportion of settleable, suspended, and floatable materials from influent wastewater within settling tanks (figure 7.4) (Ziajahromi et al. 2017).

Several recent studies have revealed that the preliminary and primary treatment stages remove the majority of MPs from wastewater. MP removal rates ranging from 50–98% after primary treatment have been recorded (Xu et al. 2021; Sun et al. 2019; Murphy et al. 2016; Talvitie et al. 2017b; Alvim et al. 2020).

SS are removed by controlling wastewater flow velocity. The settling time of SS must be shorter than the travel time of water in the sedimentation tank (Zhang et al. 2020). As wastewater enters the tank, its velocity is reduced to approximately 30–60 $cm \cdot min^{-1}$ (1–2 $ft \cdot min^{-1}$) (Spellman and Drinan 2003) so that many SS sink to the bottom. Target pollutants for removal are SS of low density ($\geq$ 1.1 $g \cdot ml^{-1}$ and $\leq$ 1.5 $g \cdot ml^{-1}$) or small particles of high density (> 1.5 $g \cdot ml^{-1}$). In the primary tank, both the settled sludge solids (termed *primary sludge*) and floating grease and scum are removed. These waste streams are pumped to separate units for disposal or further treatment (Spellman and Drinan 2003).

It is expected that wastewater enriched with MPs of common polymer types (PE, PP, PS, etc.) will undergo substantial MP removal during primary treatment. Several MPs, such as polybutylene terephthalate (PBT) (1.31 $g \cdot cm^{-1}$), PET (1.38 $g \cdot cm^{-3}$), PVC (1.38 $g \cdot cm^{-3}$) and polytetrafluoroethylene (PTFE) (~ 2.2 $g \cdot cm^{-3}$), have a density greater than wastewater (Zhang et al. 2020; Xu et al. 2021); therefore, it is likely that primary sedimentation will remove MPs composed of these polymers. However, it is more difficult to remove PE (density 0.91 $g \cdot cm^{-3}$) and PP (0.85 $g \cdot cm^{-3}$) as their density is less than water.

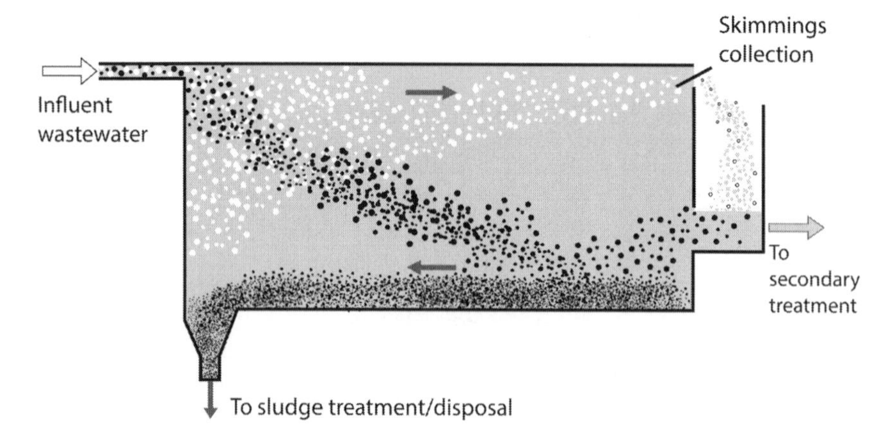

Figure 7.4 Primary Settling Tank at WWTP

Flotation

A flotation step may be included with primary sedimentation. The air flotation techniques most commonly utilized in WWTPs are electro-flotation, dispersed air flotation, and dissolved air flotation (DAF) (Rubio et al. 2002).

DAF has been used extensively in treatment of various types of municipal and industrial wastewaters (Palaniandy et al. 2017). The design function of DAF is to separate suspended particles from liquids by lifting them to the surface of the liquid.

The flotation system consists of four main components, that is, an air supply, pressurizing pump, retention tank, and flotation chamber (figure 7.5). In this unit operation, micro-bubbles are produced via aeration in a tank. A compressed air system may be used. A stream of bubbles (size approximately 20–70 μm) is generated and passes upward through the wastewater (Sol et al. 2020). Bubbles are dispersed within the tank and adhere to fine suspended particles, forming a bubble-solid complex which has an overall lower density less than that of water. As these agglomerations rise to the surface, they collide with other air bubbles and suspended particles and flocculate (figure 7.6). This ultimately forms a stable, floating foam that rises against the force of gravity. MP particles tend to partition into the floating foam for removal (Rubio et al. 2002; Xu et al. 2021). Contaminants such as oil and small SSs also attach to the bubbles and float to the surface; these

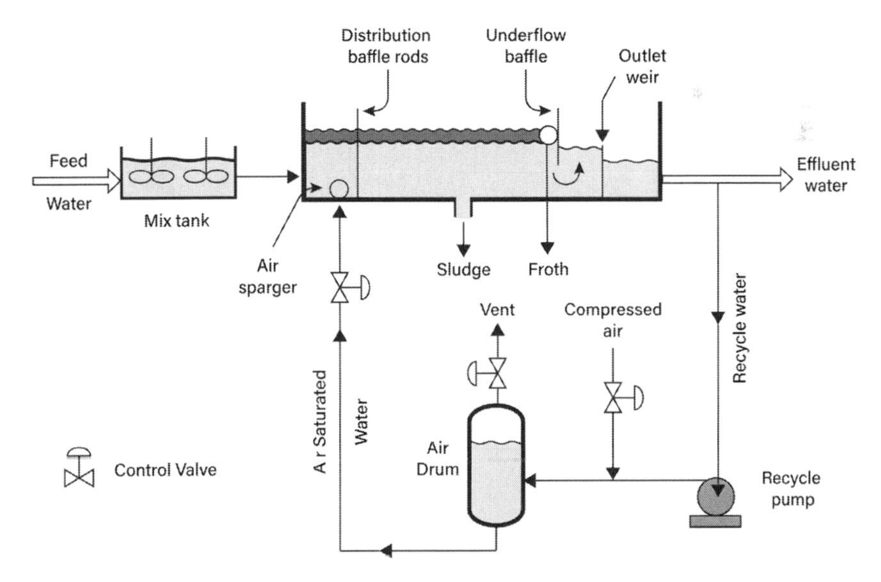

Figure 7.5 Schematic Showing Air Flotation Technique during Primary Wastewater Treatment

Source: Adapted from Mbeychok, https://en.wikipedia.org/wiki/Dissolved_air_flotation#/media/File:DAF_Unit.png https://creativecommons.org/licenses/by/4.0/

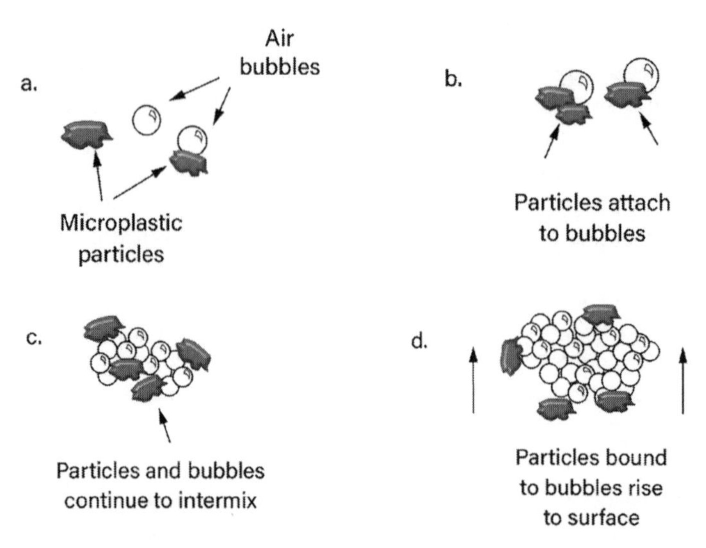

Figure 7.6 Removal of MPs by Air Flotation

are all removed via skimming (Sol et al. 2020; Palaniandy et al. 2017; Wang et al. 2005). Sludge removal is accomplished on either a continuous or batch basis.

Various researchers have reported that low-density plastic particles such as PE and PP can be effectively removed by selective flotation. Air flotation technology may be effective for removal of moderate-density MPs such as PS and PA as well (Ngo et al. 2019; Xu et al. 2021). In many cases DAF serves as an alternative process to sedimentation and offers advantages such as better final water quality, rapid startup, high rates of operation, and thicker sludge. In facilities where primary sedimentation and air flotation are combined, both heavy and light MPs can be removed, resulting in high overall removal of MPs (Xu et al. 2021).

It must be noted that the presence of oil, grease, surfactants, and other contaminants in wastewater will modify the surface physicochemical properties of MPs (Xu et al. 2021). Such contaminants will also alter the specific gravity of the suspension and consequently affect removal efficiency of MPs.

DAF is also used during the tertiary treatment of wastewater.

MP Removal in Secondary Treatment

The key design function of secondary treatment is the biological decomposition of dissolved and suspended organic matter in wastewater. Decomposition is carried out by a vast consortium of heterotrophic aerobic microorganisms under controlled conditions. Removal of carbonaceous wastes also occurs via adsorption and sedimentation. A number of biological treatment processes and systems are available (table 7.3). The two most common are activated sludge ("suspended growth") and biofilm ("fixed film") processes.

According to published literature, the biological processes taking place during secondary treatment are capable of removing from 2–55% of MPs from wastewater (Lv et al. 2019; Yang et al. 2019; Murphy et al. 2016; Alvim et al. 2020). Removal mechanisms and quantities of MPs recovered differ markedly among the different biological processes.

MP Removal in the Activated Sludge Process

The activated sludge process is the most widely applied biological method for the treatment of urban wastewater. Dissolved and suspended biodegradable organic matter and other SSs, including MPs, can be effectively removed from wastewater by activated sludge treatment (Xu et al. 2021). The process promotes microbial activity within a suspended growth reactor, typically operated in continuous mode. The main components of the process include an aeration tank and a sedimentation tank (figure 7.7).

The suspension is aerated and mixed thoroughly using physical mixers or air spargers fed by compressors. Microbial biomass attaches to and reacts with organics within the influent waste stream. As organic matter is oxidized by the microbial consortia, extracellular polymeric substances (EPS) secreted

Table 7.3 Types of Secondary Treatment Processes for Wastewater

Process	*Description*	*Examples*
Suspended Growth	Biomass is actively mixed with wastewater. Can be operated in smaller space than trickling filters that treat the same amount of water.	Activated sludge Granular activated sludge
Fixed Film (Attached Growth)	Microbial biomass grows on a physical support and wastewater passes over its surface. System allows for ready oxygen incorporation. Higher removal rates for organics and suspended solids than suspended growth systems.	Trickling filters Rotating biological contactors Bio-towers Moving bed biofilm reactors (MBBR) Integrated fixed-film activated sludge process Constructed wetlands
Innovative Technologies	Equipment designs which remove a specific nutrient during microbial oxidation. Filtration (membrane) combined with oxidation.	Membrane bioreactors Anaerobic ammonium oxidation reactor (ANAMMOX) Short-cut nitrification and denitrification

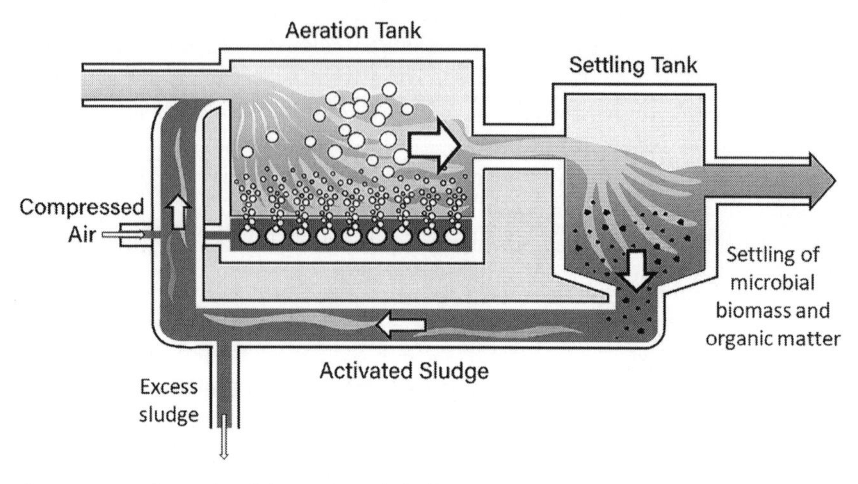

Figure 7.7 Schematic of the Activated Sludge Process
Source: Adapted from US EPA (2004).

by the cells adsorb MPs. These viscous gel-like structures, which are species-specific, are composed of long-chain polysaccharides (figure 7.8). Biofilms rapidly form on surfaces. EPS and biofilms play an important role in activated sludge, as they serve to maintain granule structure. They also serve as a defense for cells against toxic substances. Biofilms also may contribute to adsorption and eventual removal of MPs (Xu et al. 2021).

Suspended biomass accumulates quickly; therefore, sedimentation must be closely managed. After the appropriate reaction time in the aeration tank, the microbial suspension is transferred to a sedimentation tank. The biomass is allowed to coagulate and settle. The clarified liquid is collected and is either discharged or treated further. The sludge (predominantly microbial biomass with some sorbed MPs) is removed from the tank bottom; a portion is recycled to the aeration tank, which allows for sustained microbial decomposition of

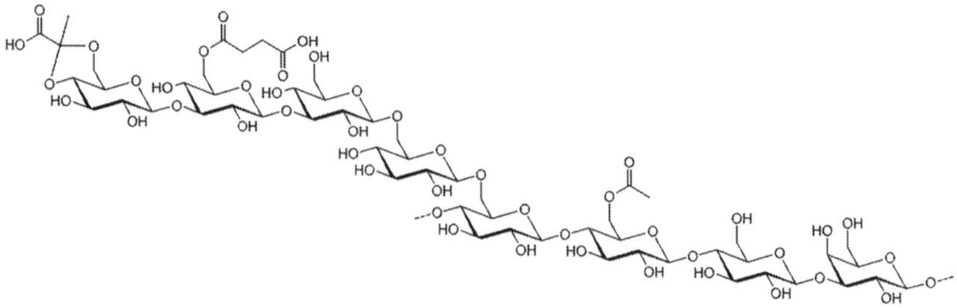

Figure 7.8 A Succinoglycan Molecule, an Example of EPS
Source: Https://en.wikipedia.org/wiki/File:Sinorhizobium_meliloti_monosuccinylated_succinoglycan_ EPS_I).svg

biochemical oxygen demand (BOD) and other contaminants. The recycled biomass is already acclimated to the waste stream constituents; therefore, no lag time occurs for the microbes to act on the newly added waste (Pichtel 2019).

The activated sludge process has been reported to remove a range of 3.6%–42.9% of MPs from wastewater (Carr et al. 2016; Lares et al. 2018; Mason et al. 2016; Zhang et al. 2020). The process appears to be more effective at removing small MPs compared to larger ones. At a WWTP in Spain, Bayo et al. (2020b) determined greater abundance of MPs with particle sizes of 400–600 μm in effluent from activated sludge. In another study (Liu et al. 2019), the MPs transferred to the sludge had an average size of 222.6 μm. In a study on seven WWTPs in Xiamen, China (Long et al. 2019), MP removal rate from secondary treatment improved with decreasing particle size. Average removal rate of large MPs (>355 μm) was 78.5% and removal rate for small MPs (43–63 μm) was 95.5%. The smaller MPs may be more readily ingested by protozoans and metazoans and therefore incorporated within sludge flocs. It is likely that they are more readily sorbed to EPS and biofilms as well. In contrast to these findings, some researchers have found that MPs with particle sizes of 1–5 mm were more readily removed than were smaller particles (Lares et al. 2018).

Success in MP removal is also a function of physical form of MPs occurring in wastewater (e.g., fiber, fragment, microbead, film). Films having a size similar to SSs (less than 20 μm) and fiber MPs are readily adsorbed by EPS and settle with sludge during sedimentation (Liu et al. 2019; Xu et al. 2019). Approximately 17% microbeads and 13% foam MPs were retained in the activated sludge (Liu et al. 2019).

Degradation of MPs by resident microbial populations is considered unlikely during secondary treatment, given the resistance of most polymers to biodegradation as well as the relatively short residence time of MPs in the reactor vessel. Therefore, the predominant removal mechanism during the activated sludge process is adsorption and aggregation within sludge flocs (figure 7.9) (Zhang et al. 2020).

A widely used subcategory of the activated sludge process is the anaerobic-anoxic-aerobic (A^2O) method (figure 7.10). The unique function of this method involves the removal of nitrogen and phosphorus from wastewater. As the name implies, the system is configured by three tanks in series which operate in sequential anaerobic, anoxic, and aerobic modes. In the anaerobic tanks, microorganisms are less active and the activated sludge process is limited. As a result, phosphorus is released. In the anoxic tank, denitrifying microbes convert nitrate ions to N_2 gas which is released into the atmosphere. Within the aerobic tank nitrifying bacteria decompose ammonia-nitrogen sent from the anoxic tank and convert it to nitrate ion. The nitrate-containing liquid is pumped into the anoxic tank.

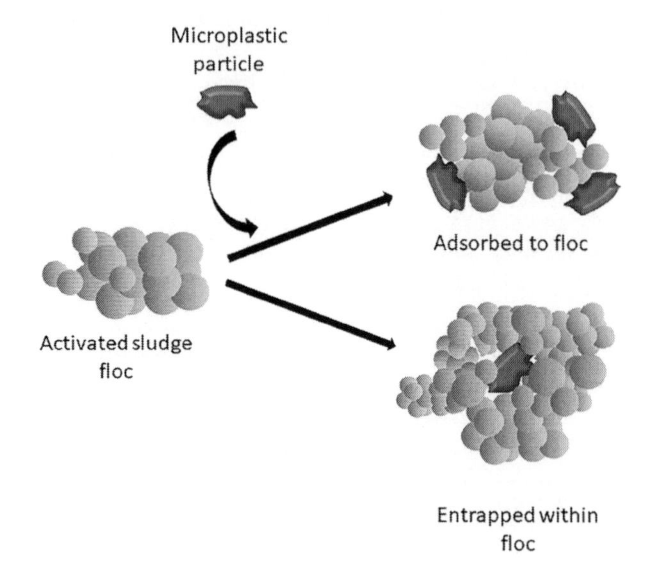

Figure 7.9 MPs Removal Mechanisms during the Activated Sludge Process

The A^2O exhibits a lower removal efficiency for MPs in wastewater compared to the activated sludge process. At a WWTP in Wuhan, China, Liu et al. (2019) determined a 16.6% MP removal with the A^2O process. A WWTP in Beijing utilized A^2O for secondary wastewater treatment, and removal rate of MPs by A^2O was 54.5% (Yang et al. 2019). The removal rate of MPs by the A^2O process may be influenced by the combined effects of MP density and morphology, resulting in a higher overall removal rate for high-density and fibrous MPs (Xu et al. 2021). MP removal rate may also be influenced by hydraulic retention time (HRT) (Hidayaturrahman and Lee 2019). The short HRTs in A^2O may inhibit the formation of biofilms on MP surfaces and limit the ability of MPs to settle and accumulate in sludge (Xu et al. 2021).

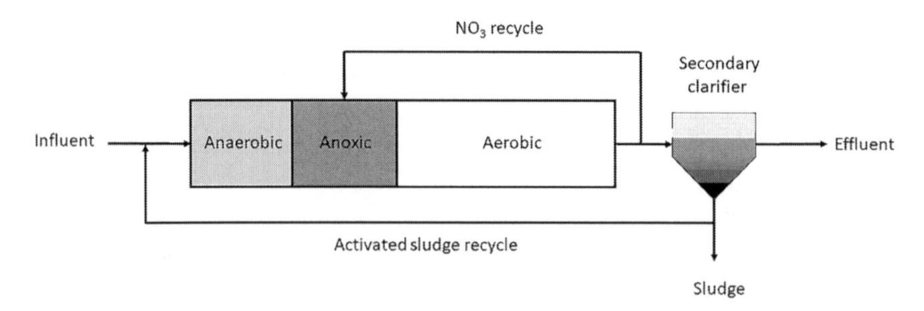

Figure 7.10 The A^2O Method

MP Removal in Biofilm-Related Process

As discussed in chapter 5, a biofilm is an actively growing layer of microbes occurring on a particle surface. Also present is an extensive accumulation of EPS. These components are key to MP removal in biofilm-related wastewater treatment.

Rotating biological contactors (RBCs) is an example of a fixed-film reactor for treatment of BOD and other contaminants in wastewater. An RBC consists of a series of closely spaced corrugated plastic disks mounted on a horizontal shaft. The disks rotate through the wastewater, where approximately half of the RBC surface is immersed. The remainder is exposed to the atmosphere, which provides oxygen to the attached microorganisms and promotes oxidation of contaminants (figure 7.11). Microbes become established on the surfaces of the disks and degrade organics occurring in the waste. The extensive microbial populations attached to the disks provide a high degree of waste treatment within a short time.

The effluent from the primary clarifier enters the reactor. The disks rotate through the wastewater at a rate of about 1.5 rpm. A microbial slime forms on the disks, which degrades the organic contaminants in the waste stream. The rotating motion of the disks through the wastewater causes excess biomass to shear off. After RBC treatment, the effluent enters a secondary clarification stage to separate suspended microbial biomass. Additional treatment of solids and liquids may be required. Clarified secondary effluents may be discharged while residual solids must be disposed of.

As in the case of the activated sludge process, adsorption is a key mechanism of MP removal in biofilm processes. The MPs, newly bound to EPS, now become a carrier for supporting biofilm growth (Zhang et al. 2020) (figure 7.12).

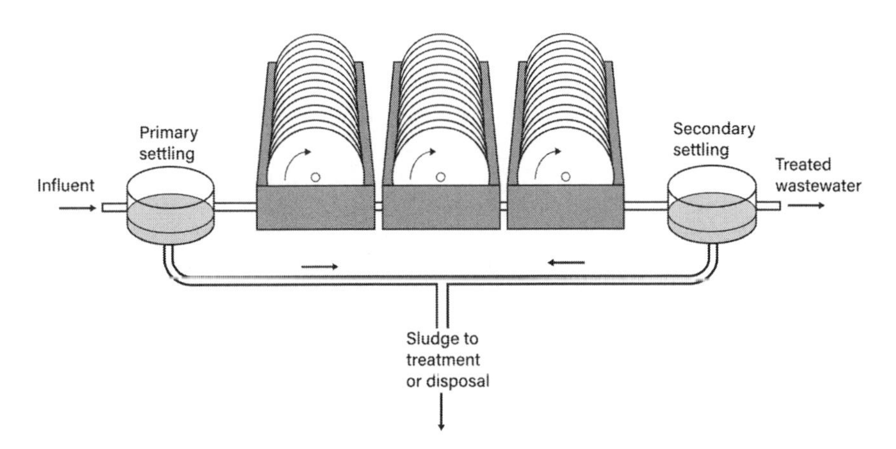

Figure 7.11 Rotating Biological Contactor in WWTP

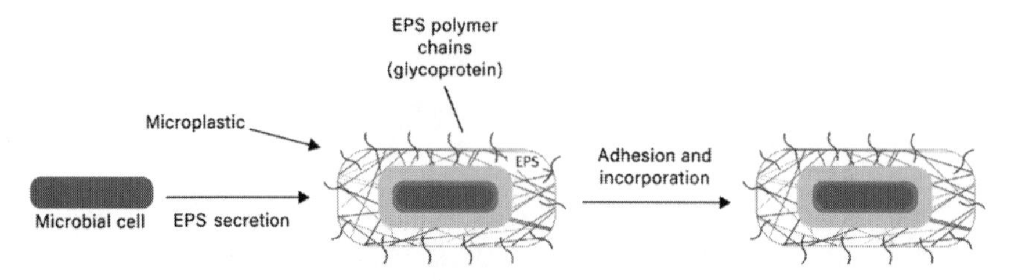

Figure 7.12 Entrapment of MPs in EPS

Source: Adapted from *Journal of Cleaner Production*, vol. 291, Xu, Z., Bai, X., & Ye, Z., Removal and generation of microplastics in wastewater treatment plants: A review. 125982 (2021) with permission from Elsevier.

Some researchers (Hidayaturrahman and Lee 2019) have observed that MP removal efficiency in biofilm-based (membrane disc) processes is moderate (79% removal). Biofilm-based processes remove mostly the smaller-sized MPs (Zhang et al. 2020). These findings indicate that microbial uptake and adsorption are the primary mechanisms of MP removal because smaller particles are more readily adsorbed than larger particles. In cases where a biofilm reactor contains fully packed media having small pore sizes, capture of MPs is expected to be greater. The media, therefore, serves an additional filtration function. Particles are removed and recovered during backwashing (Zhang et al. 2020).

Membrane Bioreactors

The membrane bioreactor (MBR) is an activated sludge process which combines a suspended growth biological reactor with solids removal via filtration within a single unit (Talvitie et al. 2017b). The filtration component is installed either in a bioreactor vessel or a separate tank. A rising flow of the contaminant-enriched water is directed across a membrane. Water passes through the membrane into a separate channel and is recovered (figure 7.13) (US EPA 2007). Only the smallest particles are permitted to pass through the membrane filter; thus, the effluent (now termed the *permeate*), is of high quality (Poerio et al. 2019). As a result of the cross-flow movement of water and waste components, those particles left behind (the *retentate*) do not accumulate on the membrane surface but are transported out of the unit for subsequent recovery and disposal (US EPA 2007).

Membranes can be configured in several ways. The two most popular configurations are hollow fibers grouped in bundles, or flat sheets (Meng et al. 2017). They can include a backwash system which reduces membrane fouling by pumping permeate back through the membrane (Wang et al. 2008). The membranes commonly used in MBR systems are microfiltration (MF) and ultrafiltration (UF) types and are often manufactured of cellulose fibers. Pore sizes are 0.1–50 μm and 0.001–0.1 μm for MF and UF membranes,

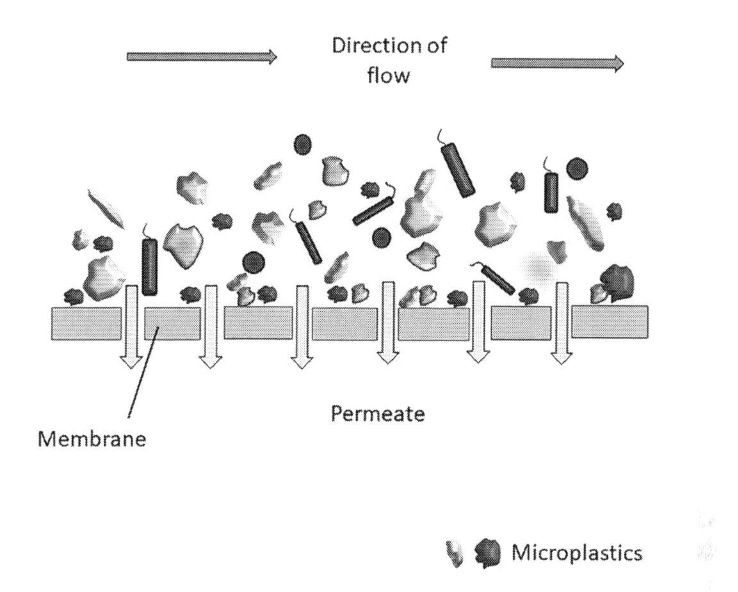

Figure 7.13 Membrane Biological Reactor

respectively. These small sizes are therefore capable of capturing a substantial proportion of MPs. Essentially all MPs should be stopped by the UF membrane due to the fine pore sizes.

To protect membranes from damage, the wastewater should undergo substantial debris removal prior to interaction with the MBR. MBR systems must be equipped with 1- to 3-mm-cutoff fine screens immediately upstream of the membranes, depending on the manufacturer. These screens may require frequent cleaning and maintenance (US EPA 2007).

Compared to other WWTP unit operations, MBR has significant advantages and is quite successful in removing MPs; in some cases, removal rates have approached 100% (Xu et al. 2019; Cristaldi et al. 2020; Sol et al. 2020; Zhang et al. 2020). Baresel et al. (2019) reported that MP removal attained 100% when applying membranes with pore sizes of 0.2 μm. MBR is considered to be more efficient than conventional activated sludge processes for MP capture (Michielssen et al. 2016). Talvitie et al. (2017a, 2017b) found that the MBR process at a WWTP in Viikinmäki, Finland, removed ~ 99.9% MPs from primary treatment effluent without using tertiary treatment. In comparison, conventional activated sludge treatment followed by a tertiary process such as disc filter, rapid sand filtration (RSF), or DAF attained 98.5% MP removal (Talvitie et al. 2017a). At a WWTP in Mikkeli, Finland, the MBR (with MF) reduced MP concentration from 57600 MP·m^{-3} to 400

MP·m^{-3}. By comparison, conventional activated sludge treatment reduced MP concentration to 1000 MP·m^{-3} (Lares et al. 2018). At WWTPs studied in the United States, secondary treatment removed 95.6% of small anthropogenic litter (SAL), one plant equipped with tertiary treatment removed 97.2% of SAL, and a plant with MBR removed 99.4% (Michielssen et al. 2016). In a study by Lv et al. (2019), an MBR reduced MP content from 280 MP·m^{-3} to 50 MP·m^{-3} in wastewater following secondary sedimentation, thus removing 82.1% of MPs from secondary effluent; the final total MP removal rate was 99.5%.

In an 18-month study (Bayo et al. 2020a), the average MP concentration in WWTP influent was 4400 MPs·m^{-3}. The main types of MPs found were fibers (1340 items·m^{-3}), films (590 items·m^{-3}), fragments (200 items·m^{-3}), and beads (20 items·m^{-3}). MBR technologies were successful in removing MPs: effluent MP concentrations were 920 MPs·m^{-3} for MBR. Removal of microfibers was difficult—they were able to bypass the MBR due to the high pressure required in the system, and were eventually released to recipient waterways. Removal efficiency of MP fragments was 98.8%; for fibers, removal efficiency was lower—57.7%.

The use of MBRs is not without its disadvantages; these include membrane costs, high energy demand, control of fouling, and low flow (Ersahin et al. 2012; Sol et al. 2020). Li et al. (2020) examined MPs removal and its effects on membrane fouling in MBRs for treating surface water for drinking purposes. Although the MBR was effective in eliminating organic matter and ammonia, removal performance was inhibited in the presence of PVC MPs. The MBR system recovered within a few days after MP addition. MPs contamination may have led to substantial membrane fouling, some of which was irreversible (Li et al. 2020).

MP Removal in Tertiary Treatment

Tertiary treatment serves as a final wastewater treatment stage for effluent prior to release to the aquatic environment. Specifically, tertiary treatment processes remove additional SS, COD, nitrogen, and phosphorus from effluent following secondary sedimentation. On rare occasions wastewater will be subjected to specialized treatment following the influx of unique pollutants, for example from an accidental chemical release upstream. More than one tertiary treatment process can be employed at a WWTP.

Processes associated with tertiary treatment include RSF and membrane filtration (for SS removal), coagulation (removal of SS and phosphorus), adsorption (organic matter), denitrification/nitrification (nitrogen removal), ozone oxidation (refractory contaminant removal), and others. Several of these processes have been shown to be effective for MPs capture and/or partial decomposition. In many cases, WWTPs which include tertiary treatment have experienced higher MP removal rates than those relying solely on primary or secondary treatment (Xu et al. 2021). This result is likely the result of efficient filtration and/or segregation of MPs by RSF, MBR, DAF, coagulation, and other mechanisms.

Rapid Sand Filtration

RSF is a popular, low-cost technology that allows for rapid and efficient removal of a range of pollutants. RSF involves passing wastewater through a bed of granular media in order to remove suspended inorganic and organic particles along with microbial biomass and floating or emulsified oils (figure 7.14) (Xu et al. 2021).

The system utilizes natural quartz sand and gravel as filter materials. Anthracite is sometimes included. The filtering process operates based on two principles, that is, mechanical straining, and chemical/physical adsorption (figure 7.15). Removal efficiency of MPs using RSF has been documented to range from 45–97% (Cristaldi et al. 2020; Talvitie et al. 2017a).

A diverse consortium of microorganisms proliferates on the surface of the granular filter material (Poghosyan et al. 2020; Gülay et al. 2016). MPs can become entrapped within EPS secreted by microbes which form biofilms and larger aggregates. These aggregates become retained within the sand layer

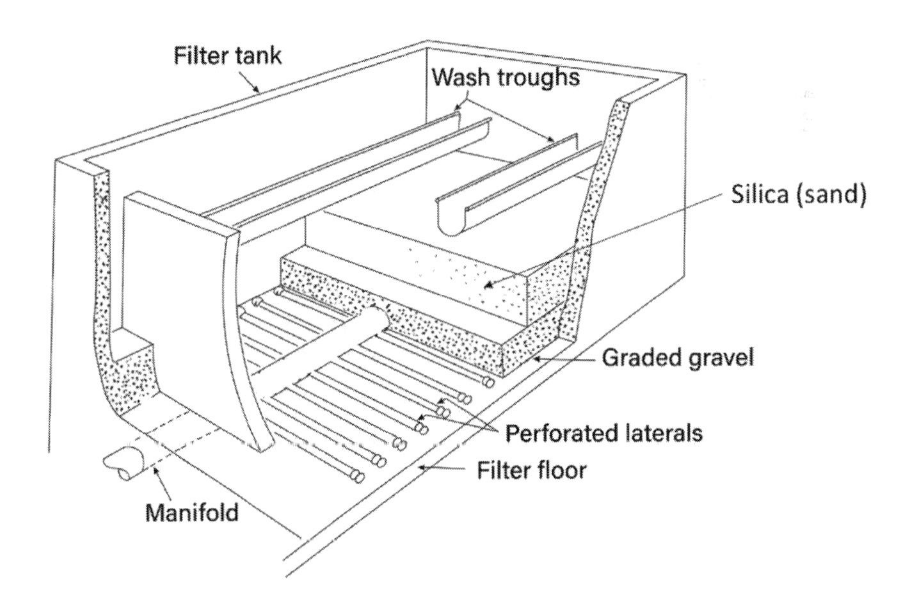

Figure 7.14 Rapid Sand Filtration Unit at a WWTP
Source: US EPA 1989.

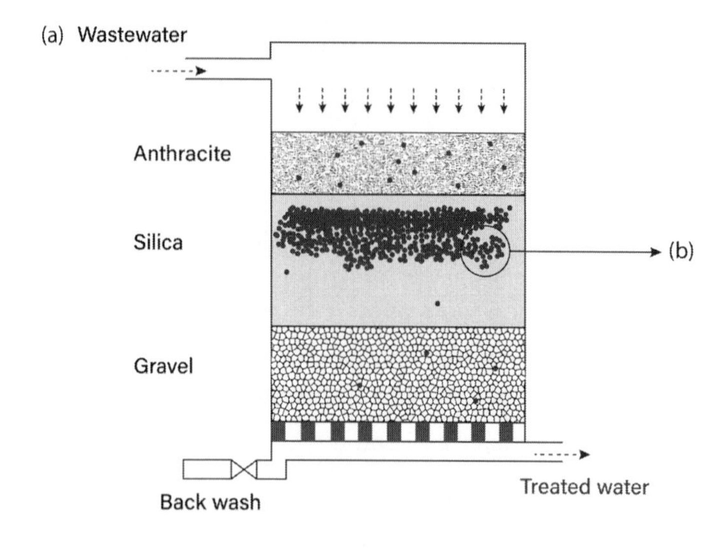

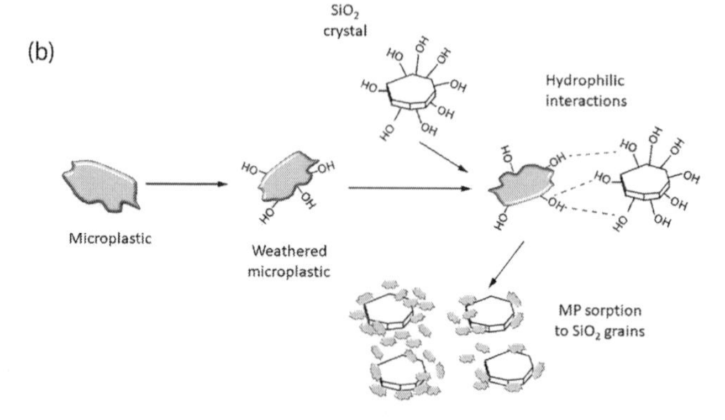

Figure 7.15 Processes Involved in RSF by Quartz
Source: Republished with permission of Elsevier Science & Technology Journals, from *Water Research*: Nano/microplastics in water and wastewater treatment processes—Origin, impact and potential solutions. Enfrin, M., Dumee, L., and Lee, J., vol. 161 (2019), permission conveyed through Copyright Clearance Center, Inc.

(Xu et al. 2021). Such bio-aggregation occurs even though the surface of the sand/gravel filter material is subject to frequent washing.

MP removal is additionally due to hydrophilic interactions between MPs and sand particles. MPs can sorb to silica surfaces via hydrophilic interaction with hydroxyl groups on weathered MP surfaces (figure 7.16). This type of MP sorption is difficult to reverse (Xu et al. 2021). With time, the adsorbed MPs saturate all available reaction sites on silica surfaces. As a consequence of microbial growth and MP sorption to silica, the removal rate of MPs may

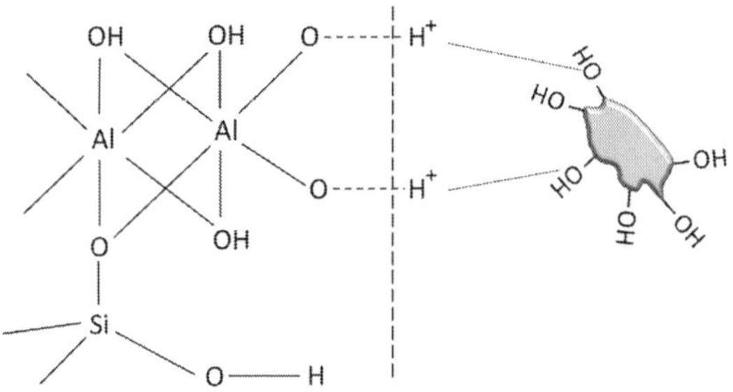

Figure 7.16 Hydrophilic Interactions between MPs and Sand Particles

be markedly reduced with long-term RSF operation, due to clogging. Such system blockages are a significant concern regarding the use of RSF in tertiary wastewater treatment (Xu et al. 2021). An additional concern with RSF is that MPs can fragment into smaller particles by mechanical action with silica (Penumadu et al. 2009; Xu et al. 2021).

In a study by Bayo et al. (2020a), the efficiency of RSF for MP removal was tested at a WWTP in Spain. Average concentration in the influent was 4400 MPs·m^{-3}, and effluent concentrations were 1080 MPs·m^{-3} for RSF. LDPE, nylon, and polyvinyl (PV) fragments were detected in the effluent. Removal efficiency of MP fragments was 95.5%. Given the small size and morphology of fibers, the RSF allowed many to pass through the system: removal efficiency was 53.8%.

Dissolved Air Flotation

DAF can be used in primary treatment or tertiary treatment of wastewater. Compared to conventional sedimentation processes in primary and secondary clarifiers, DAF is less sensitive to flow variations and has high removal efficiency with short retention times (5–10 minutes), especially for low-density particles which are less likely to settle (Sol et al. 2020). In a study at a WWTP in Hämeenlinna, Finland (Talvitie et al. 2017a), a 95% MP removal efficiency was achieved with DAF: MP concentration decreased from 2000 to 100 MP m^{-3}.

Reverse Osmosis

Reverse osmosis (RO) is sometimes included as a tertiary process in municipal and industrial water treatment systems. In RO, high pressure (10–100 bar) is applied to the suspension which forces wastewater through a semi-permeable membrane. Salts, heavy metals, and other impurities are retained by the membrane as a more concentrated solution. Ziajahromi et al. (2017) reported

90% removal efficiency of MPs from a WWTP that employed RO in Sydney, Australia. Microfibers were, however, observed in effluent samples after RO. This may have been due to the occurrence of larger-size pores within the membrane and the presence of unidentified membrane imperfections.

The primary disadvantages of RO technology include membrane fouling and high energy demand (Ahmed et al. 2017; Sol et al. 2020).

Coagulation

Coagulation/flocculation is employed worldwide in facilities that treat drinking water and municipal and industrial wastewater. Contaminants targeted for removal are NOM (humic compounds, microbial, and other biomass), and soil mineral fines (clays) (Chen et al. 2020). However, removal of MPs is certainly feasible using this technology.

Coagulation methods are designed to remove particles from wastewater via destabilizing (i.e., neutralizing) the repulsive forces that keep particles suspended in water. When the repulsive forces are neutralized, suspended particles form agglomerations that can settle out of suspension for separation from water (Moussa et al. 2017). In order to neutralize particle (i.e., MP) surface charges, ions having opposite charge are attracted to the surface, forming an electric double layer (figure 7.17). Within the electric double layer an inner region (the *Stern layer*) occurs, where oppositely charged ions are bound tightly to the particle surface. Beyond this is an outer layer where ions move freely due to diffusion (*ion diffuse layer* or *slipping plane*) (Vepsalainen 2012). Simple inorganic or organic chemicals are applied to wastewater which alter the properties of the electrical double layer and reduce the repulsive energy between particles. This causes micro- and colloidal particles to become destabilized and to aggregate.

Coagulation/flocculation results from four possible mechanisms depending on variables such as chemical and physical properties of the suspension, coagulant properties, and pollutant characteristics (Vepsalainen 2012; Crittenden et al. 2012; Bratby 2016).

Compression of the electrical double layer

The thickness of the electric double layer has a direct impact on the extent of repulsion or attraction between particles. With decreased double-layer thickness, repulsive forces are reduced and particles will adhere to each other. In order to compress the electrical double layer, ions having opposite charge are added to the solution either by metal salts in the case of chemical coagulation, or by oxidation of the anode in the case of electrochemical coagulation (see below). The increased concentration of counterions in the solution causes the double layer to compress; this allows particles to come together.

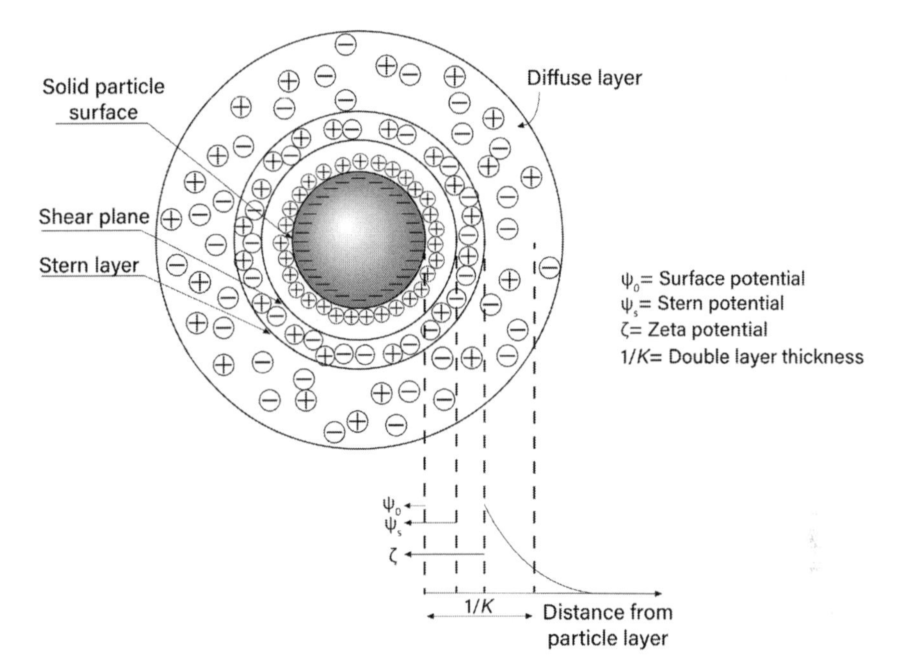

Figure 7.17 The Electrical Double Layer

Adsorption/charge neutralization

Adsorption of counterions to particle surfaces will neutralize particle surface charge; repulsive forces are therefore suppressed and van der Waals attraction dominates. Particles now approach each other and coagulate (Mollah et al. 2004; Vepsalainen 2012; Ghernaout et al. 2011).

Inter-particle bridging

Metal coagulants polymerize and can create bridges between colloidal particles. This is especially pronounced when polymers have high molecular weights and occur as long chains. The bridging effects result in the formation of larger entities (Ghernaout et al. 2011; Vepsalainen 2012).

Entrapment of particles in precipitate

When high rates of metal salts are used as coagulants, the metals react with water, forming insoluble metal hydrates that precipitate and create a sludge blanket. The newly formed precipitates entrap colloidal particles (Ghernaout et al. 2011; Vepsalainen 2012). This mechanism is known as *sweep coagulation.*

Coagulation/flocculation often involves the use of simple salts, for example $FeCl_3$ and $AlCl_3$, to neutralize charges on dissolved and suspended materials; however, a number of other coagulants have been employed to bind particles for removal. Complexes between salts and contaminants (MPs) can form via ligand exchange, which results in formation of strong bonds between MP particles (Chorghe et al. 2017). Agglomeration of particles results in the formation of flocs, which are eventually separated by gravity (Hu et al. 2012; Lee et al. 2012; Shirasaki et al. 2016). The MP particles are ultimately removed as insoluble aggregates via the combined effects of the above four mechanisms (figure 7.18) (Dempsey et al. 1984; Dennett et al. 1996; Vilgé-Ritter et al. 1999).

Coagulation has been evaluated for the removal of MPs from drinking water and wastewater, and Fe and Al salts have been found to be effective for agglomeration of MPs. In a laboratory study, Fe-based coagulants were evaluated for removal of PE MPs (Ma et al. 2019a). PE removal was limited during coagulation (less than 15%) even with a high dosage of $FeCl_3 \cdot 6H_2O$. This effect may have been due to: (1) the low density of PE (0.92–0.97 $g \cdot m^{-3}$), which resulted in limited settling during coagulation; and (2) average size of Fe-based flocs at pH 7.0 was about 630 μm, which was smaller than that of PE particles (which were at the mm scale). Smaller particles were more readily adsorbed and were removed efficiently. Several advanced drinking water treatment plants (ADWTP) in China were analyzed by Wang et al. (2020a); a coagulation treatment step resulted in reductions of 40–55% in MPs in treated water. At three DWTPs in the Czech Republic, it was reported (Pivokonsky et al. 2018) that coagulation-flocculation followed by sand filtration removed

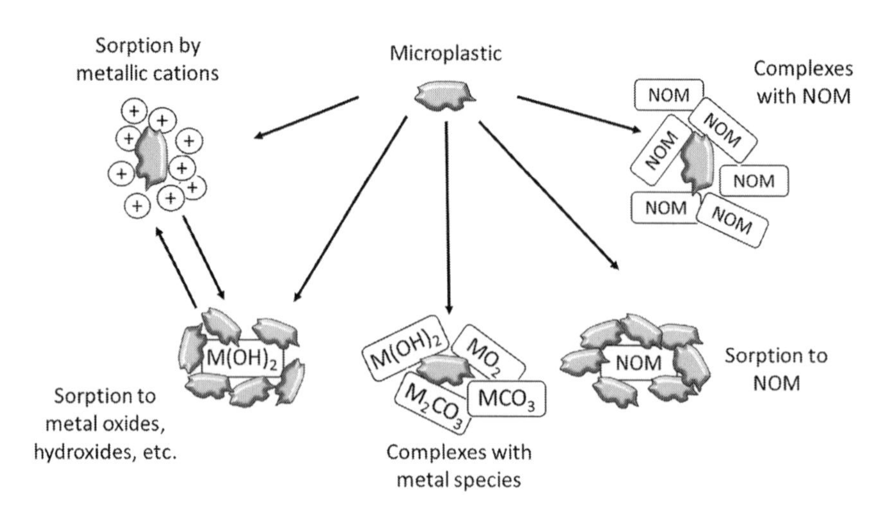

Figure 7.18 Capture of MPs by Coagulation/Flocculation

70–82% of MPs. Differences in MP removal percentage might be related to facility treatment technologies. One plant utilized conventional sand filtration and the others used a two-stage separation system (sedimentation + sand filtration, or flotation + sand filtration) with additional granular activated carbon filtration.

Aluminum-based reagents were evaluated as coagulants to remove MPs at three WWTPs in South Korea. Removal efficiencies ranged between 47% and 82% (Hidayaturrahman and Lee 2019). Aluminum salts are considered more effective than $FeCl_3$ in MP removal (Ma et al. 2019b); however, if high quantities of $AlCl_3$ are used at a water/wastewater treatment facility, there is the risk of adversely affecting receiving water bodies (and resident biota), and also human health (Sol et al. 2020; Li et al. 2020).

Polyaluminium chloride (PAC) has a polymeric structure with the general formula $(Al_n(OH)_mCl_{(3n-m)})_x$ and is completely soluble in water. Upon hydrolysis several mono- and polymeric species are formed; the $Al_{13}O_4(OH)_{247}^+$ is a particularly important cation (Contec.com 2021). Removal of PS and PE MPs by coagulation using PAC was found to be superior to that by $FeCl_3$ (Zhou et al. 2021).

Process variables such as pH and reagent concentration must be addressed in order to optimize coagulation/flocculation. Solution pH affects floc properties due to the potential for hydrolysis of the coagulant (Duan and Gregory 2003; Sillanpää et al. 2017). In one study (Ma et al. 2019a), PE particles were more readily removed under low pH conditions. In contrast, removal efficiency of MPs under alkaline pH conditions was greater than that in acidic conditions; recovery of PS and PE were greatest at an initial pH of 9 (Zhou et al. 2021). Properties of water including ionic strength, NOM, and turbidity levels may impart only minimal effects on floc characteristics (Ma et al. 2019a).

Aluminum sulfate ($Al_2(SO_4)_3 \cdot 18\ H_2O$) (alum) was tested as a coagulant for treatment of MPs and microfibers composed of PE, rayon, or polyester (Skaf 2020). Solutions containing microspheres, with an initial turbidity of 16 NTU were treated. The alum-treated water had turbidity levels < 1.0 NTU. Sweep flocculation was suggested as the dominant mechanism for microsphere removal. Although removal of several types of MPs by coagulation/flocculation has proven successful to some degree, little work has been demonstrated on the effectiveness of removal of microfibers. In this study, stability of PE microfibers in water was strongly influenced by addition of nonionic and anionic surfactants: the smallest fibers were dispersed in solution and effectively removed via coagulation.

Polyacrylamide (PAM) is a polymer-based material used as a water treatment additive to coagulate and remove suspended particles (figure 7.19). It has also been used in soil erosion control to promote aggregation. Studies have demonstrated enhanced coagulation performance in water with PAM addition (Aguilar et al. 2003; Aboulhassan et al. 2006; Lee and Westerhoff 2006). Both cationic and anionic forms of PAM have been evaluated for effectiveness in PE MP removal from water. In one study (Ma et al. 2018), removal efficiency of PE was greater with anionic PAM than with cationic species. Recovery was found to be more effective with smaller PE particle sizes.

Aluminum sulfate, PAC, and PAM have raised public health concerns due to the residual aluminum and acrylamide which remain in treated water and sludge. Several sources have recommended that cationic PAM not be used in water treatment due to aquatic toxicity (Duggan et al. 2019).

Herbort et al. (2018) suggested coagulation/flocculation of MPs based on alkoxy-silyl bond formation via sol-gel reactions as a sustainable approach for MP removal. Complex organic-SiO_x molecules were generated to act as adhesion reagents between PE and PP MP particles. In laboratory studies, the MPs adhered to form large three-dimensional agglomerates which could be removed using cost-efficient filtration methods such as RSF. The agglomeration is carried out independently of the type, size, and amount of the MP as well as environmental variables (e.g., pH, temperature, pressure). It has been suggested that this coagulation process can be transferred to decentralized systems (e.g., industrial processes) in addition to WWTPs (Herbort et al. 2018).

Microalgae cultures have been proposed for MP coagulation treatment of wastewater due to their added ability to remove heavy metals as well as some toxic organic compounds (Abdel-Raouf et al. 2012). Microalgae furthermore incorporate inorganic nitrogen and phosphorus present in wastewater for growth. Microalgae have the potential for MPs removal, from water, as various species produce substantial quantities of EPS. Recent work (Cunha et al. 2019) evaluated the interactions of MPs and EPS secreted by marine and freshwater microalgae. The EPS formed by certain microalgal species formed a dense and viscous mesh. Two marine microalgae, *Tetraselmis* sp. and *Gloeocapsa* sp., had good EPS production which resulted in MPs aggregation. Aggregation capabilities were

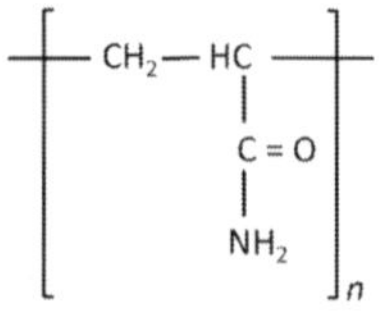

Figure 7.19 Generic Structure of PAM Monomer

effective for both low- and high-density MPs. Effectiveness of coagulation was partly limited by MP particle size. Certain species, for example *Microcystis panniformis*, produced small-sized EPS which did not permit aggregation of MPs larger than 25 μm.

Electrochemical techniques, for example electrocoagulation (EC), are relatively low-cost methods of treatment that do not require addition of chemicals or use of microorganisms. EC technology is applied for treatment of wastewater containing food wastes, dyes, suspended particles, organic matter from landfill leachates, synthetic detergent effluents, mine wastes, and other solutions (WOC 2021). In this process, an electric current is passed through the wastewater; an electrochemical reaction takes place which releases metal ions into water and subsequently generates flocs via electrolysis (figure 7.20) (Garcia-Segura et al. 2017; Perren et al. 2018).

The basic EC technology relies on an electrolytic cell with anode and cathode electrodes immersed into the wastewater and connected to a DC power source. A variety of anode and cathode geometries are possible including plates, balls, fluidized bed spheres, wire mesh, rods, and tubes (WOC 2021). Iron and aluminum electrodes are the most commonly used metals for EC cells as they are readily available, inexpensive, and proven reliable. The anode generates the coagulant by liberating metal cations when a current is passed through the cell (Moussa et al. 2017). Metal

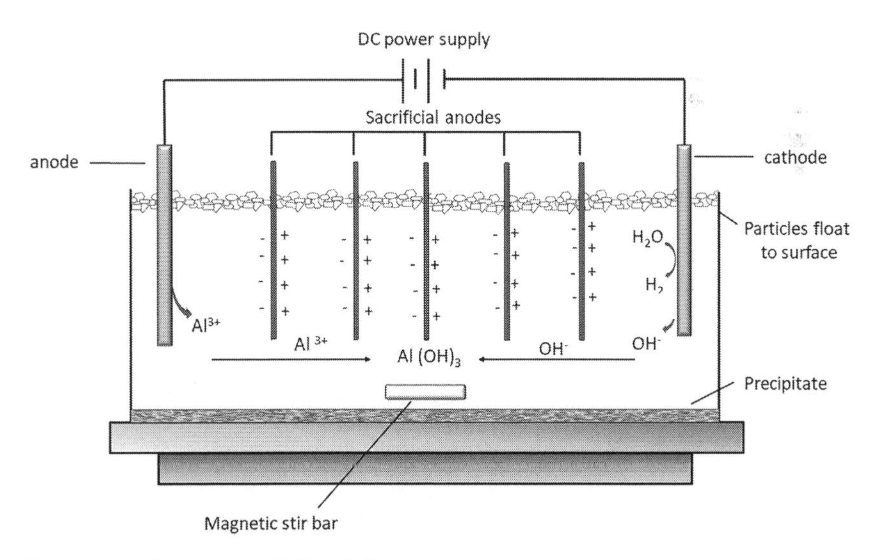

Figure 7.20 Electrocoagulation Cell
Source: Adapted from Perren et al. (2018). Reproduced with kind permission of American Chemical Society (ACS). https://pubs.acs.org/doi/10.1021/acsomega.7b02037. Further permissions related to the material excerpted should be directed to the ACS.

hydroxide coagulants are generated in situ by the reaction of metal ions (e.g., Fe^{2+} and Al^{3+}) released from sacrificial electrodes. Using the example of an iron electrode, reactions at the anode are as follows (Moussa et al. 2017):

$$Fe_{(s)} \rightarrow Fe^{2+}_{(aq)} + 2e^{-} \quad (7.1)$$

$$4Fe^{2+}_{(aq)} + 10H_2O + O_{2(aq)} \rightarrow 4Fe(OH)_{3(s)} + 8H^{+} \quad (7.2)$$

$$Fe^{2+}_{(aq)} + 2OH^{-} \rightarrow Fe(OH)_{2(s)} \quad (7.3)$$

At the cathode, hydrogen gas is generated along with OH^- ions, which leads to an increase in pH of the solution (Eq. 7.4):

$$2H_2O_{(l)} + 2e^{-} \rightarrow H_{2(g)} + 2OH^{-} \quad (7.4)$$

The newly-generated metal hydroxide coagulants destabilize surface charges on suspended microparticles. The particles can now migrate sufficiently close for interacting via van der Waals forces (Akbal and Camci 2011). The coagulant concurrently forms a sludge blanket which traps SS particles. The H_2 gas which is generated during the electrolysis process lifts the sludge to the water surface (Moussa et al. 2017).

Polyethylene MPs were effectively removed from artificial wastewater using EC in a stirred-tank batch reactor (Perren et al. 2018)–removal efficiencies were in excess of 90% over pH values ranging from 3 to 10. Optimum removal (99%) occurred at pH 7.5. A current density of 11 $A \cdot m^{-2}$, which was considered energy efficient, was most efficient for achieving the highest removal rate.

Benefits of electrochemical processes include being environmentally benign, minimization of sludge production, energy efficiency, cost-effectiveness, and adaptability to automation (Zeboudji et al. 2013; Perren et al. 2018). The primary advantage of EC over chemical coagulation/flocculation is that the latter relies upon added reagents such as metal salts and polyelectrolytes, while in EC the coagulants are generated in situ via electrolytic oxidation of a suitable anode. This ultimately results in the generation of significantly less sludge (Moussa et al. 2017).

CLOSING STATEMENTS

A number of studies have demonstrated that various technologies can effectively remove MPs from raw water and wastewater, and that most are recovered in the sludge (Magnusson and Noren 2014; Michielssen et al. 2016; Talvitie et al. 2015, 2017a; Carr et al. 2016; Mintenig et al. 2017; Murphy et al. 2016; Lares et al. 2018). The reported reduction in MPs quantities in WWTP outflows is quite significant—a preponderance of data for such plants reveals a removal efficiency greater than 90%; even at these rates, however, vast quantities of MPs continue to be discharged into waterways, even with the use of tertiary treatment (Talvitie 2017a; Sutton et al. 2016; Habib et al. 2020; Browne et al. 2011; Cristaldi et al. 2020; Ziajahromi 2016). In order to optimize the wastewater treatment process for improving MP removal, additional research is needed to explore the behavior of MPs within various unit operations.

REFERENCES

Abdel-Raouf, N., Al-Homaidan, A., and Ibreheem, I. (2012). Microalgae and wastewater treatment. *Saudi Journal of Biological Sciences* 19(3):257–275. DOI: 10.1016/j.sjbs.2012.04.005.

Aboulhassan, M., Souabi, S., Yaacoubi, A., and Baudu, M. (2006). Improvement of paint effluents coagulation using natural and synthetic coagulant aids. *Journal of Hazardous Materials* 138(1):40–45.

Aguilar, M., Saez, J., Llorens, M., Soler, A., and Ortuno, J. (2003). Microscopic observation of particle reduction in slaughterhouse wastewater by coagulation–flocculation using ferric sulphate as coagulant and different coagulant aids. *Water Research* 37(9):2233–2241.

Ahmed, M.B., Zhou, J., Ngo, H.H., Guo, W., Thomaidis, N., and Xu, J. (2017). Progress in the biological and chemical treatment technologies for emerging contaminant removal from wastewater: A critical review. *Journal of Hazardous Materials* 323:274–298.

Akarsu, C., Kumbur, H., Gökdağ, K., Kıdeyş, A.E., and Sanchez-Vidal, A. (2020). Microplastics composition and load from three wastewater treatment plants discharging into Mersin Bay, north eastern Mediterranean Sea. *Marine Pollution Bulletin* 150:110776.

Akbal, F., and Camcı, S. (2011). Copper, chromium and nickel removal from metal plating wastewater by electrocoagulation. *Desalination* 269(1–3):214–222.

Alvim, C.B., Mendoza-Roca, J.A., and Bes-Piá, A. (2020). Wastewater treatment plant as microplastics release source–Quantification and identification techniques. *Journal of Environmental Management* 255:109739.

Baresel, C., Harding, M., and Fang, J. (2019). Ultrafiltration/granulated active carbon-biofilter: Efficient removal of a broad range of micropollutants. *Applied Sciences* 9(4):710.

Bayo, J., Lopez-Castellanos, J., and Olmos, S. (2020a). Membrane bioreactor and rapid sand filtration for the removal of microplastics in an urban wastewater treatment plant. *Marine Pollution Bulletin* 156:111211.

Bayo, J., Olmos, S., and Lopez-Castellanos, J. (2020b). Microplastics in an urban wastewater treatment plant: The influence of physicochemical parameters and environmental factors. *Chemosphere* 238:124593.

Bilgin, M., Yurtsever, M., and Karadagli, F. (2020). Microplastic removal by aerated grit chambers versus settling tanks of a municipal wastewater treatment plant. *Journal of Water Process Engineering* 38:101604.

Blair, R.M., Waldron, S., and Gauchotte-Lindsay, C. (2019). Average daily flow of microplastics through a tertiary wastewater treatment plant over a ten-month period. *Water Research* 163:114909.

Bratby, J. (2016). *Coagulation and Flocculation in Water and Wastewater Treatment–Third Edition*. London: IWA Publishing.

Browne, M.A., Crump, P., Niven, S., Teuten, E., Tonkin, A., Galloway, T., and Thompson, R. (2011). Accumulation of microplastic on shorelines woldwide: Sources and sinks. *Environmental Science & Technology* 45(21):9175–9179.

Burns, E.E., and Boxall, A.B. (2018). Microplastics in the aquatic environment: Evidence for or against adverse impacts and major knowledge gaps. *Environmental Toxicology and Chemistry* 37(11):2776–2796.

Carr, S.A., Liu, J., and Tesoro, A. (2016). Transport and fate of microplastic particles in wastewater treatment plants. *Water Research* 91:174–182.

Chen, G., Feng, Q., and Wang, J. (2020). Mini-review of microplastics in the atmosphere and their risks to humans. *Science of the Total Environment* 703:135504.

Chorghe, D., Sari, M., and Chellam, S. (2017). Boron removal from hydraulic fracturing wastewater by aluminum and iron coagulation: Mechanisms and limitations. *Water Research* 126:481–487.

Chukwulozie, O.P., Segun, O.B., Lekwuwa, C.J., and Chiagoro, U.D. (2018). Design of sewage treatment plant for CBN housing estate Trans-Ekulu Enugu Nigeria. *Advances in Research* 13(3):1–18.

Conley, K., Clum, A., Deepe, J., Lane, H., and Beckingham, B. (2019). Wastewater treatment plants as a source of microplastics to an urban estuary: Removal efficiencies and loading per capita over one year. *Water Research* X3:100030.

Contec.com (2021). Poly Aluminium Chloride – PAC. http://pac-contec.com/index.php/pac-poly-aluminium-chloride.

Cristaldi, A., Fiore, M., Zuccarello, P., Conti, G., Grasso, A., Grasso, A., Nicolosi, I., Copat, C., and Ferrante, M. (2020). Efficiency of wastewater treatment plants (WWTPs) for microplastic removal: A systematic review. *International Journal of Environmental Research and Public Health* 17(21):8014.

Crites, R., and Tchobanoglous, G. (1998). *Small and Decentralized Wastewater Management Systems*. Boston (MA): WCB McGraw-Hill, pp. 703–760.

Crittenden, J.C., Trussell, R., Hand, D., Howe, K., and Tchobanoglous, G. (2012). *MWH's Water Treatment: Principles and Design*. New York: John Wiley & Sons.

Cunha, C., Faria, M., Nogueira, N., Ferreira, A., and Cordeiro, N. (2019). Marine vs freshwater microalgae exopolymers as biosolutions to microplastics pollution. *Environmental Pollution* 249:372–380.

Dempsey, B.A., and O'Melia, C.R. (1984). Removal of naturally occurring compounds by coagulation and sedimentation. *Critical Reviews in Environmental Control* 14(4):311–331.

Dennett, K.E., Amirtharajah, A., Moran, T.F., and Gould, J.P. (1996). Coagulation: Its effect on organic matter. *Journal of the American Water Works Association* 88(4):129–142.

Dris, R., Gasperi, J., Rocher, V., Saad, M., Renault, N., and Tassin, B. (2015). Microplastic contamination in an urban area: A case study in Greater Paris. *Environmental Chemistry* 12(5):592–599.

Duan, J., and Gregory, J. (2003). Coagulation by hydrolysing metal salts. *Advances in Colloid and Interface Science* 100:475–502.

Dubaish, F., and Liebezeit, G. (2013). Suspended microplastics and black carbon particles in the Jade system, southern North Sea. *Water, Air, & Soil Pollution* 224(2):1–8.

Duggan, K.L., Morris, M., Bhatia, S., Khachan, M., and Lewis, K. (2019). Effects of cationic polyacrylamide and cationic starch on aquatic life. *Journal of Hazardous, Toxic, and Radioactive Waste* 23(4):04019022.

Edo, C., González-Pleiter, M., Leganés, F., Fernández-Piñas, F., and Rosal, R. (2020). Fate of microplastics in wastewater treatment plants and their environmental dispersion with effluent and sludge. *Environmental Pollution* 259:113837.

Enfrin, M., Dumee, L., and Lee, J. (2019). Nano/microplastics in water and wastewater treatment processes–Origin, impact and potential solutions. *Water Research* 161:621–638.

Ersahin, M.E., Ozgun, H., Dereli, R., Ozturk, I., Roest, K., and Lier, J. (2012). A review on dynamic membrane filtration: Materials, applications and future perspectives. *Bioresource Technology* 122:196–206.

Garcia-Segura, Eiband, M., de Melo, J., and Martinez-Huitle, C. (2017). Electrocoagulation and advanced electrocoagulation processes: A general review about the fundamentals, emerging applications and its association with other technologies. *Journal of Electroanalytical Chemistry* 801:267–299.

Ghernaout, D. (2011). A review of electrocoagulation as a promising coagulation process for improved organic and inorganic matters removal by electrophoresis and electroflotation. *Desalination and Water Treatment* 28(1–3):287–320.

Gies, E.A., LeNoble, J.L., Noël, M., Etemadifar, A., Bishay, F., Hall, E.R., and Ross, P.S. (2018). Retention of microplastics in a major secondary wastewater treatment plant in Vancouver, Canada. *Marine Pollution Bulletin* 133:553–561.

Gülay, A., Musovic, S., Albrechtsen, H., Al-Soud, W., Sorensen, S., and Smets, B. (2016). Ecological patterns, diversity and core taxa of microbial communities in groundwater-fed rapid gravity filters. *The ISME Journal* 10(9):2209–2222.

Gündoğdu, S., Çevik, C., Güzel, E., and Kilercioğlu, S. (2018). Microplastics in municipal wastewater treatment plants in Turkey: A comparison of the influent and secondary effluent concentrations. *Environmental Monitoring and Assessment* 190(11):1–10.

Habib, R.Z., Thiemann, T., and Al Kendi, R. (2020). Microplastics and wastewater treatment plants—A review. *Journal of Water Resource and Protection* 12(1):1.

Hammouda, O., Gaber, A., and Abdel-Raouf, N. (1995). Microalgae and wastewater treatment. *Ecotoxicology and Environmental Safety* 31(3):205–210.

Herbort, A.F., and Sturm, M. (2018). Alkoxy-silyl induced agglomeration: A new approach for the sustainable removal of microplastic from aquatic systems. *Journal of Polymers and the Environment* 26(11):4258–4270.

Hidayaturrahman, H., and Lee, T.-G. (2019). A study on characteristics of microplastic in wastewater of South Korea: Identification, quantification, and fate of microplastics during treatment process. *Marine Pollution Bulletin* 146:696–702.

Hou, L., Kumar, D., Yoo, C.G., Gitsov, I., and Majumder, E.L.-W. (2021). Conversion and removal strategies for microplastics in wastewater treatment plants and landfills. *Chemical Engineering Journal* 406:126715.

Hu, C., Liu, H., Chen, G., and Qu, J. (2012). Effect of aluminum speciation on arsenic removal during coagulation process. *Separation and Purification Technology* 86:35–40.

Jia, Y., Zhou, M., Chen, Y., Hu, Y., and Luo, J. (2020). Insight into short-cut of simultaneous nitrification and denitrification process in moving bed biofilm reactor: Effects of carbon to nitrogen ratio. *Chemical Engineering Journal* 400:125905.

Katrivesis, F., Karela, A., Papadakis, V., and Paraskeva, C. (2019). Revisiting of coagulation-flocculation processes in the production of potable water. *Journal of Water Process Engineering* 27:193–204.

Lares, M. (2018). Occurrence, identification and removal of microplastic particles and fibers in conventional activated sludge process and advanced MBR technology. *Water Research* 133:236–246.

Lee, K.E., Morad, N., Teng, T., and Poh, B. (2012). Development, characterization and the application of hybrid materials in coagulation/flocculation of wastewater: A review. *Chemical Engineering Journal* 203:370–386.

Lee, W., and Westerhoff, P. (2006). Dissolved organic nitrogen removal during water treatment by aluminum sulfate and cationic polymer coagulation. *Water Research* 40(20):3767–3774.

Leslie, H., Brandsma, S., Velzen, M., and Vethaak, A. (2017). Microplastics en route: Field measurements in the Dutch river delta and Amsterdam canals, wastewater treatment plants, North Sea sediments and biota. *Environment International* 101:133–142.

Li, L., Liu, D., Song, K., and Zhou, Y. (2020). Performance evaluation of MBR in treating microplastics polyvinylchloride contaminated polluted surface water. *Marine Pollution Bulletin* 150:110724.

Li, X., Mei, Q., Chen, L., Zhang, H., Dong, B., Dai, X., He, C., and Zhou, J. (2019). Enhancement in adsorption potential of microplastics in sewage sludge for metal pollutants after the wastewater treatment process. *Water Research* 157:228–237.

Liu, X., Yuan, W., Di, M., Li, Z., and Wang, J. (2019). Transfer and fate of microplastics during the conventional activated sludge process in one wastewater treatment plant of China. *Chemical Engineering Journal* 362:176–182.

Long, Z., Pan, Z., Wang, W., Ren, J., Yu, X., Lin, L., Lin, H., Chen, H., and Jin, X. (2019). Microplastic abundance, characteristics, and removal in wastewater treatment plants in a coastal city of China. *Water Research* 155:255–265.

Lv, X., Dong, Q., Zuo, Z., Liu, Y., Huang, X., and Wu, W.-M. (2019). Microplastics in a municipal wastewater treatment plant: Fate, dynamic distribution, removal efficiencies, and control strategies. *Journal of Cleaner Production* 225:579–586.

Ma, B., Li, W., Liu, R., Liu, G., Sun, J., Liu, H., Qu, J., and Meer, W. (2018). Multiple dynamic Al-based floc layers on ultrafiltration membrane surfaces for humic acid and reservoir water fouling reduction. *Water Research* 139:291–300.

Ma, B., Xue, W., Ding, Y., Hu, C., Liu, H., and Qu, J. (2019a). Removal characteristics of microplastics by Fe-based coagulants during drinking water treatment. *Journal of Environmental Sciences* 78:267–275.

Ma, B., Xue, W., Hu, C., Liu, H., Qu, J., and Li, L. (2019b). Characteristics of microplastic removal via coagulation and ultrafiltration during drinking water treatment. *Chemical Engineering Journal* 359:159–167.

Magni, S., Binelli, A., Pittura, L., Avio, C.G., Della Torre, C., Parenti, C.C., Gorbi, S., and Regoli, F. (2019). The fate of microplastics in an Italian wastewater treatment plant. *Science of the Total Environment* 652:602–610.

Magnusson, K., and Norén, F. (2014). *Screening of Microplastic Particles in and Down-Stream a Wastewater Treatment Plant*. Stockholm (Sweden): Swedish Environmental Research Institute.

Mason, S.A., Garneau, D., Sutton, R., Chu, Y., Ehmann, K., Barnes, J., Fink, P., Papazissimos, D., and Rogers, D. (2016). Microplastic pollution is widely detected in US municipal wastewater treatment plant effluent. *Environmental Pollution* 218:1045–1054.

McCormick, A., Hoellein, T., Mason, S., Schluep, J., and Kelly, J. (2014). Microplastic is an abundant and distinct microbial habitat in an urban river. *Environmental Science & Technology* 48(20):11863–11871.

Meng, F., Zhang, S., Oh, Y., Zhou, Z., Shin, H., and Chae, S. (2017). Fouling in membrane bioreactors: An updated review. *Water Research* 114:151–180.

Michielssen, M.R., Michielssen, E., Ni, J., and Duhaime, M. (2016). Fate of microplastics and other small anthropogenic litter (SAL) in wastewater treatment plants depends on unit processes employed. *Environmental Science: Water Research & Technology* 2(6):1064–1073.

Mintenig, S., Veen, I., Loder, M., Primpke, S., and Gerdts, G. (2017). Identification of microplastic in effluents of waste water treatment plants using focal plane array-based micro-Fourier-transform infrared imaging. *Water Research* 108:365–372.

Mollah, M.Y., Morkovsky, P., Gomes, J., Kesmez, M., Parga, J., and Cocke, D. (2004). Fundamentals, present and future perspectives of electrocoagulation. *Journal of Hazardous Materials* 114(1–3):199–210.

Morling, S. (2019). Swedish experience and excellence in wastewater treatment demonstrated especially in phosphorus removal. *Journal of Water Resource and Protection* 11(3):333–347.

Moussa, D.T., El-Naas, M., Nasser, M., and Al-Marri, M. (2017). A comprehensive review of electrocoagulation for water treatment: Potentials and challenges. *Journal of Environmental Management* 186:24–41.

Murphy, F., Ewins, C., Carbonnier, F., and Quinn, B. (2016). Wastewater treatment works (WwTW) as a source of microplastics in the aquatic environment. *Environmental Science & Technology* 50(11):5800–5808.

Ngo, P.L., Pramanik, B., Shah, K., and Roychand, R. (2019). Pathway, classification and removal efficiency of microplastics in wastewater treatment plants. *Environmental Pollution* 255:113326.

Padervand, M., Lichtfousa, E., Robert, D., and Wang, C. (2020). Removal of microplastics from the environment: A review. *Environmental Chemistry Letters* 18(3):807–828.

Palaniandy, P., Adlan, M., Aziz, H., and Murshed, M. (2017). Dissolved air flotation (DAF) for wastewater treatment. In: *Waste Treatment in the Service and Utility Industries*, pp. 145–182.

Park, H.-J., Oh, M., Kim, P., Kim, G., Jeong, D., Ju, B., Lee, W., Chung, H., and Kang, H. (2020). National reconnaissance survey of microplastics in municipal wastewater treatment plants in Korea. *Environmental Science & Technology* 54(3):1503–1512.

Penumadu, D., Dutta, A.K., Luo, X., and Thomas, K.G. (2009). Nano and neutron science applications for geomechanics. *KSCE Journal of Civil Engineering* 13:233–242. DOI: 10.1007/s12205-009-0233-2.

Perren, W., Wojtasik, A., and Cai, Q. (2018). Removal of microbeads from wastewater using electrocoagulation. *ACS Omega* 3(3):3357–3364.

Pichtel, J. (2019). *Fundamentals of Site Remediation for Metal and Hydrocarbon-Contaminated Soils*. Lanham (MD): Bernan Press.

Pivokonsky, M., Cermakkova, L., Novotna, K., Peer, P., Cajthaml, T., and Janda, V. (2018). Occurrence of microplastics in raw and treated drinking water. *Science of the Total Environment* 643:1644–1651.

Pivokonský, M., Pivokonská, L., Novotná, K., Čermáková, L., and Klimtová, M. (2020). Occurrence and fate of microplastics at two different drinking water treatment plants within a river catchment. *Science of the Total Environment* 741:140236.

Poerio, T., Piacentini, E., and Mazzei, R. (2019). Membrane processes for microplastic removal. *Molecules* 24(22):4148.

Poghosyan, L., Koch, H., Frank, J., Kessel, M., Cremers, G., Alen, T., Jetten, M., Camp, H., and Lucker, S. (2020). Metagenomic profiling of ammonia-and methane-oxidizing microorganisms in two sequential rapid sand filters. *Water Research* 185:116288.

Prata, J.C. (2018). Microplastics in wastewater: State of the knowledge on sources, fate and solutions. *Marine Pollution Bulletin* 129(1):262–265.

Radityaningrum, A.D., Trihadiningrum, Y., Soedjono, E.S., and Herumurti, W. (2021). Microplastic contamination in water supply and the removal efficiencies of the treatment plants: A case of Surabaya City, Indonesia. *Journal of Water Process Engineering* 43:102195.

Rajala, K., Gronfors, O., Hesampour, M., and Mikola, A. (2020). Removal of microplastics from secondary wastewater treatment plant effluent by coagulation/flocculation with iron, aluminum and polyamine-based chemicals. *Water Research* 183:116045.

Raju, S., Carbery, M., Kuttykattil, A., Senthirajah, K., Lundmark, A., Rogers, Z., Suresh, A., Evans, G., and Palanisami, T. (2020). Improved methodology to determine the fate and transport of microplastics in a secondary wastewater treatment plant. *Water Research* 173:115549.

Rubio, J., Souza, M., and Smith, R. (2002). Overview of flotation as a wastewater treatment technique. *Minerals Engineering* 15(3):139–155.

Sarkar, D.J., Sarkar, S.D., Das, B.K., Praharaj, J.K., Mahajan, D.K., Purokait, B., and Kumar, S. (2021). Microplastics removal efficiency of drinking water treatment plant with pulse clarifier. *Journal of Hazardous Materials* 413:125347.

Schneiderman, E. (2015). Discharging microbeads to our waters: An examination of wastewater treatment plants in New York. New York State Office of the Attorney General Environmental Protection Bureau.

Shirasaki, N., Matsushita, T., Matsui, Y., and Marubayashi, T. (2016). Effect of aluminum hydrolyte species on human enterovirus removal from water during the coagulation process. *Chemical Engineering Journal* 284:786–793.

Sillanpää, M., and Sainio, P. (2017). Release of polyester and cotton fibers from textiles in machine washings. *Environmental Science and Pollution Research* 24(23):19313–19321.

Simon, M., van Alst, N., and Vollertsen, J. (2018). Quantification of microplastic mass and removal rates at wastewater treatment plants applying Focal Plane Array (FPA)-based Fourier Transform Infrared (FT-IR) imaging. *Water Research* 142:1–9.

Skaf, D.W., Vito, P., Javaz, R., and Kyle, K. (2020). Removal of micron-sized microplastic particles from simulated drinking water via alum coagulation. *Chemical Engineering Journal* 386:123807.

Sol, D., Laca, A., Laca, A., and Diaz, M. (2020). Approaching the environmental problem of microplastics: Importance of WWTP treatments. *Science of the Total Environment* 740:140016.

Spellman, F.R., and Drinan, J. (2003). *Wastewater Treatment Plant Operations Made Easy: A Practical Guide for Licensure*. Lancaster (PA): DEStech Publications, Inc.

Sun, J., Dai, X., Wang, Q., Loosdrecgt, M., and Ni, B. (2019). Microplastics in wastewater treatment plants: Detection, occurrence and removal. *Water Research* 152:21–37.

Sutton, R., Mason, S., Stanek, S., Willis-Norton, E., Wren, I., and Box, C. (2016). Microplastic contamination in the San Francisco Bay, California, USA. *Marine Pollution Bulletin* 109(1):230–235.

Talvitie, J., and Heinonen, M. (2014). HELCOM 2014–Base project 2012–2014: Preliminary study on synthetic microfibers and particles at a municipal wastewater treatment plant. Online verfügbar unter. http://helcom. fi/helcom-at-work/projects/completed-projects/base/components/microplastics, Zugriff am 11, 2016.

Talvitie, J., Heinonen, M., Paakkonen, J., Vahtera, E., Mikola, A., Setala, O., and Vahala, R. (2015). Do wastewater treatment plants act as a potential point source of microplastics? Preliminary study in the coastal Gulf of Finland, Baltic Sea. *Water Science and Technology* 72(9):1495–1504.

Talvitie, J., Mikola, A., Setala, O., Heinonen, M., and Koistinen, A. (2017a). How well is microlitter purified from wastewater?–A detailed study on the stepwise removal of microlitter in a tertiary level wastewater treatment plant. *Water Research* 109:164–172.

Talvitie, J., Mikola, A., Koistinen, A., and Setala, O. (2017b). Solutions to microplastic pollution–Removal of microplastics from wastewater effluent with advanced wastewater treatment technologies. *Water Research* 123:401–407.

US EPA. (2003). *Screening and Grit Removal. Wastewater Technology Fact Sheet. EPA 832-F-03-011*. Washington (DC): Office of Water.

US EPA. (2004). *Primer for Municipal Wastewater Treatment Systems. EPA 832-R-04-001*. Washington (DC): Office of Water.

US EPA. (2007). *Membrane Bioreactors. Wastewater Management Fact Sheet*. Washington (DC): EPA.

Vepsäläinen, M., and Sillanpää, M. (2020). Electrocoagulation in the treatment of industrial waters and wastewaters. In: *Advanced Water Treatment*. Amsterdam (Netherlands): Elsevier, pp. 1–78.

Vilgé-Ritter, A., Rose, J., Masion, A., Bottero, J.-Y., and Lainé, J.-M. (1999). Chemistry and structure of aggregates formed with Fe-salts and natural organic matter. *Colloids and Surfaces A: Physicochemical and Engineering Aspects* 147(3):297–308.

Wang, F., Wang, B., Duan, L., Zhang, Y., Zhou, Y., Sui, Q., and Yu, G. (2020). Occurrence and distribution of microplastics in domestic, industrial, agricultural and aquacultural wastewater sources: A case study in Changzhou, China. *Water Research* 182:115956.

Wang, J., Zheng, L., and Li, J. (2018). A critical review on the sources and instruments of marine microplastics and prospects on the relevant management in China. *Waste Management & Research* 36(10):898–911.

Wang, L.K., Fahey, E., and Wu, Z. (2005). *Dissolved Air flotation: Physicochemical Treatment Processes*. Berlin (Germany): Springer, pp. 431–500.

Wang, Z., Lin, T., and Chen, W. (2020). Occurrence and removal of microplastics in an advanced drinking water treatment plant (ADWTP). *Science of the Total Environment* 700:134520.

Wang, Z., Wu, Z., Mai, S., Yang, C., Wang, X., An, Y., and Zhou, Z. (2008). Research and applications of membrane bioreactors in China: Progress and prospect. *Separation and Purification Technology* 62(2):249–263.

Wiśniowska, E., Moraczewska-Majkut, K., and Nocoń, W. (2018). Efficiency of microplastics removal in selected wastewater treatment plants: Preliminary studies. *Desalination and Water Treatment* 134:316–323.

WOC (World of Chemicals). (2021). Electrocoagulation – New technology for wastewater treatment [accessed 2021 May 19]. https://www.worldofchemicals.com/18/chemistry-articles/electrocoagulation-new-technology-for-the-wastewater-treatment.html.

Xu, X., Jian, Y., Xue, Y., Hou, Q., and Wang, L. (2019). Microplastics in the wastewater treatment plants (WWTPs): Occurrence and removal. *Chemosphere* 235:1089–1096.

Xu, Z., Bai, X., and Ye, Z. (2021). Removal and generation of microplastics in wastewater treatment plants: A review. *Journal of Cleaner Production* 291:125982.

Xue, J., Peldszus, S., Van Dyke, M.I., and Huck, P.M. (2021). Removal of polystyrene microplastic spheres by alum-based coagulation-flocculation-sedimentation (CFS) treatment of surface waters. *Chemical Engineering Journal* 422:130023.

Yang, L., Li, K., Cui, S., Kang, Y., An, L., and Lei, K. (2019). Removal of microplastics in municipal sewage from China's largest water reclamation plant. *Water Research* 155:175–181.

Ziajahromi, S., Neale, P., and Leusch, F. (2016). Wastewater treatment plant effluent as a source of microplastics: Review of the fate, chemical interactions and potential risks to aquatic organisms. *Water Science and Technology* 74(10):2253–2269.

Ziajahromi, S., Neale, P., Rintoul, L., and Leusch, F. (2017). Wastewater treatment plants as a pathway for microplastics: Development of a new approach to sample wastewater-based microplastics. *Water Research* 112:93–99.

Chapter 8

Recovery of Microplastics from Sediment

INTRODUCTION

Coastal and benthic sediments have been identified as long-term sinks for microplastics (MPs) (Clark et al. 2016; Woodall et al. 2014; Zalasiewicz et al. 2016; Coppock et al. 2017; Alomar et al. 2016). In marine environments worldwide, the predominant fraction (>90%) of MPs are estimated to accumulate in sediments (Van Melkebeke et al. 2020). Coastal sediments on Kamilo Beach on the island of Hawai'i were documented to contain 3.3% MPs (w/w) (Carson et al. 2011). Sediments from Marine Protected Areas in the Mediterranean Sea contained up to 0.90 ± 0.10 MPs·g^{-1} (Alomar et al. 2016). Claessens et al. (2011) reported an average concentration of 97.2 particles·kg^{-1} dry sediment for the Belgian coast. At 14 sites in the Canadian Arctic, Adams et al. (2021) found that concentrations of MPs occurring as fragments, foams, films, and spheres ranged from 0 to 1.6 particles·g^{-1} dw. Concentrations of microfibers ranged from 0.4 to 3.2 fibers·g^{-1} dw. Microfibers comprised 82% of all MPs, followed by fragments at 15%.

High-density plastics such as PET and PVC will likely settle rapidly and become incorporated into the sediment. Low-density particles are expected to float on the water surface (Suaria and Aliani 2014) or become suspended in the water column (Fossi et al. 2012). Low-density MPs can, however, settle due to a gradual increase in density (Van Cauwenberghe et al. 2015). Studies have determined increased particle density over time among polymer types which are typically buoyant in water (Ye and Andrady 1991; Imhof et al. 2012). The buoyancy of plastics is affected over the short- and long-term by biofouling, mineral adsorption, and incorporation of MPs into fecal pellets, which increase particle density, resulting in accumulation on lake bottoms, shorelines, and the seafloor (Corcoran et al. 2015; Cole et al. 2016; Long et al. 2015; Coppock et al. 2017). Biofouling may also impart polarity to MPs, which consequently affects sinking behavior.

It must be noted that only a small percentage of plastic products consist of pure polymer; copolymers are commonly used in finished plastic products, and a wide range of additives may occur in the plastic formulation (see chapter 2), which ultimately modify particle density from that of the pure polymer.

PRACTICAL ISSUES IN PARTICLE SEPARATION

Once MP-enriched sediment is collected, it must be processed so that the target synthetic MPs are recovered from the natural matrix. Separation techniques have long been available for the beneficiation of mineral ores. In addition, technologies are available in the recycling industry for segregation of different components of a waste stream (e.g., aluminum, ferrous). Technologies are under study for separation of plastics by polymer type (Hori et al. 2009; Tsunekawa et al. 2005). Based upon successful utilization in various industries, a range of methods are being evaluated for possible removal of MPs from sediment mixtures, both in the field and at laboratory scale. These include application of gravity methods, flotation, electrostatic separation, and others (Pita and Castilho 2016).

As stated in chapter 1, MPs are defined as having a size less than 5 mm. Marine sediment particles vary in size across a wide range depending on geographic location; however, they tend to occur primarily in either the sand or mud fraction (NGDC 1976). The sand fraction includes particles between 63 μm and 2 mm, while the mud fraction are those particles smaller than 63 μm (Blott and Pye 2012). Given this overlap of size ranges, it follows that relying on particle size (for example, via sieving) does not serve as an adequate criterion for MP separation and removal from sediment. Particle density may, however, serve as a useful basis for remediation of MP-affected sediment. Sediment particles are characterized as having a density of 2.65 $g{\cdot}cm^{-3}$; in contrast, the density of many common plastic types, for example, HDPE, LDPE, PP, polyamide (PA), and PS typically occurs below 1.0 $g{\cdot}cm^{-3}$ (see table 2.2). In addition, sediment particles are predominantly hydrophilic (Giese and Van Oss 2002; Borysenko et al. 2009) while most plastics, especially fresh particles, tend to be nonpolar and therefore exhibit relatively hydrophobic character (Angu et al. 2000; van Oss 2006). Other factors such as particle shape affect sinking behavior, which influences the performance of many separation processes (Rhodes 2008). Based on these variables, several techniques may be applicable for separation of MPs from sediment.

Technologies showing promise include separation by density, air flotation, and electrostatic methods, as they require differences in particle density, polarity, size, and/or shape for effective separation.

Large-scale remediation of historic MP contamination is not feasible in most cases. In such circumstances, MPs are highly dispersed and the scale is too great. Additionally, ecological damage would be substantial and the costs of sediment removal excessive (UNEP 2015). However, in the case of large recent spills that occurred over a limited area (for example, nurdle losses from a facility release or container ship accident; see chapter 3), many technologies are possible. This chapter provides a review of techniques suitable for separation of MPs from sediment over a limited area and depth.

DENSITY SEPARATION

The concept of *gravity concentration* involves creating conditions where particles of different densities, sizes, and shapes move relative to each other while under the influence of gravity (Ambrós 2020). A range of devices has been designed to promote the separation of particles such as minerals by gravity. Many can be adapted for MPs removal from sediment.

When a heterogeneous mixture of particles having different densities (e.g., sediment minerals plus MPs) is introduced into a liquid of intermediate density (e.g., a salt solution), the low-density particles will float, while those of higher density will sink. The density of the separation medium—typically a salt solution—can be adjusted to support the flotation of all, or certain plastic formulations, to the surface. The floating layer (MPs) is subsequently collected by simple and effective methods such as vacuuming (Crawford and Quinn 2017); alternatively, the heavier sediment can be removed via a spigot or similar device at the base of the reaction vessel.

A number of methods and systems are available, with varying levels of efficiency, for separation of MPs from sediment using the principle of density flotation. Recent research has evaluated a range of salt solutions, mixing times, and settling times for effective operation.

Prior to introduction into a reactor vessel, the MP-sediment mixture may first have to undergo simple preparation; for example, large stones and roots should be removed by rough sieving. Sediment particles that are agglomerated should be separated into individual particles ("liberation") by crushing or grinding ("comminution"). The crushing step will incur additional cost for equipment and maintenance.

Separation Solutions

Filtered seawater (density 1.023 g·ml^{-1}; Browne et al. 2011; Costa et al. 2010; Ivar do Sul et al. 2009; Ryan and Moloney 1990; Imhof et al. 2011), distilled water (density 1.0 g·ml^{-1}; Zhang et al. 2018) and freshwater (McDermid and McMullen 2004; Alomar et al. 2016) have been documented for MP separation by density; however, sodium chloride (NaCl) is the most frequently used media for separation (Thompson et al. 2004; Pagter et al. 2018). Saturated NaCl has a density of 1.202 g·ml^{-1}; therefore, particles with a density below this value will float to the surface, while heavier particles (e.g., sediment) will sink. The benefits of NaCl include its low cost and ready availability; furthermore, it is environmentally benign (depending on the recipient ecosystem, e.g., seawater versus freshwater). A drawback to using NaCl is that its density does not permit separation of all polymer types—only those particles having a density below this value, that is, PE, foam PS and some PP blends (which tend to be the most common MP types identified in the aquatic environment) will be segregated. If high-density plastics such as PET and PVC are prevalent in the affected sediment, the separating solution can be replaced by a higher-density solution. Densities of most of the common polymers appear in table 2.2.

A number of salt solutions of variable densities have been evaluated for extraction of MPs from sediment (table 8.1) (Hanvey et al. 2016; Horton et al. 2016; Coppock et al. 2017). If the ultimate project goal is to remove all plastic particles, it is recommended to use a solution having a density greater than 1.4 g·ml^{-1}. Several high-density extracting solutions have been successfully used including zinc chloride (1.5 g·ml^{-1}) (Imhof 2012; Liebezeit et al. 2012), sodium polytungstate (Cooper and Corcoran 2010; Zhao et al. 2015) and sodium iodide (1.8 g·ml^{-1}) (Claessens et al. 2013; Dekiff et al. 2014; Nuelle 2014). If the density of the separation solution is excessively high, however, there is the risk that unwanted particles (i.e., sediment), may become entrained, thus limiting effectiveness.

Technologies for Density Separation

The Munich Plastic Sediment Separator (MPSS) was developed by Imhof et al. (2012) (figure 8.1). The MPSS is divided into three main compartments, that is, sediment container, standpipe, and dividing chamber. A rotor driven by an electric motor is mounted at the base of the sediment container to provide constant stirring of the mixture. A $ZnCl_2$ solution (density 1.6–1.7 g·ml^{-1}) as separation fluid permitted extraction of plastic particles ranging from large fragments to small MPs (< 1 mm). Recovery rates were reported at 100% for large MPs (1–5 mm) and 95.5% for particle sizes < 1 mm. A suspension high in sediment will likely alter the density of the separation fluid;

Table 8.1 Densities of Selected Salt Solutions for MP Separation

Name	*Chemical Formula or Composition*	*Density* $g{\cdot}ml^{-1}$	*Reference*
Distilled water	H_2O	1.000	Zhang et al. (2018)
Seawater	H_2O; Cl^-, Na^+, SO_4^{2-}, Mg^{2+}, Ca^{2+}, K^+; dissolved inorganics and organics; other	1.020–1.029	Browne et al. (2011), Costa et al. (2010), Ivar do Sul et al. (2009), Ryan and Moloney (1990)
Sodium chloride	NaCl	1.202	Li et al. (2018), Nuelle et al. (2014), Hidalgo-Ruz et al. (2012)
Sodium iodide	NaI	1.6	Li et al. (2018), Nuelle et al. (2014), Hidalgo-Ruz et al. (2012)
Calcium chloride	$CaCl_2$	1.4	Stolte et al. (2015)
Zinc chloride	$ZnCl_2$	1.5–1.8	Mohamed Nor and Obbard (2014), Quinn et al. (2017), Imhof et al. (2012)
Sodium polytungstate	$Na_2WO_4{\cdot}9O_3{\cdot}H_2O$	3.1	Corcoran et al. (2009)
Lithium metatungstate	Li_2O_4W	2.95	Masura et al. (2015)
Zinc bromide	$ZnBr_2$	1.9	Quinn et al. (2017), Imhof et al. (2012)

therefore, density must be monitored and adjusted as necessary. The MPSS should be effective in both batch and continuous mode of operation.

Coppock et al. (2017) tested the utility of NaCl, sodium iodide (NaI), and $ZnCl_2$ for MP separation using a modified version of the MPSS. Different natural sediment types were used, ranging from fine silt/clay (mean 10.2 µm) to coarse sand (149.3 µm), and the sediment was spiked with known quantities of MPs (PE, PVC, nylon). Zinc chloride (density 1.5 $g{\cdot}cm^{-3}$) was found to be the most effective solution as it allowed fine sediment to settle while supporting flotation of dense polymers. The unit consistently separated MPs from sediment in a single step with an efficiency of 95.8% (range 70–100%). The method is reported to be inexpensive, reproducible, and portable; it can therefore be employed both in the laboratory and in the field.

Zhang et al. (2020) used sodium dihydrogen phosphate (NaH_2PO_4) solution, heated to increase its density, for separation of PE, PP, PS, PVC, PET, PMMA, and POM. The densities of saturated NaH_2PO_4 solutions were 1.4, 1.46, and 1.51 $g{\cdot}ml^{-1}$ at 20, 30, and 40°C, respectively, while the density of NaCl remained at 1.18 $g{\cdot}ml^{-1}$. The NaH_2PO_4 solution exhibited higher recovery rates for MPs

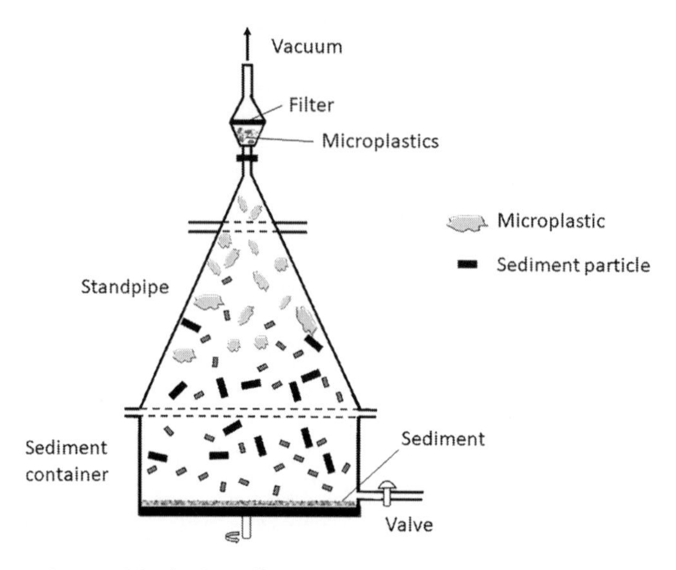

Figure 8.1 The Munich Plastic Sediment Separator (MPSS)
Source: Republished with permission of American Society of Limnology and Oceanography from: A novel, highly efficient method for the separation and quantification of plastic particles in sediments of aquatic environments. Imhof et al. *Limnology and Oceanography: Methods,* 10(7), 524–537. (2012), permission conveyed through Copyright Clearance Center, Inc.

compared to the NaCl solution—three high-density plastics (PVC, PET, and POM) that could not be recovered by NaCl were removed with NaH_2PO_4.

In batch studies of density separation, settling times of the suspensions have varied between 30 seconds and 6 hours (Besley et al. 2017; Zhang et al. 2020). Short settling times are preferred in terms of efficiency of operation; however, there is the risk of reduced purity of the MP product, as lightweight colloids such as clays and fine silts may remain suspended.

Some of the most effective reagents for density separation are also the most expensive. In order to control costs, reuse of the salt solution is beneficial. Kedzierinski et al. (2016) found that NaI solutions are recyclable without change in density; the total loss was 35.9% after ten cycles of use. During storage, chemical reactions may occur but are reversible. Rodrigues et al. (2020) showed that $ZnCl_2$ solution could be reused at least five times, maintaining an efficiency above 95%.

Practical Issues with Density Separation

To date, density separation methods have been troubled by several practical drawbacks, including variable extraction efficiencies, incompatibility with very fine sediments, particle degradation from flotation media, and complexity (Coppock et al. 2017). The salts which comprise the heavier separation liquids (e.g., metatungstate) tend to be expensive; some are also considered to be environmental and/or public health hazards. According to the published literature, removal of fine MPs has proven more difficult than for larger particles. In addition, recovery of microfibers using density separation is less effective compared to that of particles. The recovery efficiency of fibers has been reported to vary over a wide range, i.e., from 0–98% (Crawford and Quinn 2017). A single extraction should not be expected to result in complete MP removal, even if the density of the solution occurs within the required range for separation. Multiple extractions may be necessary.

In several reports of pilot-scale work using density separation, MP particles were often larger than 1 mm in size, spherical in shape, and mixed with relatively coarse sediments. Therefore, reports from some studies may not accurately represent separation efficiency from actual contaminated environments. Lastly, certain synthetic polymers such as fluoropolymers (density 1.7–2.28 $g{\cdot}ml^{-1}$), epoxide resins (1.85–2.0), melamine (1.45–2.0), and phenolic resins (1.17–2.0) are denser than common polymers such as PVC (Kedzierski et al. 2017). Thus, these polymers cannot be recovered with most of the salts mentioned above. To recover the heaviest MPs particles, a denser solution like sodium polytungstate (density 3.7 $g{\cdot}ml^{-1}$) may be necessary (Corcoran et al. 2009).

Some researchers initially apply reduction methods such as elutriation or froth flotation before using density separation in order to improve overall efficiency of separation (Claessens et al. 2013; Nuelle et al. 2014).

ELUTRIATION

The technique of *elutriation* occurs via application of a fluid stream (i.e., liquid or gas) counter-current to the direction of sedimentation (figure 8.2). Elutriation separates MPs from sediment based primarily on differences in density, but also on variations in particle size and shape.

Fluidization of the MP/sediment mixture is induced by creating an upward flow of fluid through the vessel containing the mixture. A salt solution is typically not required. Depending on the system in use, recovery rates with elutriation may exceed 90% (Van Melkebeke 2019). Separation can be optimized by varying the design and setup of the elutriation system. Elutriation is often applied upstream of density separation or other separation processes to reduce sample volume (Van Melkebeke 2019).

FROTH FLOTATION

Froth flotation technology involves the physical separation of fine particles from a solids/water suspension based on the ability of air bubbles to adhere to particle

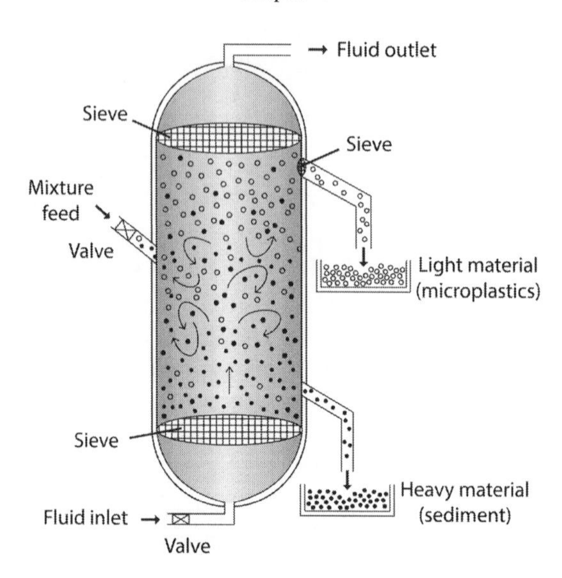

Figure 8.2 Diagram of an Elutriation Apparatus
Source: Adapted from *Microplastic Pollutants*, Crawford, M.B., and Quinn, B., Chapter 9, Microplastic separation techniques, p. 214 (2017), with permission from Elsevier.

surfaces. Froth flotation systems have been applied in many industries. The technology has been employed in the mining industry since the early twentieth century; it was originally developed for the treatment of sulfide minerals, non-metallic minerals, and fuel minerals (e.g., fine coal particles). Flotation technology has since been adapted to solid waste processing for recycling, in deinking recycled newsprint, for wastewater treatment, and de-oiling petroleum refinery effluents (Kawatra 2011; Wills and Finch 2016; Wang et al. 2010). Froth flotation is also used in the recycling industry to separate plastic particles from a mixed waste stream, and to separate mixtures of different polymer types (Drelich et al. 1998; Fraunholcz 2004; Marques et al. 2000; Carvalho et al. 2010).

Flotation focuses upon the differences in surface chemical properties of solid particles; specifically, separation is a function of the relative hydrophilicity or hydrophobicity of particles. These differences can be inherent to particles or activated by addition of reagents. To a lesser extent, separation is influenced by particle physical properties such as bulk density, particle size, shape, and even surface roughness (Shent et al. 1999; Imhof et al. 2012). Other physical factors include mechanical variables in the reaction chamber such as rate of air input and bubble size (Wills and Finch 2016).

The Reactor Vessel

The flotation system comprises a single vessel which is relatively straightforward in design (figure 8.3). Several types are in common use in various

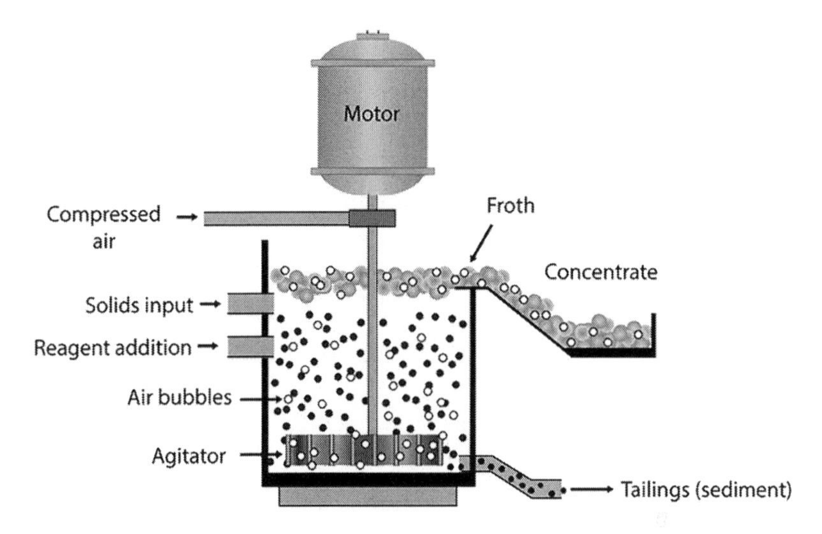

Figure 8.3 Froth Flotation Apparatus

industries and include electrolytic flotation, dissolved-air flotation, dispersed-air flotation, mechanical cells, and flotation columns (Van Melkebeke 2019). Air is pumped into an impeller, or agitator, which generates turbulence within the suspension. The purpose of agitation is to encourage collision and contact between particles and bubbles.

The sediment/MP suspension is introduced into a reactor vessel and is agitated. Those particles that attach to air bubbles are lifted to the surface of the slurry (figure 8.4.). Many MP particles, particularly nurdles and other primary particles, tend to possess nonpolar surfaces (see chapter 3) and are thus hydrophobic or not readily wetted; in contrast, sediment particles often have strongly polar surfaces which are hydrophilic, that is, they are wettable.

Hydrophobic particles tend to adhere to the bubbles. The attachment of the target product to air bubbles (termed *true flotation*) is the crucial mechanism in froth flotation (Wills and Finch 2016). If the density of the bubble-particle aggregate is less than that of the slurry, the aggregate rises to the surface and is collected in the froth phase. The process is more effective with low-density particles, as is the case for PE or PS MPs (Kawatra 2011). The particle and bubble must remain attached as they migrate upward into the froth layer. The product (termed *concentrate* or *launder*) is transported to the overflow. Particles having low hydrophobicity (i.e., high wettability) do not interact markedly with air bubbles and sink to the bottom of the vessel. This unwanted sediment is removed by drainage from the base of the vessel.

In order for froth flotation to be effective, the target particles must be of a fine size and separate from other particles. Particle sizes most effectively treated by flotation are typically less than 100 μm (Nihill et al. 1998).

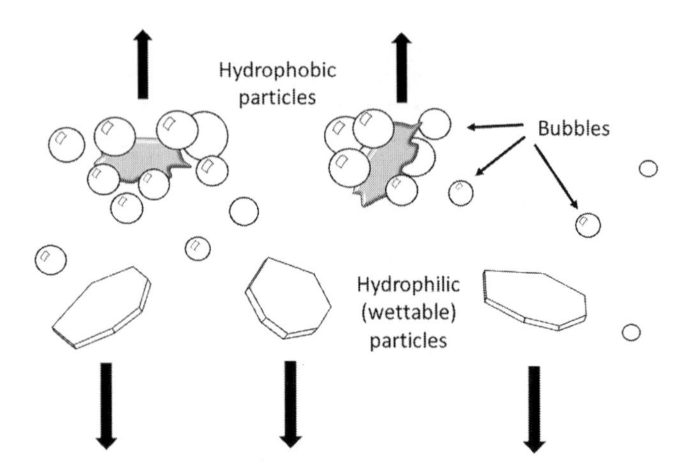

Figure 8.4 Reaction of MPs and Sediment Particles with Air Bubbles

If particles are excessively large, the density of the bubble-particle aggregate will exceed that of the slurry and the aggregates will sink.

A primary function of the froth phase is to allow for separation (i.e., *selectivity*) of differing particle types; another function is to *retain* the collected particles and transport them to the overflow. These goals are attained by allowing entrained particles to drain while preferentially retaining the particles attached to bubbles in the froth. This results in an increased concentrate.

Particle recovery or rejection by flotation is a function of three mechanisms (Wills and Finch 2016; Kawatra 2011):

1. attachment of MP particles to air bubbles (i.e., *true flotation*);
2. entrapment of particles within the froth (aggregation); and
3. entrainment of particles in water that passes through the froth.

Relevant Particle Properties

Mineral colloids commonly possess strong ionic surface bonding and are polar; they will react strongly with water molecules and are thus naturally hydrophilic. An exposed quartz surface has open Si– and O– bonds, which rapidly hydrolyze to form Si–OH groups. These, in turn, adhere to the dipoles of water molecules via hydrogen bonds. The interaction of a mineral surface with water is shown in figure 8.5 for quartz, a hydrophilic mineral.

Flotation Reagents

A suite of reagents is available for the flotation process, which serve a number of unique functions. Ultimately, however, they work collectively to

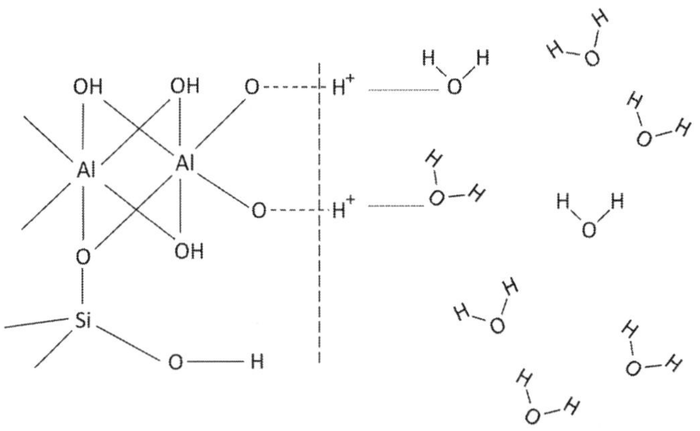

Figure 8.5 Attraction of Exposed Silica Crystal Edges to H_2O Molecules via Hydrogen Bonding

provide the most effective separation and concentration of MPs. Reagents are classified based on function and include collectors, frothers, regulators, and depressants. Hundreds are used extensively in flotation (Bulatovic 2007).

Collectors

Collectors are a group of organic surfactants whose purpose is to form a hydrophobic coating on a particle surface in the flotation pulp, thereby creating optimal conditions for attachment of hydrophobic particles to air bubbles. This coating enhances the separability of the hydrophobic and hydrophilic particles. Collectors can be divided into distinct groups based on their ability to dissociate in water. Ionizing collectors are heteropolar organic molecules, both cationic and anionic. Anionic collectors can be subdivided into oxhydryl and sulfhydryl collectors (figure 8.6). The primary amines (i.e., $R\text{–}NH_2$) are the most important flotation collectors. Non-ionizing collectors cannot dissociate to form ions or are insoluble in water. They are hydrocarbon oils derived from crude oil or coal.

Frothers

As stated above, the target product (MPs) attaches to air bubbles only if they are hydrophobic to some degree. Once the bubble-particle aggregate has reached the surface of the slurry, the bubbles will support the MP particles as long as the froth remains stable; otherwise, bubbles will burst and release the particles. Controlling the stability of the froth phase is important to achieving adequate particle separation (Wills and Finch 2016). To attain the desired

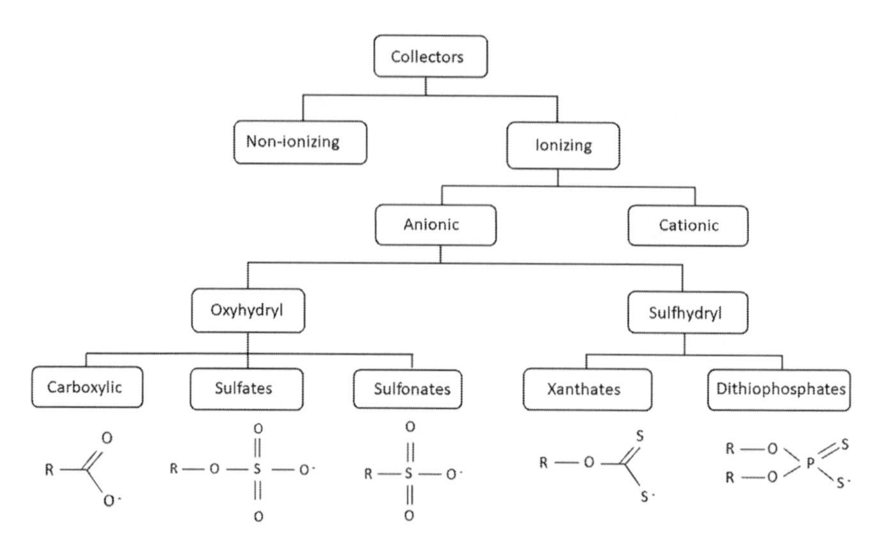

Figure 8.6 Classification of Flotation Collectors
Source: Erica Bilodeau. Https://en.wikipedia.org/wiki/Froth_flotation#/media/File:Collectors.png Creative Commons Attribution-Share Alike 4.0 International license.

conditions, it may be necessary to introduce *frothers*. Frothers comprise a group of surfactant compounds which are active at the air-water interface. They are short-chain compounds which consist of a polar head and hydrocarbon tail. Water readily bonds to the polar head of the frother molecule by hydrogen bonding, but the nonpolar tail will not react with water. This causes the nonpolar groups to be oriented toward the air phase, while the polar group is in contact with water (figure 8.7).

Frothers help generate bubbles, support production of froth, and stabilize fine bubbles. In a stable froth, bubbles do not burst at the slurry surface, which enables collected particles to overflow as the float concentrate.

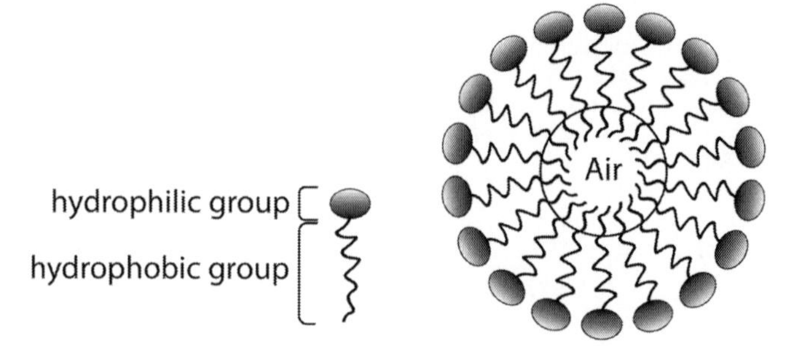

Figure 8.7 Frother Molecule Showing Polar Head and Hydrophobic Tail; Orientation of Frother Molecules on Bubble Surfaces
Source: Adapted from Pancholi et al. (2018). Creative Commons License 3.0.

Frothers additionally reduce the velocity of bubble rise in the froth (Klimpel and Isherwood 1991; Wills and Finch 2016), which increases residence time of bubbles in the slurry thus increasing the number of collisions with particles.

Several different classifications of frothers occur, based on their properties and behaviors in solution and pulp. Common classification methods are based on pH sensitivity, solubility, frothing collecting ability, and others (Khoshdast and Abbas Sam 2011). Some major types of frothers appear in table 8.2.

Table 8.2 Examples of Frother Reagents

pH category	*Group*	*Subgroup*	*Chemical Formula*
Acidic	Phenols	Phenol	C_6H_5OH
		Toluol	$C_6H_5CH_3$
		Naphthalene	$C_{10}H_8$
		Xylenols (dimethylphenols)	$(CH_3)_2C_6H_3OH$
		Cresols	$CH_3C_6H_4OH$
	Alkyl sulfates	Aromatic hydrocarbon with aliphatic radical	various
Neutral	Aliphatic alcohols	*n*-propanol	$CH_3(CH_2)_2OH$
		n-butanol	$CH_3(CH_2)_3OH$
		n-pentanol	$CH_3(CH_2)_4OH$
		n-hexanol	$CH_3(CH_2)_5OH$
		n-heptanol	$CH_3(CH_2)_6OH$
		n-octanol	$CH_3(CH_2)_7OH$
		2-butanol	$CH_3CHOH(CH_2)CH_3$
		tert-butanol (~ *t*-butanol)	$CH_3C(CH_3)_2OH$
		Isoamyl alcohol	$(CH_3)_2CHCH_2CH_2OH$
		MIBC (~ methyl isobutyl carbinol)	$(CH_3)_2CHCH_2CHOHCH_3$
		4-heptanol (~ 2-ethyl hexanol)	$CH_3(CH_2)_2CH(CH_2)_2OHCH_3$
		Diacetone alcohol	$C_6H_{12}O_2$
		n-propanol	$CH_3(CH_2)_2OH$
	Cyclic alcohols	α-terpineol	$C_{10}H_{18}O$
		β-terpineol	$C_{10}H_{18}O$
		γ-terpineol	$C_{10}H_{18}O$
		Borneol	$C_{10}H_{10}O$
		Fenchyl alcohol	$C_{10}H_{18}O$
	Alkoxy paraffins	1,1,3-triethoxybutane	$C_{10}H_{22}O_3$
	Polyglycol ethers	Dow Chemical DF-200	$CH_3(C_3H_6O)_3OH$
Basic	Pyridine base	Pyridine sulfotrioxide	$C_5H_5NSO_3$

Source: Khoshdast and Abbas Sam (2011), Bulatovic (2007), Kawatra (2011).

The froth must persist long enough for removal from the unit. The froth should not, however, be so stable as to become persistent foam. A froth that is too stable is difficult to pump through the flotation unit and other unit operations. Dense froth can adversely affect operations downstream, such as thickening of the concentrate (Wills and Finch 2016).

The froth layer is discharged from the unit by flowing over the discharge lip of the unit by gravity, or is removed by mechanical froth scrapers (Kawatra 2011), producing a concentrate of MP particles. The sediment minerals that do not float within the froth are referred to as *tailings* and are removed via pumping.

Separation of MPs from sediment is never complete: some proportion of unwanted material (*gangue*) will inevitably be recovered in the froth. As described above, true flotation is a selective process between hydrophobic and hydrophilic particles. Sediment and soil contain varying quantities of organic matter (e.g., humic acid, fulvic acid, or humin), much of which is hydrophobic and can therefore attach to bubbles and be lifted into the froth. Also, in limited cases, certain mineral types (e.g., kaolinite) may have few charged surfaces and can be carried into the launder. In contrast, entrainment is non-selective; therefore, entrained sediment particles can be brought to the surface in the froth. If particles are sufficiently coarse, they settle rapidly enough so they are not carried into the froth by entrainment. Finer particulates, however, settle more slowly and thus have more opportunity to be entrapped in the froth. Clay particles in particular, which measure < 2 μm, are readily entrained. Lastly, undesirable particles of sediment can be carried into the froth by being physically attached to floatable MP particles. The recovery of bound particles can be reduced by rejecting them along with the least-buoyant liberated particles. This will, however, sacrifice recovery of the target MP particles.

Given that MPs are generally more hydrophobic than are sediment particles, froth flotation presents a promising MP separation technique (figure 8.8). As stated in chapter 3, however, MPs possess diverse chemical and physical properties, some of which are inherent (e.g., structure of the polymer, presence of additives) and others which accrue over time. The latter includes biofouling, chemical oxidation of surfaces, and other forms of weathering. All these aspects add a level of complexity to optimizing froth flotation and require further investigation.

JIGGING

Jig separation is another gravity concentration method that is applied for the separation of solids based primarily on differences in density. *Jigging* occurs

Figure 8.8 Froth Flotation Apparatus Separating Plastic Particles
Source: Argonne National Laboratory's Flickr page: Froth Flotation Plant Uploaded using F2ComButton, CC BY-SA 2.0, https://commons.wikimedia.org/w/index.php?curid =6463805.

in an open vessel filled with water and equipped with a pulsating bed. A screen is positioned near the top, where the solids mixture is introduced. The pulsating movement of water through the bed temporarily agitates all particles upward. During a pause in pulsation, dense materials collect on the screen first, which is subsequently overlain by a zone of lighter material. Each layer can, in theory, be removed and treated separately (Ambrós 2020). In other designs, heavy particles fall through the screen and settle to the bottom of the jig vessel while only lighter particles are retained by the screen (Van Melkebeke 2019). A spigot is located at the base of the hutch compartment to drain the sediment.

Jigging technology has been employed for centuries by the mining industry for ore recovery and coal beneficiation (Lyman 1992). Its widespread use arises from its high separation efficiency, high throughput rate, and cost-effectiveness (Tsunekawa et al. 2012). More recently, jigging has been applied in recycling industries (Ambrós 2020), for example, in metals recycling to recover ferrous and non-ferrous metals from automobile waste (Kuwayama et al. 2011; Mori et al. 1994), steelmaking slag (Sripriya and Murty 2005), and electric cable waste (Pita and Castilho 2018). Separation of plastics by polymer type (PE, PC, PVC) for recycling has also been studied, particularly in Japan (Hori et al. 2009; Ito et al. Tsunekawa et al. 2012).

Numerous variables guide the extent of particle stratification during jigging, which ultimately determines yield and composition of the final product. These include: (a) design and operation of the jigging vessel; (b) jigging cycle and pulsation parameters; and (c) feed characteristics (Ambrós 2020).

Design and Operation of the Jigging Vessel

The industrial jig is classified based on a range of factors including vessel geometry, position of the jig screen (fixed or mobile), bed pulsation mechanism, bed transport, and method of removal of the heavy product (Ambrós 2020).

The jigging vessel may be a single chamber or divided into two compartments (figure 8.9). In the former configuration, a mixture of sediment/MPs is introduced at the top of the vessel. A fixed sieve or screen (punch plate) supports the mixed bed of solids. The screen may be loaded with non-cohesive heavy particles termed *ragging material*. The ragging should be of a density intermediate between the densities of the particles being separated. Particles present on the screen are exposed to a pulse of water or air pumped in from below. The pulsating movement of liquid through the bed of solids agitates all particles and forces them upward. During a pause in pulsation, heavy (sediment) particles settle quickly back onto the screen, while the lighter (MP) particles are the last to settle and form the top layer. Sequential cycles of expansion and sedimentation ultimately result in stratification of particles (primarily by density, but also by size and to a lesser extent by shape) (figure 8.10).

A two-compartment apparatus is the predominant design in jigging. In the second compartment, a device directs fluid pulsation which is responsible for agitating the bed (figure 8.9b). A common pulsation device is a plunger; alternatively, the pulsating motion of water is achieved by introducing and

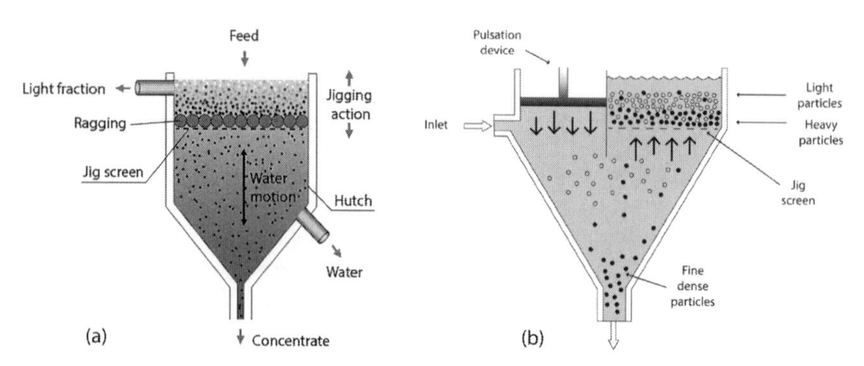

Figure 8.9 General Configuration of the Jigging Apparatus: (a) Single Compartment and (b) Two-Compartment

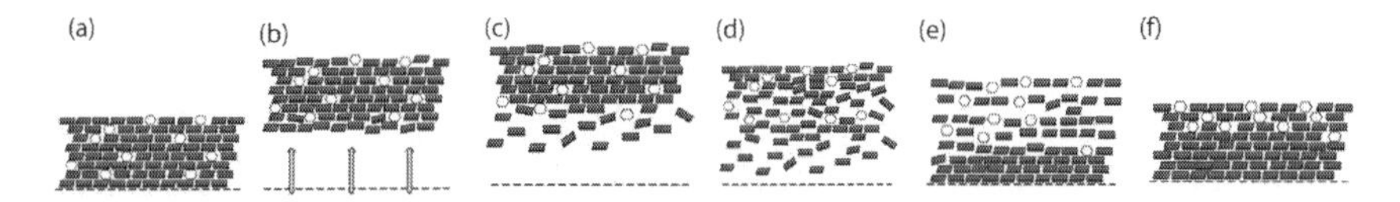

Figure 8.10 One Cycle of Expansion and Sedimentation of Mixed Particles

releasing low-pressure compressed air (Lyman 1992). Air-pulsated jigs utilize electronically controlled air valves of quick actuation for controlling water movement, thus providing a wide variation of pulse patterns. Air-pulsated jigs have been popular in mineral processing, particularly in coal beneficiation, due to their high capacity and versatility (Rao and Gouricharan 2016). In still other jigs the fluid level is stationary, and the bed is agitated and lifted via mechanical movement of the sieve.

The jigging bed is often set at a slight incline toward the outlet. When the light fraction, including MPs, migrates toward the discharge end, it is removed with the overflow.

The Jigging Cycle

Jigging is considered a rather complex gravity operation because of its continuously varying hydrodynamics (Michaud 2016). Stratification of particles occurs in the jig bed as a function of three combined mechanisms (Michaud 2016):

1) The entire bed of particles is lifted as a whole in the upward flow, which then spreads or distends. At the end of the pulse, when the water reverses its flow and moves downward, differential acceleration of particles occurs;
2) Particles experience so-called *hindered-settling* conditions (see box below) during downward flow. The bed collapses back onto the screen and segregation takes place;
3) Interstitial water movement occurs during the final stage of the downward flow before the next cycle.

HINDERED SETTLING

Hindered settling is defined as sedimentation of particles at a slower speed, resulting from interactions with nearby particles (crowding, collisions, agglomeration), compared to the speed of settling of a single particle.

Settling of particles in water is significantly affected by concentration (specific gravity) of the suspension. Individual particles or those low in concentration settle freely, unaffected by hydrodynamic influences and forces of other particles. In contrast, with a significant concentration of solids in suspension, changes occur in fluid density, particle interactions, and upwelling of fluid caused by downward particle motion (Hawksley 1954), and the falling rate of particles will consequently decrease. As a result, particle segregation is enhanced overall.

Feed Characteristics

The characteristics of the feed is an important variable which affects particle separation. Key feed characteristics which control particle stratification in jigs are particle density, size and shape, feed rate, and bed depth (Ambrós 2020).

Particle Density

The key design function of jigging is segregation of particles by density; success of this technology takes advantage of differences in density between light and dense constituents. In cases of significant differences in density (for example, lightweight plastics mixed with mineral sediment), distinct layers of particles may form during stratification. For relatively small differences in density, the profile may include a lower stratum rich in dense particles, an upper layer containing mostly the light fraction, and an intermediate layer between the two where different components are commingled.

Particle Size

Particle size has been documented to affect stratification in hydraulic jigs (Mukherjee and Mishra 2006; Olajide and Cho 1987; Pita and Castilho 2016). It is generally concluded that the finer the feed size, the less precise is overall particle separation.

Particle Shape

Lamellar (plate-like) particles tend to experience more intense drag in fluidized beds (Coulson and Richardson 1998). Pita and Castilho (2016) tested separation of plastics (PS, PMMA, PET, and PVC) in a Denver jig. Separation was greater when the light component was rich in lamellar particles, which enabled its placement at the upper layer of the bed. In contrast, the efficiency of separation decreased when the dense fraction was enriched with lamellar particles (Ambrós 2020).

Bed Depth and Feed Rate

Depth of the particle bed must be optimized in order to achieve successful particle segregation in conventional jigging. During its lateral migration along the screen of a hydraulic jig, the velocity of the particle bed varies with depth. The mean longitudinal velocity at the top of the bed can be as much as 10 times greater than that in the bottom zone (Osborne 1988). If bed thickness is excessive or if the incline of the jig screen is too great, some particles (particularly fines) may not have sufficient time to migrate to their equilibrium position. Shallow beds are advantageous when the desired product is the light component in a mixture containing a large proportion of heavy material (Feil et al. 2012). A laboratory-scale Denver jig was used to study optimization of coal beneficiation (Kumar and Venugopal 2017). Bed thickness was found to be the most significant parameter affecting the yield of clean coal.

The operator must address the proper feed rate which determines the minimum residence time for achieving optimal separation. If the feed rate is too high, particle stratification may be impeded; some target particles are lost by overflow from the jig, and excessive loading may cause heavier particles to exit the jig before attaining the desired segregation.

Jigs for Plastic-Plastic Separation

Jigging has been employed for recycling of several types of solid wastes. Due to differences in value of various waste components as well as in differences in density, dedicated jigging systems have been developed; some prototypes have been formulated solely for separating plastics. Hori et al. (2009) developed the RETAC jig, where the water pulsation and jigging cycles were optimized for the separation of plastic particles with densities close to that of water (Hori et al. 2009; Tsunekawa et al. 2005). Particles of PE and PVC of sizes 0.5–3 mm (densities of 1.1–1.4 $g \cdot ml^{-1}$) were tested. The jig separation experiments included different amplitudes, frequencies, and patterns of pulsation. High-grade PE and PVC products of over 99.8% grade were obtained with this technology.

Jigging offers a number of advantages for separation of lightweight MPs from dense mineral fractions. First, there are few moving parts other than a piston and/or a compressed air pump, and a gate for removal of heavy material. It follows that the system has only modest energy requirements and is relatively inexpensive to operate. Lastly, only water is needed to effect separation of particles (Lyman 1992). One disadvantage of jigging technology is that it becomes less efficient with decrease in particle size.

ELECTROSTATIC SEPARATION

Electrostatic separation involves the separation of particles based on their attraction or repulsion under the influence of an externally applied electrical field. Minerals possess a wide variety of conductivities, which differ markedly from those of plastics. Individual particles (mineral sediment and MPs) can be electrostatically charged for separation. Several mechanisms are available for imparting a charge to mixed particles.

Techniques for charging include ion bombardment, contact or frictional charging, and conductance charging. Ion bombardment is considered among the most effective methods of charging particles for electrostatic separation. During the process, a potential is applied to an electrode (either a point source or thin wire commonly made of tungsten), and a low-energy electrical discharge is generated around it. Gas molecules in proximity to the discharge electrode become ionized and conductive. This gas ionization is referred to as a *corona discharge*. The charged ions flow in the electric field between the electrode and earth (figure 8.11).

A positive or negative corona discharge can be achieved; however, a negative corona is preferred as it produces a more intense corona before arcing might occur (Gupta and Yan 2016). Using this scenario, both conductor (e.g., mineral particles) and nonconductor particles (MPs) are deluged by the negatively charged ions of atmospheric gas. These negatively charged gaseous ions attach to any available positive charge on the particle surface; therefore, all particles become negatively charged regardless of whether they are conductors or nonconductors. When this avalanche of gaseous ions is stopped, conductor particles readily lose their acquired charge and experience an opposite (positive) electrostatic force, repelling them from any nearby conducting surface. In contrast, nonconducting particles (i.e., MPs) remain partly coated with ions of charge opposite in electrical polarity to that of the conducting surface. As a result, these particles exhibit an electrostatic force which cause them to attach to the conducting surface. If this force is greater than the force of gravity and any other forces which could possibly separate the nonconducting particle from the surface, the particle is said to be *pinned* to the conducting surface (Gupta and Yan 2016).

At the industrial scale, electrostatic separators that rely upon charging technique via ion bombardment are the most widely applied; drum-type separators are especially common (Van Melkebeke 2019). This separator consists of a conductive rotating drum at ground potential linked with one or more high-voltage ionizing electrodes. Dry mixtures can be fed by vibratory, belt, gravity, or other methods. As shown in figure 8.12, conducive particles are released from the drum first and directed to a collection vessel, while

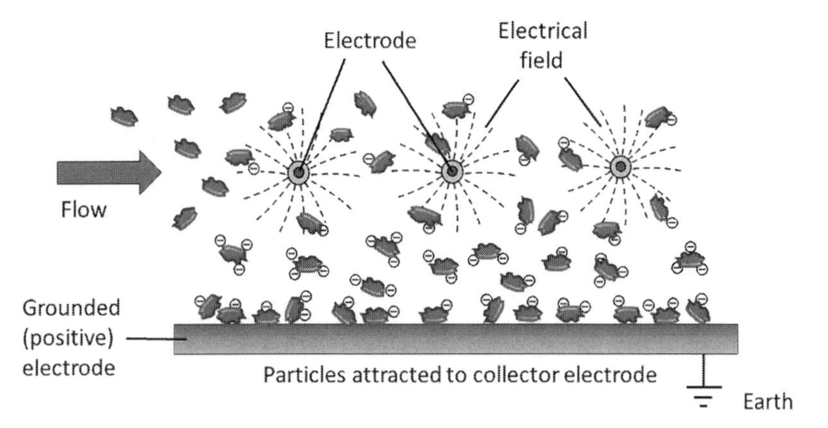

Figure 8.11 Charging of Particles Using a Low-Energy Electrical Discharge
Source: Adapted from Afshari et al. (2020). Https://creativecommons.org/licenses/by/4.0.

nonconductive particles are carried farther, separated from the conductive particles by one or more splitters, and captured separately.

A so-called scraper is often installed in the section devoted to nonconductor product collection. The scraper, essentially an alternating-current electrode device, neutralizes the charge on the pinned nonconducting particles.

Thus far little research has been published regarding the effectiveness of the electrostatic technique for MP separation from environmental media. Enders et al. (2020) evaluated the effectiveness of the Korona-Walzen-Scheider (KWS) system, designed for materials recycling, for MP separation from beach sediments and soil. When beach sediments were spiked with MPs, recovery rates of MPs were strongly dependent on MP size. Recovery of particles > 2 mm was 99–100% and recovery of fibers was approximately 80%. Recovery of MPs of size range 63–450 µm was 60–95%, while recovery of 20 µm particles was 45%. The KWS treatment was not very efficient for separating MPs from soils due to high levels of fine particulates.

Electrostatic separation technology has been applied not only for separation of MPs from different solid matrices, but also for segregation of MPs by type, which may be relevant for particle degradation by chemical and biological technologies (see chapters 9–11). Different plastic resins can be isolated via simple modifications to the charging unit and by adjusting the position of the divider plate (Hamos 2021). Felsing (2018) tested six commercial polymers of four different size ranges (2–5 mm, 0.63–2 mm, 200–630 µm, and 63–200 µm) using a Korona-Walzen-Scheider system (Hamos GmbH). Recovery rates of MPs were 90–100%. High-density PE particles were found to have a greater ability to accumulate a positive charge compared to PET, which

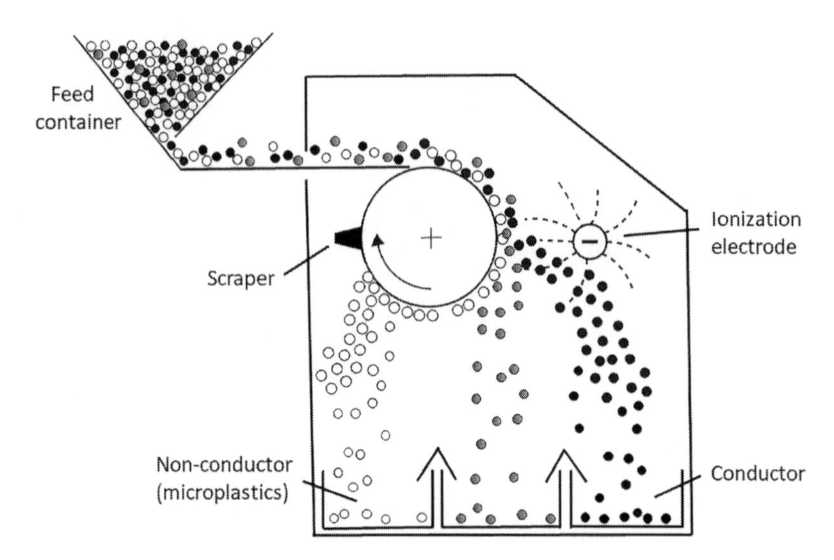

Figure 8.12 Operation of Electrostatic Separation Unit
Source: Adapted from Suponik et al. (2021). Https://creativecommons.org/licenses/by/4.0.

caused them to be more attracted to the negative electrode (drum) of the electrostatic separator (Lyskawinski et al. 2022). In a study of electrostatic separation of high-impact polystyrene (HIPS)/ABS and HIPS/ABS-PC MPs from IT waste, a recovery rate of approximately 90% and a purity of about 99% for both components of the granular mixture was recorded (Calin et al. 2022).

CLOSING STATEMENTS

This chapter provides a brief overview of several technologies which are in use, or have the potential for, separation of MPs from sediment. Some of the methods presented are based upon analytical separation techniques used in MPs research while others are proven industrial techniques for separation of plastics and/or mining industry wastes. Challenges are inherent in all methods for recovering and removing MPs, including complexity, high cost, poor extraction efficiency, and difficulty handling fine sediments (Claessens et al. 2013; Imhof et al. 2012; Hidalgo-Ruz et al. 2012; Fries et al. 2013; Lusher et al. 2016; Masura et al. 2015; Coppock et al. 2017). Relevant characteristics of the technologies discussed in this chapter are shown in table 8.3.

Table 8.3. Relevant Characteristics of Methods of Separation of MPs from Sediment

Separation Technique	*Large-Scale Operation*	*Continuous Operation*	*Complexity*	*Separation Variables*			
				Size	*Density*	*Polarity*	*Shape*
Density separation	Yes	No	Low	No	Yes	No	No
Elutriation	Yes	Yes	Low-high	Yes	Yes	No	Yes
Froth flotation	Yes	Yes	High	No	Yes	Yes	No
Jigging	Yes	Yes	High	No	Yes	No	No
Electrostatic separation	Yes	Yes	High	No	No	No	No

Source: Adapted from Van Melkebeke (2019) with kind permission of the author.

REFERENCES

Adams, J.K., Dean, B.Y., Athey, S.N., Jantunen, L.M., Bernstein, S., Stern, G., Diamond, M.L., and Finkelstein, S.A. (2021). Anthropogenic particles (including microfibers and microplastics) in marine sediments of the Canadian Arctic. *Science of the Total Environment* 784:147155.

Afshari, A., Ekberg, L., Forejt, L., Mo, J., Rahimi, S., Siegel, J., Chen, W., Wargocki, P., Zurami, S., and Zhang, J. (2020). Electrostatic precipitators as an indoor air cleaner—A literature review. *Sustainability* 12:8774. DOI: 10.3390/su12218774.

Agassant, J.-F., Binetruy, C., Charrier, P., Janin, C., Krawzczak, P., Lacrampe, M.-F., Cam, J.-B.L., Pinsolle, F., Villoutreix, G., and Zitoune, R. (2015). Plastiques et composites. In: *Base documentaire*. Saint-Denis (France): Editions T.I.

Alomar, C., Estarellas, F., and Deudero, S. (2016). Microplastics in the Mediterranean Sea: Deposition in coastal shallow sediments, spatial variation and preferential grain size. *Marine Environmental Research* 115:1–10.

Ambrós, W.M. (2020). Jigging: A review of fundamentals and future directions. *Minerals* 10(11):998.

Angu, E., Drelich, J., Laskowski, J., and Mittal, K. (2000). *Apparent and Microscopic Contact Angles*. Utrecht (Netherlands): CRC Press, p. 293.

Besley, A., Vijver, M.G., Behrens, P., and Bosker, T. (2017). A standardized method for sampling and extraction methods for quantifying microplastics in beach sand. *Marine Pollution Bulletin* 114(1):77–83.

Blott, S.J., and Pye, K. (2012). Particle size scales and classification of sediment types based on particle size distributions: Review and recommended procedures. *Sedimentology* 59(7):2071–2096.

Borysenko, A., Clennell, B., Sedev, R., Burgar, I., Ralston, J., Raven, M., Dewhurst, D., and Liu, K. (2009). Experimental investigations of the wettability of clays and shales. *Journal of Geophysical Research: Solid Earth* 114(B7).

Browne, M.A., Crump, P., Niven, S.J., Teuten, E., Tonkin, A., Galloway, T., and Thompson, R. (2011). Accumulation of microplastic on shorelines woldwide: Sources and sinks. *Environmental Science & Technology* 45(21):9175–9179.

Bulatovic, S.M. (2007). *Handbook of Flotation Reagents: Chemistry, Theory and Practice: Volume 1: Flotation of Sulfide Ores*. Amsterdam (Netherlands): Elsevier.

Carson, H.S., Colbert, S.L., Kaylor, M.J., and McDermid, K.J. (2011). Small plastic debris changes water movement and heat transfer through beach sediments. *Marine Pollution Bulletin* 62(8):1708–1713.

Claessens, M., Van Cauwenberghe, L., Vandegehuchte, M.B., and Janssen, C.R. (2013). New techniques for the detection of microplastics in sediments and field collected organisms. *Marine Pollution Bulletin* 70(1–2):227–233.

Clark, J.R., Cole, M., Lindeque, P.K., Fileman, E., Blackford, J., Lewis, C., Lenton, T.M., and Galloway, T.S. (2016). Marine microplastic debris: A targeted plan for understanding and quantifying interactions with marine life. *Frontiers in Ecology and the Environment* 14(6):317–324.

Cole, M., Lindeque, P.K., Fileman, E., Clark, J., Lewis, C., Halsband, C., and Galloway, T.S. (2016). Microplastics alter the properties and sinking rates of zooplankton faecal pellets. *Environmental Science & Technology* 50(6):3239–3246.

Cooper, D.A., and Corcoran, P.L. (2010). Effects of mechanical and chemical processes on the degradation of plastic beach debris on the island of Kauai, Hawaii. *Marine Pollution Bulletin* 60(5):650–654.

Coppock, R.L., Cole, M., Lindeque, P.K., Queirós, A.M., and Galloway, T.S. (2017). A small-scale, portable method for extracting microplastics from marine sediments. *Environmental Pollution* 230:829–837.

Corcoran, P.L., Biesinger, M.C., and Grifi, M. (2009). Plastics and beaches: A degrading relationship. *Marine Pollution Bulletin* 58(1):80–84.

Corcoran, P.L., Norris, T., Ceccanese, T., Walzak, M.J., Helm, P.A., and Marvin, C.H. (2015). Hidden plastics of Lake Ontario, Canada, and their potential preservation in the sediment record. *Environmental Pollution* 204:17–25.

Costa, M.F., Do Sul, J.A.I., Silva-Cavalcanti, J.S., Araújo, M.C.B., Spengler, Â., and Tourinho, P.S. (2010). On the importance of size of plastic fragments and pellets on the strandline: A snapshot of a Brazilian beach. *Environmental Monitoring and Assessment* 168(1):299–304.

Coulson, J., and Richardson, J. (1998). *Chemical Engineering, Vol. 2: Particle Technology and Separation Processes*. Oxford: Butterworth–Heinemann.

Cozar, A., Echevarría, F., Gonzalez-Gordillo, J.I., Irigoien, X., Úbeda, B., Hernandez-Leon, S., Palma, A.T., Navarro, S., García-de-Lomas, J., Ruiz, A., Fernandez-de-Puelles, M.L., and Duarte, C.M. (2014). Plastic debris in the open ocean. *Proceedings of the National Academy of Sciences* 111:10239-10244. DOI: 10.1073/pnas.1314705111.

Crawford, C.B., and Quinn, B. (2016). *Microplastic Pollutants*. Amsterdam (Netherlands): Elsevier Limited.

Crespo, E. (2016). Modeling segregation and dispersion in jigging beds in terms of the bed porosity distribution. *Minerals Engineering* 85:38–48.

Dekiff, J.H., Remy, D., Klasmeier, J., and Fries, E. (2014). Occurrence and spatial distribution of microplastics in sediments from Norderney. *Environmental Pollution* 186:248–256.

do Sul, J.A.I., Spengler, Â., and Costa, M.F. (2009). Here, there and everywhere. Small plastic fragments and pellets on beaches of Fernando de Noronha (Equatorial Western Atlantic). *Marine Pollution Bulletin* 58(8):1236–1238.

Drelich, J., Payne, T., Kim, J.H., and Miller, J.D. (1998). Selective froth flotation of PVC from PVC/PET mixtures for the plastics recycling industry. *Polymer Engineering and Science* 38(9):1378–1386.

Enders, K., Tagg, A.S., and Labrenz, M. (2020). Evaluation of electrostatic separation of microplastics from mineral-rich environmental samples. *Frontiers in Environmental Science* 8:112.

Feil, N.F., Sampaio, C.H., and Wotruba, H. (2012). Influence of jig frequency on the separation of coal from the Bonito seam—Santa Catarina, Brazil. *Fuel Processing Technology* 96:22–26.

Felsing, S., Kochleus, C., Buchinger, S., Brennholt, N., Stock, F., and Reifferscheid, G. (2018). A new approach in separating microplastics from environmental samples based on their electrostatic behavior. *Environmental Pollution* 234:20–28.

Fossi, M.C., Panti, C., Guerranti, C., Coppola, D., Giannetti, M., Marsili, L., and Minutoli, R. (2012). Are baleen whales exposed to the threat of microplastics? A case study of the Mediterranean fin whale (*Balaenoptera physalus*). *Marine Pollution Bulletin* 64(11):2374–2379.

Fraunholcz, N. (2004). Separation of waste plastics by froth flotation—A review, part I. *Minerals Engineering* 17(2):261–268.

Fries, E., Dekiff, J.H., Willmeyer, J., Nuelle, M.-T., Ebert, M., and Remy, D. (2013). Identification of polymer types and additives in marine microplastic particles using pyrolysis-GC/MS and scanning electron microscopy. *Environmental Science: Processes & Impacts* 15(10):1949–1956.

Giese, R.F., and Van Oss, C.J. (2002). *Reaction Kinetics and Catalysis Letters*, 1st ed. Oxford: CRC Press 77:393–394.

Gupta, A., and Yan, D.S. (2016). *Mineral Processing Design and Operations: An Introduction*. Amsterdam (Netherlands): Elsevier.

Hamos. (2021). Hamos EKS electrostatic plastic/plastic separators [accessed 2021 October 15]. https://www.hamos.com/products/electrostatic-separators/plastic_plastic-separators,35,eng,39.

Hanvey, J.S., Lewis, P.J., Lavers, J.L., Crosbie, N.D., Pozo, K., and Clarke, B.O. (2017). A review of analytical techniques for quantifying microplastics in sediments. *Analytical Methods* 9(9):1369–1383.

Hawksley, P. (1954). Survey of the relative motion of particles and fluids. *British Journal of Applied Physics* 5(S3):S1.

Hidalgo-Ruz, V., Gutow, L., Thompson, R.C., and Thiel, M. (2012). Microplastics in the marine environment: A review of the methods used for identification and quantification. *Environmental Science & Technology* 46(6):3060–3075.

Hori, K., Tsunekawa, M., Ueda, M., Hiroyoshi, N., Ito, M., and Okada, H. (2009). Development of a new gravity separator for plastics—A hybrid-jig. *Materials Transactions* 0911160935.

Horton, A.A., Svendsen, C., Williams, R.J., Spurgeon, D.J., and Lahive, E. (2017). Large microplastic particles in sediments of tributaries of the River

Thames, UK–Abundance, sources and methods for effective quantification. *Marine Pollution Bulletin* 114(1):218–226.

Imhof, H.K., Schmid, J., Niessner, R., Ivleva, N.P., and Laforsch, C. (2012). A novel, highly efficient method for the separation and quantification of plastic particles in sediments of aquatic environments. *Limnology and Oceanography: Methods* 10(7):524–537.

Kawatra, S. (2011). Fundamental principles of froth flotation. https://www.researchgate.net/publication/281944301_Fundamental_principles_of_froth_flotation.

Kedzierski, M., Le Tilly, V., César, G., Sire, O., and Bruzaud, S. (2017). Efficient microplastics extraction from sand. A cost effective methodology based on sodium iodide recycling. *Marine Pollution Bulletin* 115(1–2):120–129.

Khoshdast, H., and Sam, A. (2011). Flotation frothers: Review of their classifications, properties and preparation. *The Open Mineral Processing Journal* 4(1):25–44.

Klimpel, R., and Isherwood, S. (1991). Some industrial implications of changing frother chemical structure. *International Journal of Mineral Processing* 33(1–4):369–381.

Kumar, S., and Venugopal, R. (2017). Performance analysis of jig for coal cleaning using 3D response surface methodology. *International Journal of Mining Science and Technology* 27(2):333–337.

Kuwayama, Y., Ito, M., Hiroyoshi, N., and Tsunekawa, M. (2011). Jig separation of crushed automobile shredded residue and its evaluation by float and sink analysis. *Journal of Material Cycles and Waste Management* 13(3):240–246.

Leslie, H. (2015). Plastic in cosmetics: Are we polluting the environment through our personal care? Plastic ingredients that contribute to marine microplastic litter. United Nations Environment Programme. ISBN: 978-92-807-3466-9.

Liebezeit, G., and Dubaish, F. (2012). Microplastics in beaches of the East Frisian islands Spiekeroog and Kachelotplate. *Bulletin of Environmental Contamination and Toxicology* 89(1):213–217.

Lobelle, D., and Cunliffe, M. (2011). Early microbial biofilm formation on marine plastic debris. *Marine Pollution Bulletin* 62(1):197–200.

Long, M., Moriceau, B., Gallinari, M., Lambert, C., Huvet, A., Raffray, J., and Soudant, P. (2015). Interactions between microplastics and phytoplankton aggregates: Impact on their respective fates. *Marine Chemistry* 175:39–46.

Lusher, A., Welden, N., Sobral, P., and Cole, M. (2020). Sampling, isolating and identifying microplastics ingested by fish and invertebrates. In: *Analysis of Nanoplastics and Microplastics in Food*, pp. 119–148.

Lyman, G. (1992). Review of jigging principles and control. *Coal Preparation* 11(3–4):145–165.Marques, G.A., and Tenorio, J.A.S. (2000). Use of froth flotation to separate PVC/PET mixtures. *Waste Management* 20(4):265–269.

Lyskawinski, W., Baranski, M., Jedryczka, C., Mikolajewicz, J., Regulski, R., Rybarczyk, D., and Sedziak, D. (2022). Analysis of triboelectrostatic separation process of mixed poly(ethylene terephthalate) and high-density polyethylene. *Energies* 15:19. DOI: 10.3390/en15010019.

Masura, J., Baker, J., Foster, G., and Arthur, C. (2015). *Laboratory Methods for the Analysis of Microplastics in the Marine Environment: Recommendations for Quantifying Synthetic Particles in Waters and Sediments*. Silver Spring, MD: National Oceanic and Atmospheric Administration, U.S. Department of Commerce Technical Memorandum NOS-OR&R-48.

McDermid, K.J., and McMullen, T.L. (2004). Quantitative analysis of small-plastic debris on beaches in the Hawaiian archipelago. *Marine Pollution Bulletin* 48(7–8):790–794.

Morét-Ferguson, S., Law, K.L., Proskurowski, G., Murphy, E.K., Peacock, E.E., and Reddy, C.M. (2010). The size, mass, and composition of plastic debris in the western North Atlantic Ocean. *Marine Pollution Bulletin* 60(10):1873–1878.

Mori, S., Nonaka, M., Matsufuji, R., Fujita, T., Futamata, M., and Hata, M. (1994). Recovery of non-ferrous metals from car scrap using the jig separator (ECHO metal jig, type-SP). In: *Ecomaterials*. Amsterdam (Netherlands): Elsevier, pp. 771–774.

Mukherjee, A., and Mishra, B. (2006). An integral assessment of the role of critical process parameters on jigging. *International Journal of Mineral Processing* 81(3):187–200.

NGCD (National Geophysical Data Center). (1976). *The NGDC Seafloor Sediment Grain Size Database. NCEI/NOAA*. Silver Spring, MD.

Nihill, D.N., Stewart, C.M., and Bowen, P. (1998). The McArthur River mine—The first years of operation. In: *AusIMM '98 – The Mining Cycle, Mount Isa, 19–23 April 1998 (The Australasian Institute of Mining and Metallurgy: Melbourne)*, pp. 73–82.

Nuelle, M.-T., Dekiff, J.H., Remy, D., and Fries, E. (2014). A new analytical approach for monitoring microplastics in marine sediments. *Environmental Pollution* 184:161–169.

Olajide, O., and Cho, E. (1987). Study of the jigging process using a laboratory-scale Baum jig. *Mining, Metallurgy & Exploration* 4(1):11–14.

Osborne, D.G. (1988). *Coal Preparation Technology*. Dordrecht (Netherlands): Springer.

Pagter, E., Frias, J., and Nash, R. (2018). Microplastics in Galway Bay: A comparison of sampling and separation methods. *Marine Pollution Bulletin* 135:932–940.

Pancholi, K., Robertson, P.K.J., Okpozo, P., Beattie, N.S., and Huo, D. (2018). Observation of stimulated emission from Rhodamine 6G-polymer aggregate adsorbed at foam interfaces. *Journal of Physics Energy* 1(1):015007.

Pita, F., and Castilho, A. (2016). Influence of shape and size of the particles on jigging separation of plastics mixture. *Waste Management* 48:89–94.

Pita, F., and Castilho, A. (2018). Separation of copper from electric cable waste based on mineral processing methods: A case study. *Minerals* 8(11):517.

Rao, D., and Gouricharan, T. (2019). *Coal Processing and Utilization*. Boca Raton (FL): CRC Press.

Rhodes, M.J. (2008). *Introduction to Particle Technology*. New York: John Wiley & Sons.

Rodrigues, M., Gonçalves, A., Gonçalves, F., and Abrantes, N. (2020). Improving cost-efficiency for MPs density separation by zinc chloride reuse. *MethodsX* 7:100785.

Rodríguez-Narvaez, O.M., Goonetilleke, A., Perez, L., and Bandala, E.R. (2021). Engineered technologies for the separation and degradation of microplastics in water: A review. *Chemical Engineering Journal* 128692.

Ruggero, F., Gori, R., and Lubello, C. (2020). Methodologies for microplastics recovery and identification in heterogeneous solid matrices: A review. *Journal of Polymers and the Environment* 28(3):739–748.

Ryan, P., and Moloney, C. (1990). Plastic and other artefacts on South African beaches: Temporal trends in abundance and composition. *South African Journal of Science* 86(7):450–452.

Shent, H., Pugh, R., and Forssberg, E. (1999). A review of plastics waste recycling and the flotation of plastics. *Resources, Conservation and Recycling* 25(2):85–109.

Sripriya, R., and Murty, C.V. (2005). Recovery of metal from slag/mixed metal generated in ferroalloy plants—A case study. *International Journal of Mineral Processing* 75(1–2):123–134.

Suaria, G., and Aliani, S. (2014). Floating debris in the Mediterranean Sea. *Marine Pollution Bulletin* 86(1–2):494–504.

Suponik, T., Franke, D.M., Nuckowski, P.M., Matusiak, P., Kowol, D., and Tora, B. (2021). Impact of grinding of printed circuit boards on the efficiency of metal recovery by means of electrostatic separation. *Minerals* 11:281. DOI: 10.3390/min11030281.

Thompson, R., Olsen, Y., Mitchell, R., Davis, A., Rowland, S., John, A.W.G., McGonigle, D., and Russell, A.E. (2004). Lost at sea: Where is all the plastic? *Science* 34:838.

Tsunekawa, M., Kobayashi, R., Hori, K., Okada, H., Abe, N., Hiroyoshi, N., and Ito, M. (2012). Newly developed discharge device for jig separation of plastics to recover higher grade bottom layer product. *International Journal of Mineral Processing* 114:27–29.

Tsunekawa, M., Naoi, B., Ogawa, S., Hori, K., Hiroyoshi, N., Ito, M., and Hirajima, T. (2005). Jig separation of plastics from scrapped copy machines. *International Journal of Mineral Processing* 76(1–2):67–74.

Van Cauwenberghe, L., Claessens, M., Vandegehuchte, M.B., and Janssen, C.R. (2015). Microplastics are taken up by mussels (*Mytilus edulis*) and lugworms (*Arenicola marina*) living in natural habitats. *Environmental Pollution* 199:10–17.

Van Melkebeke, M. (2019). Exploration and Optimization of Separation Techniques for the Removal of Microplastics from Marine Sediments During Dredging Operations [Master's Thesis]. Ghent (Belgium): Ghent University.

Van Melkebeke, M., Janssen, C., and De Meester, S. (2020). Characteristics and sinking behavior of typical microplastics including the potential effect of biofouling: Implications for remediation. *Environmental Science & Technology* 54(14):8668–8680.

Van Oss, C.J. (2006). *Interfacial Forces in Aqueous Media*. Boca Raton (FL): CRC Press.

Wang, L.K., Shammas, N.K., Selke, W.A., and Aulenbach, D.B., editors (2010). *Flotation Technology*. Totowa (NJ): Humana Press.

Wills, B.A., and Finch, J. (2015). *Wills' Mineral Processing Technology: An Introduction to the Practical Aspects of Ore Treatment and Mineral Recovery*. Oxford: Butterworth-Heinemann.

Woodall, L.C., Robinson, L.F., Rogers, A.D., Narayanaswamy, B.E., and Paterson, G.L. (2015). Deep-sea litter: A comparison of seamounts, banks and a ridge in the Atlantic and Indian Oceans reveals both environmental and anthropogenic factors impact accumulation and composition. *Frontiers in Marine Science* 2:3.

Woodall, L.C., Sanchez-Vidal, A., Canals, M., Paterson, G.L., Coppock, R., Sleight, V., Calafat, A., Rogers, A.D., Narayanaswamy, B.E., and Thompson, R.C. (2014). The deep sea is a major sink for microplastic debris. *Royal Society Open Science* 1(4):140317.

Ye, S., and Andrady, A.L. (1991). Fouling of floating plastic debris under Biscayne Bay exposure conditions. *Marine Pollution Bulletin* 22(12):608–613.

Zalasiewicz, J., Waters, C.N., Do Sul, J.A.I., Corcoran, P.L., Barnosky, A.D., Cearreta, A., Edgeworth, M., Gałuszka, A., Jeandel, C., Leinfelder, R., McNeill, J.R., and Leinfelder, R. (2016). The geological cycle of plastics and their use as a stratigraphic indicator of the Anthropocene. *Anthropocene* 13:4–17.

Zhang, X., Yu, K., Zhang, H., Liu, Y., He, J., Liu, X., and Jiang, J. (2020). A novel heating-assisted density separation method for extracting microplastics from sediments. *Chemosphere* 256:127039.

Zhao, S., Zhu, L., and Li, D. (2015). Characterization of small plastic debris on tourism beaches around the South China Sea. *Regional Studies in Marine Science* 1:55–62.

Part III

DECOMPOSITION OF MICROPLASTICS

As revealed in previous chapters, technologies are available for removal of microplastics (MPs) from water, wastewater, and sediment. Both drinking water and wastewater treatment facilities are equipped with unit operations designed to capture and remove common pollutants from the treatment train; fortunately, these operations are also capable of recovering large volumes of MPs. Methods used in mineral processing and other industries are available for MP recovery from loose sediments such as sand.

MPs which are removed from water and solids cannot be recycled due to challenges of incompatible polymer types as well as contamination by mineral and organic matter. Incineration of wastes, including plastics, poses issues of air pollution, ash disposal, and public opposition. Direct disposal of large volumes of MPs to land only perpetuates the cycle of MP releases to the environment. In order to effectively mitigate pollution by MPs, it may be beneficial to apply one or more technologies for their ultimate destruction following removal from environmental media.

All synthetic polymers are composed of long chains of molecules; degradation, in this context, is defined as any chemical, physical, or biological process which results in breakage of the large chains into short units of lower MW, and, ideally, to carbon dioxide and water. Mineralization of certain polymers by abiotic and biotic pathways has been demonstrated under laboratory and/or pilot-scale conditions. Most methods, however, are capable of only partial decomposition and require extensive time periods that are impractical for large-scale water treatment. Other recent technologies, however, have shown remarkable potential for degradation of MPs.

The ideal degradative technology would rapidly convert MPs to simple and innocuous forms with minimal inputs of energy and other resources. This concept is especially challenging, as synthetic polymers are formulated to be inherently recalcitrant to chemical/physical and biological processes. With this reality in mind, a number of innovative techniques focused on degradation of MP wastes are being researched in various laboratories. Technologies under consideration embrace thermal, mechanical, chemical, catalytic, photo-oxidative, and biological degradation. Chapter 9 addresses treatment of MPs via conventional and advanced oxidation processes including ozonation, Fenton, and wet air oxidation, and chapter 10 presents heterogeneous photocatalysis. Chapter 11 focuses on microbial-assisted decomposition of MPs.

Chapter 9

Chemical Technologies for Microplastic Destruction

INTRODUCTION

For decades, chemical oxidation processes have been applied for treatment of drinking water by reducing concentrations of residual organics, removal of ammonia, and controlling odors. Chemical oxidation technologies have also been successfully employed to treat many types of industrial effluents, including those containing recalcitrant organic compounds such as textile dyeing and finishing effluents (Hassaan and Nemr 2017).

Chemical oxidation can be divided into two categories: conventional chemical treatments and advanced oxidation processes (Oller et al. 2011; Buthiyappan et al. 2016). Conventional chemical oxidation by H_2O_2 alone is not sufficient to decompose high concentrations of certain refractory organic contaminants such as highly chlorinated aromatic compounds, due to low rates of reaction at practical H_2O_2 concentrations. However, ozone, UV irradiation, and transition metals (e.g., iron salts) can activate H_2O_2 to form hydroxyl radicals which are strong oxidants (Eqs. 9.1–9.3) (Neyens and Baeyens 2003):

Ozone and hydrogen peroxide

$$O_3 + H_2O_2 \rightarrow {}^{\bullet}OH + O_2 + HO_2^{\bullet} \tag{9.1}$$

UV light and hydrogen peroxide

$$H_2O_2 + h\upsilon \rightarrow 2{}^{\bullet}OH \tag{9.2}$$

Iron salts and hydrogen peroxide

$$Fe^{2+} + H_2O_2 \rightarrow Fe^{3+} + {}^{\bullet}OH + OH^- \quad (9.3)$$

These processes hold promise for microplastics (MPs) decomposition as well.

ADVANCED OXIDATION PROCESSES

Advanced oxidation processes (AOPs) are a family of intensive chemical reactions which involve generation of oxidizing agents that transform hazardous organic and inorganic pollutants to less hazardous compounds. AOPs rely on reactive oxygen species such as hydroxyl radicals ($^{\bullet}OH$), superoxide radical anions ($O_2^{\bullet -}$), and perhydroxyl radicals ($HO_2^{\bullet}$) to degrade and mineralize organic pollutants including microplastics.

AOP technologies were introduced in the 1980s for treatment of drinking water. More recently, municipal effluents are treated by AOPs in order to remove recalcitrant organic pollutants including pharmaceutical ingredients, personal care products and endocrine-disrupting compounds, among others. AOP processes have been studied and applied for treatment of varied industrial and commercial wastewaters including those from pulp and paper, textile, agrochemical (e.g., pesticides), and metal plating industries. Other hazardous wastewaters treated by AOP technologies include hospital and slaughterhouse effluent (Fatta-Kassinos et al. 2010).

The goal in most AOPs is generation of the hydroxyl radical ($^{\bullet}OH$) or other high-reactivity species such as the sulfate radical ($SO_4^{\bullet -}$). These react with many organic substances at rates close to the diffusion-controlled limit (Machulek et al. 2013). The $^{\bullet}OH$ radical is one of the most powerful known oxidants; it reacts rapidly and relatively non-selectively with organic compounds. Key reaction mechanisms of $^{\bullet}OH$ with organics include (Munter 2001; Siitonen 2007; Sillanpää and Matilainen 2015; Neyens and Baeyes 2003; Karat 2013):

1) addition of $^{\bullet}OH$ radicals to double bonds and aromatic rings;
2) hydrogen removal from an aliphatic carbon, which yields carbon-centered radicals that initiate radical chain oxidation;
3) the $^{\bullet}OH$ radical obtains an electron from an organic substituent; and
4) the $^{\bullet}OH$ radical transfers its unpaired electron to another substrate (carbonates, bicarbonates etc.).

Table 9.1 Examples of Non-Photochemical and Photochemical Oxidation Processes for Possible Microplastics Destruction

Non-Photochemical	*Common Reagents*	*Photochemical*	*Common Reagents*
Alkaline ozonation	O_3/OH^-	Water photolysis in vacuum; UV.	–
Ozonation with hydrogen peroxide	O_3/H_2O_2	UV/ hydrogen peroxide	UV/H_2O_2
Fenton and related processes	Fe^{3+}/H_2O_2	UV/ozone	UV/O_3
Electrochemical oxidation	Various electrode types, e.g., C, Pt, PbO_2, mixed metal oxides.	Photo-Fenton and related processes	$Fe^{3+}/H_2O_2/O_3$
γ-Radiolysis and electron-beam treatment	–	Sono-photo Fenton	Fe^{3+}/H_2O_2
Oxidation in sub/and supercritical water	–	UV/periodate	UV/IO_4^-
Zero-valent iron	Fe^o	Heterogenous photocatalysis	Ag, TiO_2, Cu_2O, ZnO, $BiVO_4$, CdS, etc./ $H_2O_2/O_2/UV$
Ferrate	K_2FeO_4, Fe (VI)		

In addition to the above, it must be noted that since the •OH radical is highly unstable it can rapidly auto-terminate (Eq. 9.4).

$$\bullet OH + HO_2\bullet \rightarrow O_2 + H_2O \quad (9.4)$$

AOPs can be divided into non-photochemical and photochemical types (table 9.1). In non-photochemical AOPs, the •OH radical is generated when oxidizing agents such as O_3 or H_2O_2 are applied to the wastewater to be treated. Photochemical AOPs rely on the same oxidizers, but in combination with UV irradiation. Various combinations of oxidants and catalysts have been formulated for oxidation and mineralization of organic pollutants, for example O_3/H_2O_2, UV/H_2O_2, UV/O_3, UV/TiO_2, Fe^{2+}/H_2O_2, and Fe^{2+}/H_2O_2 + hυ (Matilainen and Sillanpää 2010).

Some AOPs are used in combination with catalysts to sustain the desired rate of reaction (Goi 2005; Karat 2013; Legrini et al. 1993; Von Sonntag 2008; Matilainen and Sillanpää 2010; Machulek et al. 2013). Catalysts are often classified as homogenous and heterogeneous. Homogeneous catalysts are dissolved within the same phase as the reactants and are thus uniformly distributed throughout the reaction medium. The most popular transition metals employed as homogeneous catalysts in AOPs are Fe^{2+}, Fe^{3+}, Zn^{2+}, Mn^{2+}, Mn^{3+}, Mn^{4+}, Ti^{2+}, Cr^{3+}, Cu^{2+}, Co^{2+}, Ni^{2+}, Cd^{2+}, and Pb^{2+} (Buthiyappan et al. 2016). In contrast, heterogeneous catalysts occur in a different phase from the reactants; the catalysis reaction takes place on the external and internal surfaces of the catalyst at gas-solid or liquid-solid interfaces (Spivey et al. 2005). Heterogeneous photocatalysis will be discussed in chapter 10.

When seeking the most effective AOP for microplastic degradation, key factors to be considered include microplastic load, process limitations (e.g., sludge production), and necessary operating conditions. Some of the most commonly applied AOPs for recalcitrant water pollutants are ozonation, electrochemical oxidation, Fenton's and photo-Fenton's reagent, wet air oxidation, and heterogeneous semiconductor photocatalysis (Tarr 2003; Parsons 2004; Fatta Kassinos et al. 2010); however, many others are available.

AOPs are employed alone or can be combined with other processes. Such integration often leads to higher removal efficiencies. In the case of microplastic-enriched water, decisions as to which technologies to combine are based on the ultimate treatment objective, properties of the wastewater stream (i.e., types and quantities of MPs to be treated), waste production, and cost. The potential of several AOP technologies for microplastics destruction are discussed below.

Ozone and Ozone AOPs

For decades ozone, O_3, has been employed in drinking water facilities for disinfection, taste and odor management, and color removal (Langlais et al. 1991; Matilainen and Sillanpää 2010). Ozone is a powerful oxidant, having a high oxidation potential ($E_0 = 2.07$ V). Ozone reacts with organic compounds via an electrophilic addition to double bonds (Litter and Quici 2010). This reaction, termed *direct oxidation*, is slow and selective (Eq. 9.5).

$$O_3 + R \rightarrow R_{oxid} \tag{9.5}$$

In addition, a rapid and nonselective reaction (*indirect ozonation*) occurs when $^{\bullet}OH$ radicals are formed as ozone decomposes in water (Eq. 9.6). The $^{\bullet}OH$ radical is a stronger oxidizing agent than ozone itself.

$$2O_3 + 2H_2O \rightarrow 2^{\bullet}OH + O_2 + 2HO_2 \tag{9.6}$$

At the municipal/industrial treatment scale, the conventional ozonation process requires efficient reactor design in order to maximize ozone mass transfer. The bubble reactor is the most common conventional contactor for water

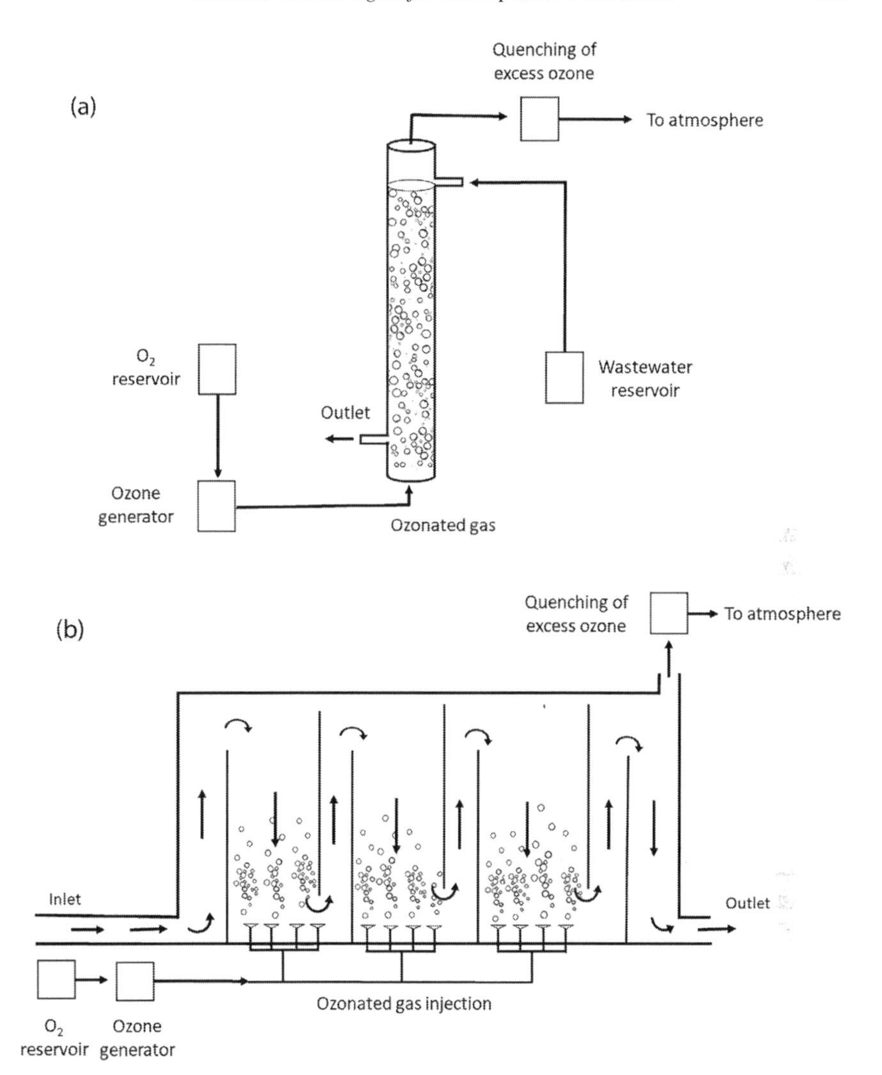

Figure 9.1 Examples of Two Types of Ozonation Reactors: (a) Bubble Column and (b) Ozonation Chamber with Multiple Compartments in Series

Source: Figure 9.1a. adapted from Graça et al. (2020). Creative Commons Attribution. (CC BY) license (http://creativecommons.org/licenses/by/4.0/).

treatment with ozone. The reactor is available in two configurations, i.e., the conventional bubble column and the ozonation chamber (figure 9.1).

In the bubble column, gas and water flow in countercurrent mode: wastewater is introduced at the top and ozone is bubbled in from the base. Solid particles (e.g., glass beads) can be included in the column to promote contact between the gas phase and wastewater. In the bubble chamber, water and ozone flow concurrently; baffles direct water movement in each compartment. Gas bubbles are produced by diffusers located at the base of the reactor. This allows the gas to be distributed throughout the entire volume of the vessel. The bubble chamber reactor is typically equipped with two to four compartments separated by baffles. The number of chambers may be adjusted based on the distribution of ozonated gas flow which is necessary to attain the required rate of chemical reaction (Suez 2021).

The ozonation reaction is enhanced by increasing the interfacial area of contact by reducing bubble size via use of small-size diffusers. Vigorous stirring, mixers, and contact towers also enhance the process. Increasing ozone solubility by increasing reaction pressure to several atmospheres also favors the process.

Standard ozonation technology offers many benefits if applied for destruction of microplastics:

1) on-site installation is relatively simple;
2) the reactor does not require large space;
3) the process does not increase the volume of wastewater;
4) sludge production is minimal;
5) no hazardous chemicals (e.g., H_2O_2) are required on-site; and
6) residual O_3 is readily decomposed to oxygen (Buthiyappan et al. 2016).

A key disadvantage of the standard ozonation method is non-optimal gas-liquid ozone mass transfer (Litter 2005; Gogate and Pandit 2004). This drawback might be overcome by including additional oxidants. To achieve maximal reaction with O_3, complex mixing techniques are often required; this may result in substantial capital equipment and maintenance costs. An additional limitation of the ozonation method is high energy consumption.

A number of modifications are available to enhance production of the •OH radical during ozonation. A simple and cost-efficient method to convert conventional ozonation to an AOP involves addition of H_2O_2 (Duguet et al. 1985; Glaze et al. 1992; von Gunten 2003). It is also possible to combine ozone with UV irradiation, or combine ozone, H_2O_2 and UV (Glaze et al. 1982, 1987; Legrini et al. 1993; Matilainen and Sillanpää 2010). Using an ozone AOP, the potential for •OH radical formation increases substantially.

O_3/H_2O_2

When ozonation is augmented by addition of hydrogen peroxide, several secondary reactions occur to improve the overall oxidation capacity of the process. Hydrogen peroxide triggers decomposition of O_3 by electron transfer. The •OH-producing reaction (Eq. 9.7) results in a cascade of other radical-producing reactions (Litter 2005):

$$O_3 + H_2O_2 \rightarrow {}^{\bullet}OH + O_2 + HO_2^{\bullet} \quad (9.7)$$

$$HO_2^{\bullet} \leftrightarrow O_2^{\bullet -} + H^+ \quad (9.8)$$

$$O_2^{\bullet -} + O_3 \rightarrow O_3^{\bullet -} + O_2 \quad (9.9)$$

$$O_3^{\bullet -} + H^+ \rightarrow HO_3^{\bullet} \quad (9.10)$$

$$HO_3^{\bullet} \rightarrow {}^{\bullet}OH + O_2 \quad (9.11)$$

$$O_3 + {}^{\bullet}OH \leftrightarrow O_2 + HO_2^{\bullet} \quad (9.12)$$

The process is rapid and can treat organic pollutants at very low concentrations (ppb) at circumneutral pH values. The optimal O_3/H_2O_2 molar ratio is ~ 2:1 (Litter and Quici 2010).

Combined ozonation and use of H_2O_2 has been found to be effective for degradation of pesticides (Litter and Quici 2010). Coupled O_3/H_2O_2 treatment of microplastics has received limited study to date.

O_3/UV

UV irradiation can be combined with ozone for MP decomposition. Irradiation of ozone in water results in the formation of H_2O_2 (Augugliaro et al. 2006) (Eq. 9.13):

$$O_3 + H_2O \xrightarrow{h\upsilon} H_2O_2 + O_2 \quad (9.13)$$

Additional reactions occur simultaneously, which yield active radicals. Photolysis of the newly formed H_2O_2 by UV-C radiation produces two hydroxyl radicals per each molecule of hydrogen peroxide (Alpert et al. 2010):

$$H_2O_2 \xrightarrow{h\upsilon} 2{}^{\bullet}OH \quad (9.14)$$

The H_2O_2 also reacts with O_3:

$$O_3 + H_2O_2 \rightarrow {}^{\bullet}OH + O_2 + HO_2^{\bullet} \quad (9.15)$$

Finally, ozone reacts with hydroxyl radicals to form superoxide radicals:

$$O_3 + {}^{\bullet}OH \rightarrow O_2 + HO_2^{\bullet} \quad (9.16)$$

Ozone and Microplastics

Microplastics can be degraded via ozone oxidation (Zhang et al. 2020). Ozone reacts with synthetic polymers by forming radical species which further oxidize polymer chains. These radicals promote the overall plastic degradation process via formation of functional groups such as hydroxyl and carbonyl groups and modified C–H groups (Amelia et al. 2021).

A number of studies are available on the effects of conventional ozone treatment for MP removal; however, little has been reported about ozone AOPs for MP degradation. At a drinking water treatment plant in China, standard ozonation resulted in fragmentation of microplastics to finer-sized particles (Wu et al. 2022).

Various studies have determined changes to surface functional groups. Ozonation of PE MPs was responsible for the appearance of carbonyl bonds and a modest loss of weight of particles (Fitri et al. 2021). During ozone treatment of PS microspheres, sample surfaces contained a large number of carboxyl and hydroxy (mainly–COOH) groups (Yang et al. 2017).

Ozonation was found effective for degradation of nano-sized PS particles (Li et al. 2022). Particle surfaces were roughened, and oxygen-containing groups were generated on surfaces, which increased hydrophilicity of the plastics, contributing to further oxidative degradation to produce formic acid, phenol, acetophenone, hydroquinone, and other products. In some reported cases, MP destruction by ozone is relatively rapid and efficient. Chen et al. (2018) measured > 90% degradation of hydrolyzed polyacrylamide with 60 min of ozone treatment at 35–45°C.

When ozone was used for treatment of 1,4-*trans*-polychloroprene (Anachkov et al. 1993), degradation was highly efficient with respect to the quantity of ozone used. Hidayaturrahman and Lee (2019) studied wastewater treatment using ozone degradation, and destruction of microbeads, fibers, sheets, and fragments was quantified. Using an average dosage of 12.6 mg L^{-1} ozone with an irradiation time of about 1 min, it was determined that 90% of microplastics (microbeads, fibers, sheets, and fragments) were decomposed.

Although ozone is a highly reactive oxidant, certain disadvantages are associated with employing ozone for MP destruction. Ozone is a non-selective oxidizing agent and can therefore be wasted if other oxidizable substrates (e.g., natural organic matter) are present. Depending on the target microplastics for destruction (e.g., PVC, PS), oxidized by-products will be formed which may be more toxic than the original polymer compound.

THE FENTON REACTION AND FENTON-RELATED AOPS

Among numerous chemical oxidation techniques, Fenton is a well-accepted process with a long history of application. The conventional Fenton reaction involves use of an oxidizing agent (typically H_2O_2) and a catalyst (metal) to generate active oxygen species that oxidize organic and inorganic solutes. A number of metals are capable of activating H_2O_2 and producing hydroxyl ($^{\bullet}OH$) radicals in water; iron is most frequently used (Voelker and Sulzberger 1996; Park and Yoon 2007; Matilainen and Sillanpää 2010).

The $^{\bullet}OH$ radical has a high standard oxidation potential (2.80 V) and exhibits high reaction rates compared with other conventional oxidants such as hydrogen peroxide, ozone, oxygen, Cl_2, or $KMnO_4$ (table 9.2). The $^{\bullet}OH$ radical reacts with numerous organic and inorganic compounds with high-rate constants (Bautista et al. 2008).

The commonly recognized mechanism of forming reactive species in the Fenton process involves reaction of H_2O_2 with Fe(II) ions in water to generate hydroxyl radicals (Eq. 9.17) (Bautista et al. 2008).

$$Fe^{2+} + H_2O_2 \rightarrow Fe^{3+} + {}^{\bullet}OH + HO^{-} \quad (9.17)$$

Table 9.2 Oxidation Potentials of Common Oxidants

Oxidant	*Oxidation Potential (V)*
Fluorine (F_2)	2.87
Hydroxyl radical ($^{\bullet}OH$)	2.80
sulfate radical ($SO_4^{\bullet-}$)	2.60
Atomic oxygen (O)	2.42
Ozone (O_3)	2.07
Hydrogen peroxide (H_2O_2)	1.77
Hydroperoxyl ($HO_2^{\bullet}$)	1.7
Potassium permanganate ($KMnO_4$)	1.67
Chlorine dioxide (ClO_2)	1.4
Hypochlorous acid (HClO)	1.49
Chlorine (Cl_2)	1.36
Oxygen (O_2)	1.23
Bromine (Br_2)	1.09

$$Fe^{3+} + H_2O_2 \rightarrow Fe^{2+} + HO_2^{\bullet} + H^+ \tag{9.18}$$

The ferrous catalyst is also regenerated from reaction of Fe^{3+} with newly formed organic radicals (Eq. 9.19) (Bautista et al. 2008):

$$R^{\bullet} + Fe^{3+} \rightarrow R^+ + Fe^{2+} \tag{9.19}$$

In addition to Eqs. 9.17–9.19, however, other reactions occur (Eqs. 9.20–9.24), which may adversely affect the oxidation process (Pignatello et al. 2006):

$$Fe^{2+} + HO^{\bullet}_2 \rightarrow Fe^{3+} + HO_2 \tag{9.20}$$

$$Fe^{2+} + {}^{\bullet}OH \rightarrow Fe^{3+} + OH^- \tag{9.21}$$

$$Fe^{3+} + HO^{\bullet}_2 \rightarrow Fe^{2+} + O_2 + H^+ \tag{9.22}$$

Equations (9.20) through (9.22) are rate-limiting steps in the Fenton process, as hydrogen peroxide is consumed (Babuponnusami and Muthukumar 2014). Other reactions may occur, involving radicals and H_2O_2:

$${}^{\bullet}OH + H_2O_2 \rightarrow HO_2^{\bullet} + H_2O \tag{9.23}$$

$$HO_2^{\bullet} + HO_2^{\bullet} \rightarrow H_2O_2 + O_2 \tag{9.24}$$

When an organic compound reacts with excess ferrous ion at low pH, hydroxyl radicals can attach to unsaturated bonds of alkenes or alkynes, and to aromatic or heterocyclic rings (Eq. 9.25) (Neyens and Baeyens 2003).

H OH H OH OH COOH •OH •OH •OH COOH (9.25)

The •OH radicals can also abstract a hydrogen atom, initiating a radical chain oxidation (Eqs. 9.26–9.28) (Walling 1975; Lipczynska-Kochany et al. 1995; Neyens and Baeyens 2003).

$$RH + {}^{\bullet}OH \rightarrow H_2O + R^{\bullet} \tag{9.26}$$

$$R^{\bullet} + H_2O_2 \rightarrow ROH + {}^{\bullet}OH \tag{9.27}$$

$$R^{\bullet} + O_2 \rightarrow ROO^{\bullet} \tag{9.28}$$

The organic free radicals produced in eq. 9.26 may then be oxidized by Fe^{3+}, reduced by Fe^{2+}, or dimerized (Eqs. 9.29–9.31) (Tang and Tassos 1997; Neyens and Baeyens 2003).

$$R^{\bullet} + Fe^{3+} \rightarrow R^{+} + Fe^{2+} \tag{9.29}$$

$$R^{\bullet} + Fe^{2+} \rightarrow R^{-} + Fe^{3+} \tag{9.30}$$

$$R^{\bullet} + R^{\bullet} \rightarrow R\text{–}R \tag{9.31}$$

The efficiency of the Fenton oxidation process is a function of several variables including solution pH, hydrogen peroxide and catalyst concentrations, and temperature, all of which guide the reduction of Fe^{3+} to Fe^{2+}.

Maximum catalytic activity of the Fenton reaction occurs at pH 2.8–3.0 regardless of the target substrate; activity declines markedly as the pH exceeds this range. At pH > 3, Fe^{3+} begins to precipitate as $Fe(OH)_3$, which decomposes H_2O_2 to O_2 and H_2O (Bautista et al. 2008; Szpyrkowicz et al. 2001). Under these conditions, fewer hydroxyl radicals are generated due to the presence of less free Fe ions. Also at high pH, auto-decomposition of hydrogen peroxide is accelerated (Eq. 9.32).

$$2H_2O_2 \rightarrow O_2 + 2H_2O \tag{9.32}$$

At pH < 3, regeneration of Fe^{2+} via reaction of Fe^{3+} with H_2O_2 (Eq. 9.18) tends to be inhibited (Kritzer and Dinjus 2001). Iron complex species such as $[Fe(H_2O)_6]^{2+}$ are formed which react more slowly with hydrogen peroxide than do other species (Xu et al. 2009). Additionally, in the presence of high concentration of H^+ ions, peroxide becomes solvated to form the stable oxonium ion, $H_3O_2^+$. Oxonium ions make hydrogen peroxide more stable and reduce its reactivity with ferrous ions (Kavitha and Palanivelu 2005; Kwon et al. 1999).

The buffer solution used in the Fenton reaction plays a role in the degradation process. An acetic acid/acetate buffer offers maximum oxidation efficiency; in contrast, phosphate and sulfate buffers do not effectively support oxidation (Benitez et al. 2001; Babuponnusami and Muthukumar 2014).

The Fenton reaction is strongly affected by concentrations of both Fe^{2+} and H_2O_2. Rate of contaminant degradation increases with increase in ferrous ion concentration (Lin and Lo 1997). Some researchers have found, however, that beyond a certain Fe^{2+} concentration, the extent of increase is negligible (Lin et al. 1999; Kang and Hwang 2000; Rivas et al. 2001). In addition, excess ferrous ions will eventually contribute to increased total dissolved solids in the effluent stream. Thus, the optimum dose of ferrous ions for mineralizing organics, including MPs, must be determined empirically (Babuponnusami and Muthukumar 2014).

It is typically observed that degradation rate of the contaminant increases with increase in dosage of hydrogen peroxide (Lin et al. 1999; Eisenhauer 1964). However, if Fenton-treated wastewater is to be released into surface water, excess H_2O_2 will be detrimental to the biota and the local environment. Another drawback to the presence of excess hydrogen peroxide is the scavenging of newly generated hydroxyl radicals (see Eq. 9.23 above) (Babuponnusami and Muthukumar 2014). Despite the numerous experimental studies and industrial applications of the Fenton reaction, there is no agreement as to a universal optimum molar ratio of $[H_2O_2]/[Fe^{2+}]$ (Ruiz 2015). It seems intuitive that the dose of H_2O_2 be based upon the stoichiometric H_2O_2 to chemical COD ratio (Lücking et al. 1998; Bautista et al. 2008); however, the substantial variability in pollutant chemistry, as well as reaction conditions, require that the $[H_2O_2]/[Fe^{2+}]$ ratio be established based on laboratory-scale studies as a function of initial pollutant concentration (Bautista et al. 2008; Ruiz 2015).

The Fenton reaction is also regulated by the reaction temperature. According to the Arrhenius equation, increased temperatures accelerate the kinetics of the process; however, higher temperatures also promote decomposition of H_2O_2 to O_2 and H_2O. Based on empirical studies, ambient temperatures allow for good efficiency of decomposition (Lin and Lo 1997; Rivas et al. 2001). If the reaction temperature is expected to rise beyond 40°C due to exothermic conditions, cooling of the system is recommended (Babuponnusami and Muthukumar 2014).

The Fenton Reactor

Fenton oxidation treatment commonly employs the use of a stirred batch reactor (figure 9.2). The solution is acidified, often with dilute sulfuric acid, to pH range 3–3.5. Iron (as Fe^{2+}) is commonly added as ferrous sulfate and H_2O_2 is fed as a concentrated aqueous solution.

Reactants are often applied in the following sequence: wastewater followed by dilute sulfuric acid, the catalyst (Fe^{2+} salt) in acidic solution, and acid or base for pH adjustment. Hydrogen peroxide is added last. The Fenton reaction takes place at atmospheric pressure and room temperature; therefore, no energy input is necessary to activate hydrogen peroxide. The optimum contact time must be established for each specific effluent (e.g., microplastics). In industry practice, the length of the oxidation stage is around 60–120 minutes in most cases (Ruiz 2015). Corrosion in the reactor vessel can be substantial; therefore,

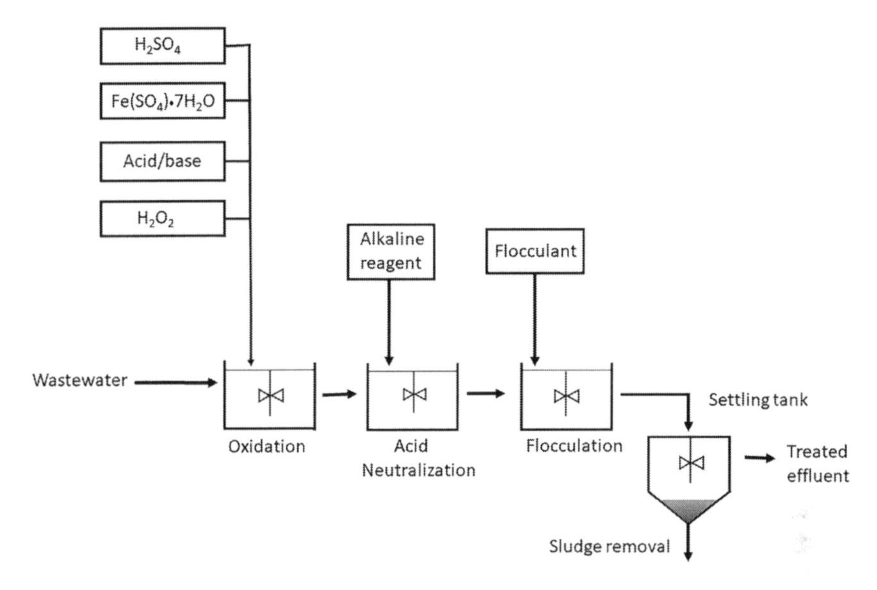

Figure 9.2 Schematic Showing the Standard Fenton Process
Source: Adapted from https://commons.wikimedia.org/wiki/File:Proceso_fenton.png.

the vessel must be coated with acid-resistant material such as glass, epoxy novolac, or polytetrafluoroethylene (PTFE) (Ozair and Al-Zahrany 2019).

The discharge from the Fenton reactor is sent to a neutralization tank where an alkaline reagent ($Ca[OH]_2$, CaO, $CaCO_3$, etc.) is added. This is followed by flocculant addition (e.g., polyaluminium chloride, various polymers) in a separate tank. The $Fe(OH)_3$ precipitate (essentially colloidal mud) and other solids are separated by settling. The effluent can be further treated using a sand-filtration stage (Bautista et al. 2008).

The conventional Fenton process offers advantages for destruction of a range of organic wastes: the necessary reagents are readily available and relatively safe to handle; the process is environmentally friendly; and reactions take place at ambient temperature and pressure.

Drawbacks exist to the standard Fenton process for organics decomposition, however. Hydrogen peroxide is known to auto-decompose to O_2 and H_2O (Eq. 9.32). Another drawback is that oxidants are wasted as a consequence of radical scavenging by hydrogen peroxide (Eq. 9.23).

The continuous loss of iron ions and formation of sludge as iron(III) compounds are other practical concerns in Fenton processes. Additional factors affecting its applicability for industrial (including microplasticsen-riched) wastewater include pH dependence, the high cost of H_2O_2, and the necessity of neutralization of the treated waste before disposal (Buthiyap-pan et al. 2016). In order to circumvent the above limitations, researchers have modified the conventional Fenton process to improve oxidation efficiency and reduce operating costs.

THE PHOTO-FENTON PROCESS

The Fenton reaction is enhanced by application of UV or UV-visible light. The photo-Fenton process utilizes irradiation with sunlight or an artificial light source, which leads to production of additional $^{\bullet}OH$ radicals, thus increasing the rate of organic contaminant degradation (Zepp et al. 1992; Fukushima et al. 2001; Moncayo-Lasso et al. 2009; Matilainen and Sillanpää 2010).

Complete mineralization of many organic compounds can be achieved: UV irradiation augments the decomposition process via photoreduction of Fe^{3+} to Fe^{2+} ions, which ultimately produces new $^{\bullet}OH$ (Eq. 9.33) (Navarro et al. 2010; Babuponnusami and Muthukumar 2014). Homogeneous photo-Fenton and photo-Fenton-like processes are effective for treating aromatic pollutants (Giroto et al. 2008; Lopez-Alvarez et al. 2012) and for treatment of dye-enriched wastewater, as they generate substantial $^{\bullet}OH$ (Elmorsi et al. 2010; Archina et al. 2015).

$$Fe(OH)^{2+} \xrightarrow{h\upsilon} Fe^{2+} + {}^{\bullet}OH \tag{9.33}$$

As is the case with standard Fenton processes, the efficiency of degradation with photo-Fenton is significantly influenced by solution pH and initial Fe^{2+} and H_2O_2 concentration. Additional factors for photo-Fenton include UV dosage and duration of contact of reagents. Optimum pH for the photo-Fenton process ranges from 2 to 5. At higher pH values, Fe^{2+} ions are unstable and convert to Fe^{3+} which form complexes with $^{\bullet}OH$ (Buthiyappan et al. 2016). Limitations of the homogeneous photo-Fenton process include processing of sludge and the narrow pH range of the treatment system (Pignatello et al. 2006; De la Cruz et al. 2012).

HOMOGENEOUS PHOTOELECTRO-FENTON PROCESS

The electro-Fenton process produces H_2O_2 electrochemically via oxygen reduction on the surface of a cathode. Further reactions occur between H_2O_2 and Fe^{2+}, as in the standard Fenton process (Matilainen and Sillanpää 2010).

Fenton and Microplastics

PVC was treated by Fenton reaction (Mackul'ak et al. 2015) with subsequent formation of *trans*-1,2-dichloroethene, *cis*-1,2-dichloroethene, trichloroethene, and tetrachloroethene. Benzene, ethylbenzene, and *o*-xylene were also identified and were believed to be products of phthalates decomposition. During photo-assisted Fenton degradation of PS microspheres mediated by broad-band UV irradiation (initial solution pH 2), IR peaks corresponded to aromatic C=C and C–O stretching of alcoholic/phenolic groups (Feng et al. 2011). These results indicate oxidative cleavage and dissolution of polymer chains of PS during the photo-assisted Fenton process. The standard Fenton process was not able to completely degrade the PS beads.

Fenton reaction with microplastics and other organic pollutants offers several advantages. No energy input is necessary to activate hydrogen peroxide, as the reaction takes place at ambient temperature and pressure. The Fenton method also requires relatively short reaction times and easy-to-handle reagents. Key disadvantages of the Fenton method are the high cost of hydrogen peroxide and the loss of the homogeneous catalyst (added as iron salt), thus requiring separation of the iron-rich mud. As shown above, a number of simple modifications to the Fenton process are available. Many are low in cost, and several prevent continuous loss of catalyst and the need of removing iron after treatment.

WET AIR OXIDATION

Wet air oxidation (WAO) technology relies upon high reaction temperatures (150–325°C) and pressures (0.0–20 MPa) in the presence of oxygen (Wakakura et al. no date; Sushma et al. 2018; Kolaczkowski et al. 1999). WAO technologies can be applied for conversion of toxic compounds to biodegradable intermediates, with the effluent subsequently treated by biological methods. In certain cases, WAO can result in complete mineralization of organic compounds to carbon dioxide and water. The latter option is costlier, as it requires more energy to bring all reactions to completion.

WAO of organic pollutants occurs via a free-radical chain-reaction mechanism. When the critical concentration of free radical is attained (i.e., the *induction period*), a fast reaction takes place (*propagation reaction*), and the contaminant is oxidized (Tungler et al. 2015).

In wet air oxidation, the reactions are thermally initiated (Rivas et al. 1998) (Eqs. 9.34–9.39):

$$RH \rightarrow R^{\bullet} + H^{\bullet} \tag{9.34}$$

$$RH + O_2 \rightarrow R^{\bullet} + HO_2^{\bullet} \tag{9.35}$$

$$R^{\bullet} + O_2 \rightarrow ROO^{\bullet} \quad (9.36)$$

$$ROO^{\bullet} + RH \rightarrow ROOH + R^{\bullet} \quad (9.37)$$

$$RH + (^{\bullet}OH, HO^{\bullet}, ROO^{\bullet}) \rightarrow R^{\bullet} + H_2O, H_2O_2, ROOH \quad (9.38)$$

$$H_2O_2 \rightarrow 2^{\bullet}OH \quad (9.39)$$

The length of the induction period is a function of oxygen concentration, temperature, type of organic compound, and catalyst concentration (Mishra et al. 1995; Rivas et al. 2001; Wu et al. 2003; Tungler et al. 2015). Solution pH also influences the induction period: it is shorter for pH values of ~4, and increases with the rise in pH (Sadana and Katzer 1974).

The WAO Reactor

A conventional wet oxidation system is a bubble column operating as a continuous process (figure 9.3). The WAO reactor is heated with N_2 under pressure. Steam or other supplemental energy is necessary for startup, and can provide trim heat if necessary. A rotary compressor and pump compress both oxygen and feed liquid to the optimum operating pressure. Once the desired temperature is attained, the pressurized, pre-heated O_2 and feed enter the chamber.

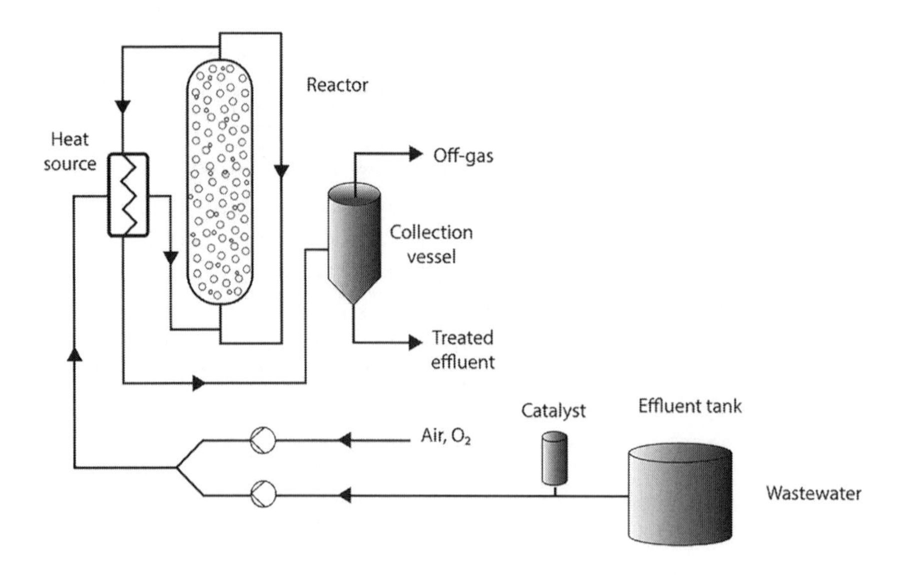

Figure 9.3 Schematic Diagram of the WAO Technology
Source: Adapted from Snipre, https://commons.wikimedia.org/wiki/File:Sch%C3%A9ma_OVH.svg.

Agitation is typically provided. Vaporized water is cooled in a condenser and returned to the reactor.

Heat exchangers recover energy from the reactor effluent; this heat is recycled for preheating the feed/air mixture entering the reactor. A separator is installed downgradient of the reactor and heat exchanger for removal of effluent and off-gases.

The oxidation reactions which take place are exothermic; therefore, sufficient energy may be generated within the reactor to permit the WAO system to operate without additional heat input at approximate COD concentrations $> 10,000\ mg \cdot L^{-1}$ (Tungler et al. 2015).

The residence time of the liquid phase in the reactor may range from 15 min to several hours at a temperature that permits the necessary oxidation reactions to proceed (Levec and Pintar 2007).

A key benefit of the WAO technology over other AOPs is that no hazardous oxidizing agents are necessary. Operating costs are moderate and pollutant discharge to the atmosphere is minimal. Drawbacks associated with WAO include high capital costs; in addition, requirements of high temperature and pressure result in safety concerns as well as possible damage to equipment (Kolaczkowski et al. 1999). During operation carboxylic acids may be generated, thus lowering pH. It may be necessary to install acid-resistant liners to prevent damage to equipment due to corrosion.

Catalytic wet oxidation (CWAO) overcomes the extreme operating conditions in WAO by incorporation of a catalyst. Catalysts include both homogeneous and heterogeneous types. Salts of iron, copper, and manganese are commonly used as homogenous catalysts in CWAO; however, many others are available. CWAO is an effective technology for treatment of highly concentrated, recalcitrant, toxic, and non-biodegradable organic pollutants in wastewater generated from various industries. The Ciba-Geigy process, developed in the late 1980s, employs a copper salt as the catalyst which is separated as copper sulfide and recycled back into the reactor (Luck 1999). Units are operated at 300°C and compressed air serves as the oxidant. The reactor is lined with titanium to withstand corrosion. Destruction efficiencies of 95–99% were documented for chemical and pharmaceutical wastewater (Luck 1996, 1999). A destruction efficiency of > 99.9% was determined for 2,8-dichlorodibenzo-*p*-dioxin. The system was not, however, capable of effectively removing ammonia from wastewater.

In the heterogeneous CWAO process, catalysts include noble metals such as platinum (Pt), palladium (Pd), ruthenium (Ru), and rhodium (Rh). The noble metals are highly effective for oxidation of organics; however, they are expensive and easily poisoned by halogens and sulfur group-containing compounds (Sushma et al. 2018). To overcome these obstacles, metal oxide catalysts such as copper, iron, manganese, and nickel oxides are used widely.

Heterogeneous catalysts have been employed for destruction of phenols, carboxylic acids, and nitrogenous compounds. This process has been used to treat municipal sewage sludge and industrial effluents such as halogenated wastewater, distillery waste, and effluents from pulp and paper mills, coke plants, and dye and textile industries (He et al. 2019; Sushma et al. 2018).

The Nippon Shokubai process was developed in the late 1980s. The heterogeneous catalysts are cylindrical pellets containing titania (TiO_2), the oxide of an element of the lanthanide series, and at least one other specialized metal (Shokubai 2018; He 2019). Combination of titania with a lanthanide oxide results in the formation of a physically stable catalyst exhibiting good oxidation activity and life span (Kolaczkowski et al. 1999). The reactor is a gas-liquid vertical monolith (Luck 1996). The Nippon Shokubai process operates within a temperature range of 160–270°C and pressure range of 0.9–8.0 MPa. The volume of oxygen is 1.0–1.5 times the theoretical oxygen demand. The process has wide applications in treatment of COD, sulfur compounds, nitrogen compounds, organic chlorides (e.g., chlorinated dibenzo dioxins) (Shokubai 2018), and others. Under typical operating conditions, destruction efficiency of > 99% was reported for organic compounds such as phenol, formaldehyde, acetic acid, and glucose (Luck 1996).

Given its successful record in degradation of numerous recalcitrant organics, CWAO holds promise for effective destruction of microplastics in water. Successful CWAO of microplastics depends on a range of parameters such as optimal reaction temperature, operating pressure, MP concentration, solution pH, and catalyst dosage.

CWAO technology has been tested for destruction of plastics of various resin types (Dubois et al. 1993; Dinsdale et al. 1999; Krisner et al. 2000; Sorensen and Bjerre 1992). WAO treatment of HDPE and PS pellets was evaluated with a Co/Bi complex as catalyst (Dubois et al. 1993). The reaction proceeded in two steps, i.e., a fast one during the first 10 min followed by a slow reaction over 4 hours. The fast reaction period, where long-chain polymers were broken down into smaller molecules such as carboxylic acids, was considered the most significant for establishing an effective oxidation rate. Oxidation rate increased slightly with initial oxygen pressure; however, temperature imparted a significantly greater effect. WAO of PE increased generation of acetic acid, which is considered a refractory intermediate of WAO (Dubois et al. 1993). Other intermediate products identified, at negligible concentrations, included acetophenone, benzaldehyde, hexanoic acid, and others. During WAO of a PE-PS mixture, a synergistic effect between both polymers occurred, with high COD reduction (99.9%) and low residual acetic acid concentration. Formation of acetic acid appeared to be inhibited by the catalyst (Dubois et al. 1993).

When wet oxidation of acrylonitrile butadiene styrene (ABS) combined with PP was studied (Wakakura et al. 1999), oxidation rate increased as a function of O_2 concentration. Air containing 50% oxygen was optimal for decomposing the ABS. Organic acids were generated but not identified; the yield of acids was controlled by adjusting pH and reaction temperature. The majority (80–90%) of nitrogen released from the ABS was in the form of NH_3; HCN was not detected. During wet air decomposition of powdered PVC (Sorensen and Bjerre 1992), products included Cl^- and CO_2, and several water-soluble compounds. A small quantity of H_2 gas was detected; its formation may have occurred via pyrolysis. At a temperature of 260°C the reaction rate increased to the point where, after 8 min, all PVC was converted to soluble products. Total liberation of Cl^- also occurred. Products of the WAO reaction were subjected to aerobic microbiological treatment. Based on decrease in COD over 12 d, the products obtained during WAO were nontoxic.

WAO of polyester at 300°C for 80 min resulted in 97% destruction of volatile solids, with 92% of the remaining total organic carbon in soluble form (Dinsdale et al. 1999). Volatile fatty acids included acetic acid (major component), maleic, glyoxylic, and benzoic acid. The soluble effluent from the WAO process had good biodegradability under both aerobic and anaerobic conditions. Krisner et al. (2000) studied WAO for decomposition of different plastics commonly encountered in municipal solid waste (PE, PP, PET, PS, and PVC). All were ground to a fine particle size. Above 270°C and with excess O_2, all plastics were destroyed within 1 h. Low concentrations of acetic and benzoic acids were formed. In order to attain 100% degradation, a second process such as biological finishing was suggested. Formation of chlorine and gaseous hydrochloric acid can be limited by adding $CaCO_3$ as a neutralizing agent.

CLOSING STATEMENTS

For treatment of wastewater containing persistent organic pollutants such as microplastics, destruction via conventional oxidation processes may not be completely effective and will incur substantial cost. Complete mineralization may require sequential application of more than one destruction technology such as pretreatment with a photochemical oxidation process or CWAO, followed by biological or electrochemical treatment of the more labile intermediates (Machulek et al. 2013; Fatta-Kassinos et al. 2010).

Advanced oxidation technologies for wastewater treatment vary widely. There is clear potential for the application of these aggressive technologies for microplastics degradation. Advantages and disadvantages of specific AOPs in treatment of organics in wastewater appears in table 9.3.

Table 9.3 Advantages and Disadvantages of Conventional AOPs

AOP type	*Advantages*	*Disadvantages*
O_3	O_3 is an excellent oxidizing agent. Short reaction times. Excess O_3 is readily decomposed. No residual sludge; does not increase volume of wastewater. Small space requirements; easy on-site installation.	O_3 must be generated electrically on-site due to its short lifetime. Pretreatment required. High energy consumption. O_3 solubility is affected by temperature change. Off-gas treatment system is necessary. High operating costs.
O_3/UV	More effective at generating $^{\bullet}OH$; more efficient than O_3 alone or UV alone.	High energy consumption. Sludge production. Not cost-effective. Turbidity interferes with penetration of light.
H_2O_2/UV	Commercially accessible; simple and economical process. Thermally stable. Reagents can be stored on-site.	Turbidity interferes with penetration of light.
O_3/H_2O_2	More efficient than O_3 or H_2O_2 alone. H_2O_2 is completely miscible with water. Good degree of flexibility, i.e., each component can be used individually or combined to oxidize different wastewater types. System can be applied at ambient atmospheric pressure and room temperature.	O_3 must be generated electrically on-site. High energy consumption. Not cost-effective.
O_3/H_2O_2/UV	Effective for degradation of complex compounds, e.g., aromatics and polyphenols.	Expensive.
Fenton-based	Rapid reaction rates. Generates strong $^{\bullet}OH$. Energy-efficient and cost-effective. Degradation of a wide range of refractory contaminants. Ferrous catalyst can be recycled by reduction of Fe^{3+}.	Highly acidic reactions pose hazards to workers and reactor vessel. Sludge production.
Electrochemical	Environmentally benign. On-site production of H_2O_2. High degradation efficiency. Relatively simple and versatile technology.	High energy consumption. Expensive.
Heterogeneous photocatalysis	Operable at ambient conditions; commercially available, inexpensive; nontoxic; photochemically stable. Certain photocatalysts take advantage of solar light. The technology can be readily scalable. Some photocatalysts (e.g. ZVI) are available in many forms and are easily handled. TiO_2 is of relatively low cost, insoluble in water, biologically and chemically inert.	Some photocatalysts are effective only under UV irradiation (more costly than using solar irradiation alone).

Source: Litter and Quici (2010), Karat (2013), Buthiyappan et al. (2016), Machulek et al. (2013).

REFERENCES

Alpert, S.M., Knappe, D.R., and Ducoste, J.J. (2010). Modeling the UV/hydrogen peroxide advanced oxidation process using computational fluid dynamics. *Water Research* 44(6):1797–1808.

Amelia, D., Fathul Karamah, E., Mahardika, M., Syafri, E., Mavinkere Rangappa, S., Siengchin, S., and Asrofi, F. (2021). Effect of advanced oxidation process for chemical structure changes of polyethylene microplastics. *Materials Today: Proceedings*. https://www.sciencedirect.com/science/article/pii/S2214785321069480.

Anachkov, M., Rakovsky, S., Stefanova, R., and Stoyanov, A. (1993). Ozone degradation of polychloroprene rubber in solution. *Polymer Degradation and Stability* 41(2):185–190.

Archina, B., Abdul Aziz, A.R., and Wan Mohd Ashri, W.D. (2015). Degradation performance and cost implication of UV-integrated advanced oxidation processes for wastewater treatments. *Reviews in Chemical Engineering* 31:263–302.

Augugliaro, V., Litter, M., Palmisano, L., and Soria, J. (2006). The combination of heterogeneous photocatalysis with chemical and physical operations: A tool for improving the photoprocess performance. *Journal of Photochemistry and Photobiology C* 7:127–144.

Babuponnusami, A., and Muthukumar, K. (2014). A review on Fenton and improvements to the Fenton process for wastewater treatment. *Journal of Environmental Chemical Engineering* 2(1):557–572.

Bautista, P., Mohedano, A., Casas, J., Zazo, J., and Rodriguez, J. (2008). An overview of the application of Fenton oxidation to industrial wastewaters treatment. *Journal of Chemical Technology & Biotechnology: International Research in Process, Environmental & Clean Technology* 83(10):1323–1338.

Benitez, F., Acero, J., Real, F., Rubio, F., and Leal, A. (2001). The role of hydroxyl radicals for the decomposition of p-hydroxy phenylacetic acid in aqueous solutions. *Water Research* 35(5):1338–1343.

Buthiyappan, A., Aziz, A.R.A., and Daud, W.M.A.W. (2015). Degradation performance and cost implication of UV-integrated advanced oxidation processes for wastewater treatments. *Reviews in Chemical Engineering* 31(3):263–302.

Buthiyappan, A., Aziz, A.R.A., and Daud, W.M.A.W. (2016). Recent advances and prospects of catalytic advanced oxidation process in treating textile effluents. *Reviews in Chemical Engineering* 32(1):1–47.

Chen, R., Qi, M., Zhang, G., and Yi, C. (2018). Comparative experiments on polymer degradation technique of produced water of polymer flooding oilfield. Paper presented at the IOP Conference Series: Earth and Environmental Science.

Chow, C.-F., Wong, W.-L., Chan, C.-S., Li, Y., Tang, Q., and Gong, C.-B. (2017). Breakdown of plastic waste into economically valuable carbon resources: Rapid and effective chemical treatment of polyvinylchloride with the Fenton catalyst. *Polymer Degradation and Stability* 146:34–41.

De la Cruz, N., Giménez, J., Esplugas, S., Grandjean, D., De Alencastro, L., and Pulgarin, C. (2012). Degradation of 32 emergent contaminants by UV and neutral photo-fenton in domestic wastewater effluent previously treated by activated sludge. *Water Research* 46(6):1947–1957.

Dinsdale, R.M., Almemark, M., Hawkes, F.R., and Hawkes, D.L. (1999). Composition and biodegradability of products of wet air oxidation of polyester. *Environmental Science & Technology* 33(22):4092–4095.

Dubois, M.A., Huard, T., and Massiani, C. (1993). Wet oxidation of non water-soluble polymers. *Environmental Technology* 14(2):195–200.

Duguet, J., Brodard, E., Dussert, B., and Mallevialle, J. (1985). Improvement in the effectiveness of ozonation of drinking water through the use of hydrogen peroxide. *Ozone: Science & Engineering* 3(7):241–258.

Eisenhauer, H.R. (1964). Oxidation of phenolic wastes. *Journal Water Pollution Control Federation*, pp. 1116–1128.

Elmorsi, T.M., Riyad, Y.M., Mohamed, Z.H., and Abd El Bary, H.M. (2010). Decolorization of Mordant red 73 azo dye in water using H_2O_2/UV and photo-Fenton treatment. *Journal of Hazardous Materials* 174(1–3):352–358.

Engberg, E., and Johansson, E. (2018). Ozonation of pharmaceutical residues in a wastewater treatment plant: Modeling the ozone demand based on a multivariate analysis of influential parameters. https://www.semanticscholar.org/paper/Ozonation-of-pharmaceutical-residues-in-a-treatment-Engberg-Johansson/104e7905d804d4bfb081531aa78cd1a2223c3ce0.

Fatta-Kassinos, D., Hapeshi, E., Malato, S., Mantzavinos, D., Rizzo, L., and Xekoukoulotakis, N.P. (2010). Removal of xenobiotic compounds from water and wastewater by advanced oxidation processes. In: *Xenobiotics in the Urban Water Cycle: Mass Flows, Environmental Processes, Mitigation and Treatment Strategies* (Vol. 16). Berlin/Heidelberg (Germany): Springer Science & Business Media, pp. 387–412.

Feng, H.-M., Zheng, J.-C., Lei, N.-Y., Yu, L., Kong, K. H.-K., Yu, H.-Q., Lau, T., and Lam, M.H. (2011). Photoassisted Fenton degradation of polystyrene. *Environmental Science & Technology* 45(2):744–750.

Fitri, A.N., Amelia, D., and Karamah, E.F. (2021). The effect of ozonation on the chemical structure of microplastics. *IOP Conference Series: Materials Science and Engineering* 1173:012017. IOP Publishing. DOI: 10.1088/1757-899X/1173/1/012017.

Fukushima, M., Tatsumi, K., and Nagao, S. (2001). Degradation characteristics of humic acid during photo-Fenton processes. *Environmental Science & Technology* 35(18):3683–3690.

Giroto, J., Teixeira, A., Nascimento, C., and Guardani, R. (2008). Photo-Fenton removal of water-soluble polymers. *Chemical Engineering and Processing: Process Intensification* 47(12):2361–2369.

Glaze, W.H., Beltran, F., Tuhkanen, T., and Kang, J.-W. (1992). Chemical models of advanced oxidation processes. *Water Quality Research Journal* 27(1):23–42.

Glaze, W.H., Kang, J.-W., and Chapin, D.H. (1987). The chemistry of water treatment processes involving ozone, hydrogen peroxide and ultraviolet radiation. *Ozone: Science & Engineering* 9(4):335–352.

Glaze, W.H., Peyton, G.R., Lin, S., Huang, R., and Burleson, J.L. (1982). Destruction of pollutants in water with ozone in combination with ultraviolet radiation. II. Natural trihalomethane precursors. *Environmental Science & Technology* 16(8):454–458.

Gogate, P.R., and Pandit, A.B. (2004). A review of imperative technologies for wastewater treatment I: Oxidation technologies at ambient conditions. *Advances in Environmental Research* 8:501–551.

Goi, A. (2005). *Advanced Oxidation Processes for Water Purification and Soil Remediation*. Tallinn (Estonia): Tallinn University Press.

Graça, C.A.L., Ribeirinho-Soares, S., Abreu-Silva, J., Ramos, I.I., Ribeiro, A.R., Castro-Silva, S.M., Segundo, M.A., Manaia, C.M., Nunes, O.C., and Silva, A.M.T. (2020). A pilot study combining ultrafiltration with ozonation for the treatment of secondary urban wastewater: Organic micropollutants, microbial load and biological effects. *Water* 12:3458.

Hassaan, M.A., and El Nemr, A. (2017). Advanced oxidation processes for textile wastewater treatment. *International Journal of Photochemistry and Photobiology* 2(3):85–93. DOI: 10.11648/j.ijpp.20170203.13.

He, S. (2019). Catalytic wet oxidation: Process and catalyst development and the application perspective. *Catalysis* 37–71.

Hidayaturrahman, H., and Lee, T.-G. (2019). A study on characteristics of microplastic in wastewater of South Korea: Identification, quantification, and fate of microplastics during treatment process. *Marine Pollution Bulletin* 146:696–702.

Kang, Y.W., and Hwang, K.-Y. (2000). Effects of reaction conditions on the oxidation efficiency in the Fenton process. *Water Research* 34(10):2786–2790.

Karat, I. (2013). Advanced oxidation processes for removal of COD from pulp and paper mill effluents: A technical, economical and environmental evaluation [*Master's Thesis*]. Stockholm (Sweden): Royal Institute of Technology.

Kavitha, V., and Palanivelu, K. (2005). Destruction of cresols by Fenton oxidation process. *Water Research* 39(13):3062–3072.

Kolaczkowski, S., Plucinski, P., Beltran, F., Rivas, F., and McLurgh, D. (1999). Wet air oxidation: A review of process technologies and aspects in reactor design. *Chemical Engineering Journal* 73(2):143–160.

Krisner, E., Ambrosio, M., and Massiani, C. (2000). Wet air oxidation of solid waste made of polymers. *Journal of Environmental Engineering* 126(3):289–292.

Kritzer, P., and Dinjus, E. (2001). An assessment of supercritical water oxidation (SCWO): Existing problems, possible solutions and new reactor concepts. *Chemical Engineering Journal* 83(3):207–214.

Kwon, B. G., Lee, D.S., Kang, N., and Yoon, J. (1999). Characteristics of *p*-chlorophenol oxidation by Fenton's reagent. *Water Research* 33(9):2110–2118.

Langlais, B., Reckhow, D.A., and Brink, D.R. (1991). Ozone in water treatment. *Application and Engineering*, 558.

Legrini, O., Oliveros, E., and Braun, A. (1993). Photochemical processes for water treatment. *Chemical Reviews* 93(2):671–698.

Levec, J., and Pintar, A. (2007). Catalytic wet-air oxidation processes: A review. *Catalysis Today* 124(3–4):172–184.

Li, Y., Li, Y., Ding, J., Song, Z., Yang, B., Zhang, C., and Guan, B. (2022). Degradation of nano-sized polystyrene plastics by ozonation or chlorination in drinking water disinfection processes. *Chemical Engineering Journal* 427:131690.

Lin, S.H., and Leu, H.G. (1999). Operating characteristics and kinetic studies of surfactant wastewater treatment by Fenton oxidation. *Water Research* 33(7):1735–1741.

Lin, S.H., and Lo, C.C. (1997). Fenton process for treatment of desizing wastewater. *Water Research* 31(8):2050–2056.

Lipczynska-Kochany, E., Sprah, G., and Harms, S. (1995). Influence of some groundwater and surface waters constituents on the degradation of 4-chlorophenol by the Fenton reaction. *Chemosphere* 30(1):9–20.

Litter, M., and Quici, N. (2010). Photochemical advanced oxidation processes for water and wastewater treatment. *Recent Patents on Engineering* 4(3):217–241.

Litter, M. I. (2005). Introduction to photochemical advanced oxidation processes for water treatment. In: *Environmental Photochemistry Part II*. Berlin/Heidelberg (Germany): Springer, pp. 325–366.

Lopez-Alvarez, B., Torres-Palma, R.A., Ferraro, F., and Peñuela, G. (2012). Solar photo-Fenton treatment of carbofuran: Analysis of mineralization, toxicity, and organic by-products. *Journal of Environmental Science and Health, Part A*. 47(13):2141–2150.

Luck, F. (1996). A review of industrial catalytic wet air oxidation processes. *Catalysis Today* 27(1–2):195–202.

Luck, F. (1999). Wet air oxidation: Past, present and future. *Catalysis Today* 53(1):81–91.

Lücking, F., Köser, H., Jank, M., and Ritter, A. (1998). Iron powder, graphite and activated carbon as catalysts for the oxidation of 4-chlorophenol with hydrogen peroxide in aqueous solution. *Water Research* 32(9):2607–2614.

Machulek, A., Jr., Oliveira, S.C., Osugi, M. E., Ferreira, V.S., Quina, F.H., Dantas, R.F., Oliveira, S.L., Casagrande, G.A., Anaissi, F.J., Silva, V.O., Cavalcante, R.P., Gozzi, F., Ramos, D.D., Rosa, A.P., Santos, A.P., de Castro, D.C., and Nogueira, A. (2013). Application of different advanced oxidation processes for the degradation of organic pollutants. In: *Organic Pollutants: Monitoring, Risk and Treatment* (Vol. 1). InTech, pp. 141–166.

Mackuľak, T., Takáčová, A., Gál, M., Marton, M., and Ryba, J. (2015). PVC degradation by Fenton reaction and biological decomposition. *Polymer Degradation and Stability* 120:226–231.

Matilainen, A., and Sillanpää, M. (2010). Removal of natural organic matter from drinking water by advanced oxidation processes. *Chemosphere* 80(4):351–365.
Martínez, F., Pariente, M. I., Melero, J. A., Botas, J. A. Catalytic wet peroxide oxidation process for the continuous treatment of polluted effluents on a pilot plant scale J. Adv. Oxid. Technol. 11(1) 65–74 (2008).
Mishra, V. (1995). Mahajani V.V., Joshi J.B. Wet air oxidation. *Industrial and Engineering Chemistry Research*. 34:2.
Moncayo-Lasso, A., Sanabria, J., Pulgarin, C., and Benítez, N. (2009). Simultaneous *E. coli* inactivation and NOM degradation in river water via photo-Fenton process at natural pH in solar CPC reactor. A new way for enhancing solar disinfection of natural water. *Chemosphere* 77(2):296–300.
Munter, R. (2001). Advanced oxidation processes–Current status and prospects. *Proceedings of the Estonian Academy of Science Chem* 50(2):59–80.
Navarro, R.R., Ichikawa, H., and Tatsumi, K. (2010). Ferrite formation from photo-Fenton treated wastewater. *Chemosphere* 80(4):404–409.
Neyens, E., and Baeyens, J. (2003). A review of classic Fenton's peroxidation as an advanced oxidation technique. *Journal of Hazardous Materials* 98(1–3):33–50.
Oller, I., Malato, S., and Sánchez-Pérez, J. (2011). Combination of advanced oxidation processes and biological treatments for wastewater decontamination—A review. *Science of the Total Environment* 409(20):4141–4166.
Ozair, G., and Al-Zahrany, S.A. (2019). Premature failure of unpromoted non-thixotroped cross-linked FRP tank from Fenton treatment system—A case study. The International Desalination Association World Congress on Desalination and Water Reuse 2019. Dubai, UAE.
Park, S., and Yoon, T.-I. (2007). The effects of iron species and mineral particles on advanced oxidation processes for the removal of humic acids. *Desalination* 208(1–3):181–191.
Parsons, S. (2004). *Advanced Oxidation Processes for Water and Wastewater Treatment*. London: IWA Publishing.
Pignatello, J.J., Oliveros, E., and MacKay, A. (2006). Advanced oxidation processes for organic contaminant destruction based on the Fenton reaction and related chemistry. *Critical Reviews in Environmental Science and Technology* 36(1):1–84.
Rivas, F., Kolaczkowski, S., Beltran, F., and McLurgh, D. (1998). Development of a model for the wet air oxidation of phenol based on a free radical mechanism. *Chemical Engineering Science* 53(14):2575–2586.
Rivas, F.J., Beltrán, F.J., Gimeno, O., and Acedo, B. (2001). Wet air oxidation of wastewater from olive oil mills. *Chemical Engineering & Technology: Industrial Chemistry - Plant Equipment - Process Engineering - Biotechnology* 24(4):415–421.
Ruiz, C.S. (2015). Fenton reactions (FS-TER-003). Technical Sheets for Effluent Treatment Plants. In: Textile Industry Fenton Reactions Series: Tertiary Treatments. Water and Environmental Engineering Group. Universidade de Coruña.
Sanada, A., and Katzer, J. (1974). Catalytic oxidation of phenol in aqueous solution. *Industrial and Engineering Chemistry* 13:127–134.

Shokubai.com. (2018). NS-LC process [accessed 2021 October 11]. https://www.shokubai.co.jp/en/products/environment/wastewater.html.

Siitonen, J.P. (2007). Paperitehtaan poistoveden otsonointi. (Ozonation of paper-mill wastewater.) Lappeenranta University of Technology, Faculty of Technology; Chemical Engineering. Lahti, Finland. Finnish.

Sillanpää, M., and Matilainen, A. (2015). NOM removal by advanced oxidation processes. In: *Natural Organic Matter in Water: Characterization and Treatment Methods*. Oxford: Butterworth-Heinemann, pp. 159–211.

Sørensen, E., and Bjerre, A.B. (1992). Combined wet oxidation and alkaline hydrolysis of polyvinylchloride. *Waste Management* 12(4):349–354.

Suez. (2021). SUEZ Degremont® Water Handbook [accessed 2021 November 17]. https://www.suezwaterhandbook.com/processes-and-technologies/oxidation-disinfection/oxidation-and-disinfection-using-ozone/selecting-ozonation-reactors.

Sushma, Kumari, M., and Saroha, A.K. (2018). Performance of various catalysts on treatment of refractory pollutants in industrial wastewater by catalytic wet air oxidation: A review. *Journal of Environmental Management* 228:169–188.

Szpyrkowicz, L., Juzzolino, C., and Kaul, S.N. (2001). A comparative study on oxidation of disperse dyes by electrochemical process, ozone, hypochlorite and Fenton reagent. *Water Research* 35(9):2129–2136.

Tang, W.Z., and Tassos, S. (1997). Oxidation kinetics and mechanisms of trihalomethanes by Fenton's reagent. *Water Research* 31(5):1117–1125.

Tarr, M.A. (2003). *Chemical Degradation Methods for Wastes and Pollutants: Environmental and Industrial Applications*. Boca Raton (FL): CRC Press.

Tungler, A., Szabados, E., and Hosseini, A.M. (2015). Wet air oxidation of aqueous wastes. *Wastewater Treatment Engineering*:153–178.

Voelker, B.M., and Sulzberger, B. (1996). Effects of fulvic acid on Fe (II) oxidation by hydrogen peroxide. *Environmental Science & Technology* 30(4):1106–1114.

Von Gunten, U. (2003). Ozonation of drinking water: Part I. Oxidation kinetics and product formation. *Water Research* 37(7):1443–1467.

Von Sonntag, C. (2008). Advanced oxidation processes: Mechanistic aspects. *Water Science and Technology* 58(5):1015–1021.

Wakakura, M., Adachi, F., Uchida, T., Okal, Y., and Ogawa, A.M.T. (1999). A Study on Treatment of Waste Fire Retardant Plastics using Wet Oxidation Method. General Topics on Recycling Processes [accessed 2022 February 21]. 1st International Symposium on Feedstock Recycling of Plastics, pp. 149–152. https://www.fsrj.org/act/7_nenkai/02-1-IFSR/ISFR.html.

Walling, C. (1975). Fenton's reagent revisited. *Accounts of Chemical Research* 8(4):125–131.

Wu, J., Zhang, Y., and Tang, Y. (2022). Fragmentation of microplastics in the drinking water treatment process – A case study in Yangtze River region, China. *Science of the Total Environment* 806(1):150545.

Wu, Q., Hu, X., and Yue, P.-l. (2003). Kinetics study on catalytic wet air oxidation of phenol. *Chemical Engineering Science* 58(3–6):923–928.

Xu, X.-R., Li, X.-Y., Li, X.-Z., and Li, H.-B. (2009). Degradation of melatonin by UV, UV/H_2O_2, Fe^{2+}/H_2O_2 and UV/Fe^{2+}/H_2O_2 processes. *Separation and Purification Technology* 68(2):261–266.

Yang, R., He, Q., Wang, C., and Sun, S. (2017). Surface modification of polystyrene microsphere using ozone treatment. Pages 130-135. Ferroelectrics Volume. 530, 2018--Issue 1: Proceedings of the Joint Symposium of the Eighth International Conference of the Chinese Society of Micro-Nano Technology and Microsystems and Nanoengineering Summit 2017 and Symposium of Materials Science and Technology 2017 (CSMNT2017 & MAN2017 & SOMST2017), Part IV.

Zepp, R.G., Faust, B.C., and Hoigne, J. (1992). Hydroxyl radical formation in aqueous reactions (pH 3-8) of iron (II) with hydrogen peroxide: the photo-Fenton reaction. *Environmental Science & Technology* 26(2):313–319.

Zhang, X., Chen, J., and Li, J. (2020). The removal of microplastics in the waste-water treatment process and their potential impact on anaerobic digestion due to pollutants association. *Chemosphere* 251:126360.

Chapter 10

Heterogeneous Photocatalytic Oxidation

INTRODUCTION

As mentioned in chapter 9, heterogeneous catalysts occur in a different phase from the reactants; catalysis reactions occur on external and internal surfaces of the catalyst at gas-solid or liquid-solid interfaces. Heterogeneous photocatalysis provides high degradation efficiency of a wide range of air and water pollutants. Recalcitrant dyes and toxic hydrocarbons such as PCBs, TCE, dichloromethane, and carbon tetrachloride have been decomposed by simple or complex heterogeneous photocatalysts to CO_2, H_2O, and other small constituents (Nguyen et al. 2018; Zhang et al. 2018; Zahran et al. 2014; Nakata and Fujishima 2012; Hsaio et al. 1983).

In photocatalytic technology, nanoscale metal oxide and metal oxide complex semiconductors may be employed for enhanced degradation of an organic contaminant. A semiconductor may be defined as a crystalline solid which is intermediate in electrical conductivity between a conductor and an insulator.

Heterogeneous photocatalytic technology has gained much popularity by virtue of its versatility, sustainability, and negligible environmental hazard (Shi et al. 2020; Wang et al. 2020; Maupin et al. 2012; Pitkäaho et al. 2013; Zhang et al. 2018). Photocatalysis offers benefits such as use of sunlight as a clean energy source and generation of many nontoxic by-products (Uheida et al. 2021).

PHOTOCATALYSIS ON MICROPLASTICS—MECHANISMS

Recent research has demonstrated that photocatalysis is a potentially viable, inexpensive, and energy-efficient method of MP elimination (Tofa et al. 2019a, 2019b; Ariza-Tarazona et al. 2020; Jiang et al. 2021). Semiconductors such as TiO_2, ZnO, Fe_2O_3, ZnS, and CdS have been evaluated for heterogeneous photocatalysis of MPs (Ibhadon and Fitzpatrick 2013). To date, application of semiconductors for MP degradation has been demonstrated at the laboratory and pilot scale only.

Effective semiconductors have a suitable band gap that, upon interaction with light, results in the generation of a suite of reactive species. When a semiconductor such as TiO_2 or ZnO is excited by light and when the energy is greater than the energy of the bandgap, separation of charge occurs in the form of free electrons. The electrons become excited and shift their position from the valence band to the conduction band. During interaction with light, several reactive species are formed (figure 10.1).

In response to photoexcitation, a "hole" is created in the valence band of the semiconductor (Eq. 10.1).

$$\text{semiconductor} \xrightarrow{h\nu} h^+_{VB} + e^-_{CB} \tag{10.1}$$

Both holes and free electrons react with O_2 and H_2O which occur in proximity to the semiconductor surface. Reactive oxygen species (ROS) including hydroxyl ($^\bullet OH$) and superoxide ($O_2^{\bullet -}$) radicals are produced. These ROS attack MPs and cause polymeric chain scission, branching, and cross-linking (figure 10.1) (Zhao et al. 2007). Changes to the polymer surface may appear as discoloration, pitting, crazing, cracking, erosion, and embrittlement (Cai et al.

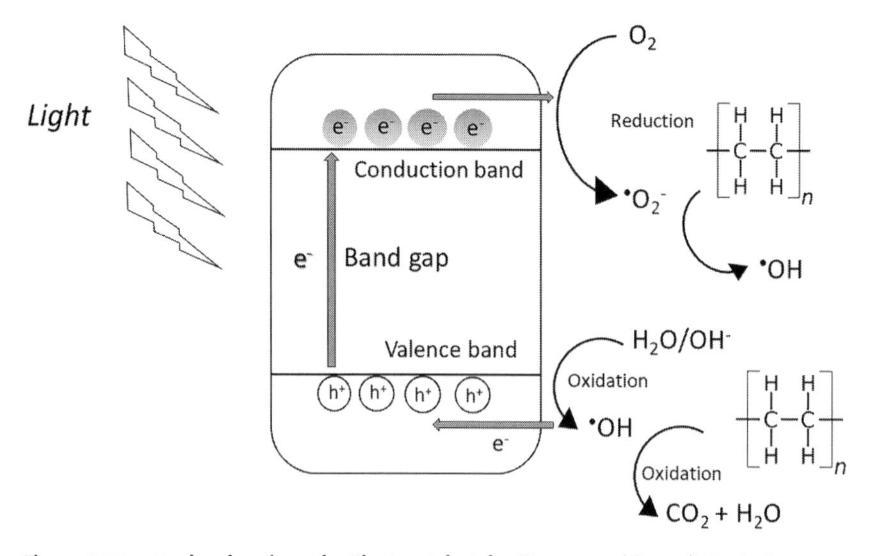

Figure 10.1 Mode of Action of a Photocatalyst for Decomposition of PE Chain

2018; Ranjan and Goel 2019). With time, the ROS migrate to deeper layers of the plastic and sustain the degradation process. In some scenarios, complete mineralization to CO_2 and H_2O has been detected (Zhao et al. 2007).

At least two possible mechanisms account for the initiation of photocatalysis of a polymer (Nabi et al. 2020). In one mechanism, the reaction is initiated at the conduction band, where the e^-_{CB} reacts with oxygen to form a superoxide radical (Eq. 10.2). The superoxide reacts with water to generate $^\bullet OOH$ (Eq. 10.3) which is converted to oxygen and hydrogen peroxide (Eq. 10.4). In the presence of light, the hydrogen peroxide is converted to the hydroxyl radical ($^\bullet OH$) (Eq. 10.5)

$$O_2 + e^-_{CB} \rightarrow O_2^{\bullet -} \tag{10.2}$$

$$O_2^{\bullet -} + H_2O \rightarrow {}^\bullet OOH + OH^- \tag{10.3}$$

$$2{}^\bullet OOH \rightarrow O_2 + H_2O_2 \tag{10.4}$$

$$H_2O_2 \xrightarrow{h\nu} 2{}^\bullet OH \tag{10.5}$$

In a second plausible mechanism, oxidation occurs at the valence band of the semiconductor.

The h^+_{VB} splits the water molecule, with the release of hydroxyl radicals (Eq. 10.6) (Kesselman et al. 1997).

$$h^+_{VB} + H_2O \rightarrow {}^\bullet OH + H^+ \tag{10.6}$$

Hydroxyl radicals ($^\bullet OH$) are electrophilic in nature and are among the most active photocatalytic oxidants. This ROS oxidizes electron-rich organic compounds non-selectively (Shang et al. 2003a, 2003b). If the polymer to be treated has a carbon-carbon backbone (for example, PE) the $^\bullet OH$ may break C–H bonds along the chain to generate polyethylene alkyl radicals ($-{}^\bullet CHCH_2-$) (Eq. 10.7) (Pham et al. 2021).

$$-(CH_2CH_2)- + {}^\bullet OH \rightarrow -({}^\bullet CHCH_2)- + H_2O \tag{10.7}$$

Subsequent reactions on the polymer chain may proceed as shown in Eqs. 10.8–10.11 (Zhao et al. 2007):

$$-({}^\bullet CHCH_2)- + O_2 \rightarrow -(CH({}^\bullet OO)CH_2)- \tag{10.8}$$

$$-(CH(^{\bullet}OO)CH_2)- + -(CH_2CH_2)- \rightarrow -(CH(OOH)CH_2)- + -(CH_2{}^{\bullet}CH)- \quad (10.9)$$

$$-(CH(OOH)CH_2)- \xrightarrow{h\nu} -(CH^{\bullet}OCH_2)- + {}^{\bullet}OH \quad (10.10)$$

$$-(CH^{\bullet}OCH_2)- \rightarrow -CHO + {}^{\bullet}CH_2CH_2- \quad (10.11)$$

The intermediates are further oxidized to CO_2 and H_2O via the action of the ROS. The overall process is shown in figure 10.1.

During photodegradation, so-called environmentally persistent free radicals (EPFRs) may also be generated, which oxidize the polymer. EPFRs can generate ROS such as $^{\bullet}OH$ which attack the polymer chain, ultimately generating monomers and various oligomers (Zhu et al. 2019). The types of EPFRs generated are a function of the chemical structure of the affected MPs (Hu et al. 2021).

End-products of MP photodecomposition vary widely and are a function of the specific polymer, type of semiconductor used, and reaction conditions (Tian et al. 2019; Nabi et al. 2020). Common products include CO_2 and soluble organic compounds, both nontoxic and some having varying levels of toxicity. When a PE-TiO_2 composite was subjected to UV irradiation for 100 h, CO_2 was the primary product, accounting for 98% of carbon release (Zhao et al. 2007). No other volatile products were detected by gas chromatography. Photocatalytic degradation of PP microplastics (MPs) by ZnO nanorods using visible light irradiation resulted in a range of by-products including formaldehyde, acetaldehyde, acetone, butanal, and several organic radicals (Uheida et al. 2020).

PHOTOCATALYSIS ON MICROPLASTICS—APPLICATIONS

Light-activated nanoparticles in aqueous suspension have been examined for MP decomposition, and both UV and visible light are employed. In other cases, nanocomposite films containing TiO_2, ZnO, Pt-ZnO, TiO_2, C-TiO_2, and C,N-TiO_2 were studied (Ariza-Tarazona et al. 2020; Tofa et al. 2019a, 2019b; Nabi et al. 2020; Uheida et al. 2021). Research has addressed the effects of irradiation (UV versus visible), duration of exposure, pH of suspension, and other variables (table 10.1). TiO_2 and ZnO are the most popular semiconductors examined for plastics photodegradation.

Table 10.1 Reactions of Selected Photocatalysts with Microplastics

Photocatalyst	*Photocatalyst Modification*	*Microplastic Type*	*Irradiation*	*Duration of Exposure (h)*	*Results*	*Reference*
TiO_2	Three TiO_2 photocatalysts synthesized by anodization.	PS nanoparticles	UV	50	A mixed structure of nanotubes and nanograss reduced the concentration of PS two-times more efficiently than photolysis with UV alone.	Domínguez-Jaimes et al. 2021
		PE film	Visible	4	Photocatalytic reaction affected the glass transition temperature and stiffness of the film material.	Luo et al. 2021
	Ag/TiO_2 modified using reduced graphene oxide	PE	UV	1, 2, 3, 4	3% Ag/TiO_2-1% RGO catalyst resulted in 76% degradation compared with 68 and 56% for Ag/TiO_2 and pure TiO_2, respectively.	Fadli et al. 2021
	C,N-TiO_2 powder	HDPE	Visible	50	Formation of free $OH^{\bullet}$ through pathways involving the photogenerated electrons plays an essential role in MP degradation.	Vital-Grappin et al. 2021
	TiO_2 films	PE MPs and PS microspheres	UV	12, 36	TiO_2 film caused near-complete mineralization (98.4%) of PS microspheres (400 nm) within 12 hours. High PE degradation after 36 h.	Nabi et al. 2020
	TiO_2– P25/β-silicon carbide alveolar foams	PS and polymethyl-methacrylate nanoparticles	UV-A	7	50% TOC conversion measured after 7 hours of treatment of PMMA. Degradation more rapid at lower flow rate and low pH (4–6). 140-nm PS particles decomposed faster than did 508-nm particles.	Allé et al. 2021

(continued)

Table 10.1 (Continued)

Photocatalyst	*Photocatalyst Modification*	*Microplastic Type*	*Irradiation*	*Duration of Exposure (h)*	*Results*	*Reference*
	Ag/TiO_2 nano-composites	PE	Visible, UV	2	Inclusion of the Ag dopant increased efficiency of microplastic decomposition. Band-gap energy of TiO_2 decreased when Ag was deposited on TiO_2.	Maulana et al. 2021
	None	Polyamide 66 microfibers	UV	105	UV-C was more effective than UV-A for degradation of microfibers (97% versus 6% mass loss, respectively).	Lee et al. 2020
	Mesoporous N-TiO_2 coating	HDPE microspheres and LDPE microplastics films	Visible	48	Less degradation occurred with microplastic films. Degradation of LDPE and HDPE controlled by a combination of variables.	Llorente-Garcia et al. 2020
	Protein-derived C,N-TiO_2	HDPE	Visible	20	Low pH facilitated plastic degradation. Low temperature caused fragmentation of MPs, thus increasing interactions with the C,N-TiO_2.	Ariza-Tarazona et al. 2020
	Au@Ni@TiO_2 micromotors	PS	UV	--	The "pushing" of the chained assemblies was effective for microplastics removal and was independent of the fuel (water or dilute peroxide solution).	Wang et al. 2019
	None	PS	UV	48	Mineralization of PS was greater in water (17%) compared to air (6%).	Tian et al. 2019
	Green synthesis and sol-gel route	HDPE	Visible	20	Low pH and low temperature imparted a combined effect on MP degradation.	Ariza-Tarazona et al. 2019

	TiO_2 NPs, and TiO_2-rGO nanocomposite	PP film	Sunlight	9 h daily	More effective photodegradation of polypropylene by TiO_2-rGO nanocomposite based on carbonyl index.	Verma et al. 2017
	Titania nanotubes Malachite green used to sensitize titania monoliths	LDPE film	Visible, UV	1080	Sensitization of nanostructures by dye increased degradation of LDPE under visible light. Over 45 days, PE films with 10% dye-sensitized titania nanotubes experienced 50% degradation under visible light	Ali et al. 2016
	polypyrrole/TiO_2 (PPy/TiO_2) nanocomposite	PE	Sunlight	240	Weight decreased 35.4%; MW decreased 54.4%.	Li et al. 2010
	Copper phthalocyanine-sensitized TiO_2	PS	Visible, UV	250	PS-(TiO_2/CuPc) showed better photocatalytic degradation than PS-TiO_2; higher charge separation efficiency of the latter photocatalyst resulted in more ROS generation. Little formation of toxic by-products.	Shang et al. 2003b
	TiO_2 in suspension, TiO_2 in PVC film	PVC	Visible, UV	720	TiO_2-blended PVC film photodegraded more rapidly than did the PVC particle/TiO_2 dispersions	Horikoshi et al. 1998
		PE, PP	Visible	200	Weight loss less than 5% in both plastics.	Ohtani et al. 1989
ZnO	ZnO NPs decorated with Fe	LDPE	Solar	120	Photocatalytic efficiency of the LDPE and Fe-ZnO film was greater compared to pure LDPE and LDPE/un-doped ZnO as shown by weight loss data.	Lam et al. 2021

(continued)

Table 10.1 (Continued)

Photocatalyst	*Photocatalyst Modification*	*Microplastic Type*	*Irradiation*	*Duration of Exposure (h)*	*Results*	*Reference*
	None	PP	UV	6	Catalyst loading, reaction temperature, and microplastic size must be addressed for efficient microplastic degradation.	Razali et al. 2020
	ZnO nanorods	PP	Visible	50	Average PP particle volume reduced by ~ 65% in comparison with fresh PP particles. Generation of formaldehyde, acetaldehyde, acetone, butanal, and various organic radicals occurred.	Uheida et al. 2021
	Pt nanoparticles deposited on ZnO nanorods and coated onto glass fibers	LDPE	Visible	175	The ZnO-Pt catalysts showed approximately 13% higher oxidation of LDPE film compared to ZnO nanorods.	Tofa et al. 2019a
	None	LDPE	Visible	175	Hydroxyl and superoxide radicals generated by ZnO initiated degradation at weak spots (e.g., chromophoric groups and defects) of the polymer to produce low-molecular-weight alkyl radicals.	Tofa et al. 2019b
Other photocatalyst types						
Au nanoparticles	-	LDPE	Solar	240	LDPE film with 1.0% Au NPs experienced 90.8% degradation efficiency.	Olajire and Mohammed 2021

MXene/ $Zn_xCd_{1-x}S$ photocatalyst	-	PET	Visible	2	PET was converted to glycolate, acetate, ethanol, and others. H_2 evolution rate was 14.17 $mmol \cdot g^{-1} \cdot h^{-1}$ in alkaline solution.	Cao et al. 2022
Cu_2O/CuO	-	PS	Visible	50	Photocatalysis promoted polymer chain scission; reduced concentration of PS up to 23% (six times greater reduction than achieved by photolysis).	Acuna-Bedoya et al. 2021
$NiAl_2O_4$	Spinels produced via both hydrothermal and precipitation methods	LDPE	Visible	5	Greatest degradation of PE by spinels prepared by hydrothermal method.	Venkataramana et al. 2021
NiO	Green synthesis using *Ananas comosus*	LDPE	Solar	240	Polymeric nanocomposites experienced 33% weight loss compared with pure LDPE (8.6%).	Olajire et al. 2021
BiOCl	Hydroxy-enriched ultrathin BiOCl	Microspheres of HDPE, Nylon-66, POM and PP; recycled HDPE	Visible	5	Surface hydroxy groups were important in enhancing photocatalytic decomposition of microplastics via the ready generation of •OH.	Jiang et al. 2021
Nb_2O_5	Atomic layers	PE, PP, PVC	Visible	40, 60, 90	PE degradation: 100% in 40 h; PP: 100% in 60 h; PVC: 100% in 90 h	Jiao et al. 2020
$CN_x\|Ni_2P$		PET, PLA, polyester microfibers	Visible	50	Photoreforming of polymers to H_2.	Uekert et al. 2019

(*continued*)

Table 10.1 (Continued)

Photocatalyst	*Photocatalyst Modification*	*Microplastic Type*	*Irradiation*	*Duration of Exposure (h)*	*Results*	*Reference*
CdS/CdO_x quantum dots		PE, PVC, PLA, PET, PUR	Visible	22	Photoreforming of polymers to H_2.	Uekert et al. 2018
TiO_2 and ZnO		PVC and polyvinylidene chloride copolymer	UV and sunlight	240-720	Photodegradation enhanced in TiO_2/water media by added oxidants such as hydrogen peroxide and potassium persulfate.	Hidaka et al. 1996

HDPE—high-density polyethylene; LDPE—low-density polyethylene; PE—polyethylene; PET—polyethylene terephthalate; PLA—polylactic acid; PMMA—polymethyl-methacrylate; POM—polyoxymethylene; PP—polypropylene; PS—polystyrene; PUR—polyurethane; PVC—polyvinyl chloride; RGO—reduced graphene oxide.

Titanium Dioxide

Titanium dioxide (TiO_2) is considered a highly effective photocatalyst by virtue of its high reactivity, good photostability, nontoxicity, and moderate cost (Zhao et al. 2007). HDPE microplastics occurring in facial scrubs were reacted with nano-TiO_2 semiconductors in water (Ariza-Tarazona et al. 2019). The nano-TiO_2 had a modest capacity (<10%) for MP destruction using visible light irradiation. It was found that low pH and low temperature imparted a combined effect on MP degradation—acidic conditions promoted the incorporation of H^+ ions into the surface, which enhanced plastic degradation and resulted in greater interaction of nanoparticles with MPs. Low temperatures caused fragmentation of particle surfaces, thus increasing surface area and interaction with TiO_2 nanoparticles (Ariza-Tarazona et al. 2019).

Photocatalytic decomposition of PE MPs and PS microspheres (400 nm) was evaluated using TiO_2 NP films under UV irradiation (Nabi et al. 2020). The TiO_2 film caused near-complete mineralization (98.4%) of PS microspheres within 12 h. PE MPs also underwent substantial photodegradation after 36 h. Intermediate products included hydroxyl, carbonyl, and carbon-hydrogen groups; CO_2 was the main end-product.

Degradation of ^{14}C-PS nanoplastic in both air and water was investigated by Tian et al. (2019) using UV light. Short-term photooxidation apparently took place only on shallow surfaces of the nanoplastics. X-ray photoelectron spectroscopy revealed that C——O groups formed on surfaces after 48 h of UV exposure. UV irradiation in air caused a significant increase in MW of PS molecules, possibly due to cross-linking of chains. In water, mineralization was significantly greater (17%) compared to that in air (3%). Several low-MW products, having condensed aromatic rings with side chains containing –OH and C=O groups or oxidized monomers with single benzene rings were detected. The hydrophilic products likely undergo further degradation or mineralization (Tian et al. 2019). There are concerns that transformation products leached from PS nano- and microplastics under photodegradation may pose health and environmental hazards.

Heterogeneous photocatalysis using UV-A irradiation of TiO_2–P25/β-silicon carbide alveolar foams was tested in a flow-through photoreactor at pH 6.3 on PS and PMMA particles in wastewater (Allé et al. 2020). A 50% TOC conversion was measured after 7 h of treatment of PMMA. Degradation was more rapid at lower flow rate and low pH (4–6). Particle size also affected the reaction rate: 140 nm PS particles decomposed faster than did 508 nm particles (Allé et al. 2020).

Photocatalytic degradation of HDPE microspheres and LDPE microplastic films was investigated using visible light and mesoporous nano-TiO_2 coating (Llorente-Garcia et al. 2020). Degradation rate was greater with smaller microplastics, and less degradation occurred with microplastic films. Degradation of MPs was influenced by a combination of factors including microplastic morphology and surface-to-volume ratio of particles (Llorente-García et al. 2020). These researchers suggest that photocatalysis of MPs in WWTPs is feasible; however, certain variables such as MP particle size and shape must be addressed for optimal design of a rapid and effective process.

Polyamide 66 microfibers were degraded partially by photocatalysis using different combinations of TiO_2 and UV irradiation (Lee et al. 2020). Extent of degradation was a function of UV wavelength. UV-C was more effective than UV-A for degradation: 97% of mass loss occurred within 48 h compared to 6% mass loss with UV-A. Concentration of TiO_2 was also important in the degradation rate of microfibers; the half-life of fibers decreased from 267 (concentration 20,000 $mg \cdot L^{-1}$ TiO_2) to 10 h (100 $mg \cdot L^{-1}$ TiO_2).

The TiO_2 semiconductor catalyst can generate substantial oxidizing agents for degradation of microplastics. However, a high recombination rate of holes and electrons occurs. In order to augment the photocatalytic capability of TiO_2 for decomposition of PE microplastics, TiO_2 was modified with Ag dopant (Fadli et al. 2021). Degradation rate of PE microplastics using two different Ag/TiO_2 nanocomposites under UV radiation was 76% and 68%, compared with 56% for pure TiO_2. When Ag/TiO_2 nanocomposites were synthesized by Maulana et al. (2021), the band gap energy of TiO_2 decreased. Efficiency of microplastic decomposition increased: degradation rate was 100% at an initial PE concentration of 100 mg/L within 120 min of irradiation. The most rapid degradation occurred with particle size range 125–150 μm, where 100% removal was noted under UV irradiation at 90 min (Maulana et al. 2021).

It is possible, in photocatalytic technology, that contact between the catalyst and microplastic particle may not be optimal. Low adsorption of plastics on the photocatalyst surface may result in low photocatalytic degradability of plastics (Sanwald et al. 2016; Zhang et al. 2016). Some researchers state that physical deposition and constant stirring are expensive and/or not satisfactory for adequate reaction with fine plastic pieces (Beladi-Mousavi et al. 2021). Self-motile photocatalysts, that is, microrobots, have been introduced which create a stirring effect with controlled remote motion which improves contact with microplastics. Autonomous light-driven microrobots are highly energy-efficient as they can be activated and navigated via a light source (e.g., the sun) and harvest "fuel" (e.g., water) from their surroundings (Dong et al. 2017; Beladi-Mousavi et al. 2021).

Zinc Oxide

ZnO has been shown effective for degradation of commercial microplastics by virtue of visible light absorption capacity and significant electron movement. In addition, as Zn is an essential element for organisms, there is minimal hazard to the local environment (Uheida et al. 2021).

When LDPE was reacted with visible light-excited ZnO photocatalysts, low-MW compounds such as peroxides, carbonyl groups, and hydroperoxides were formed (Tofa et al. 2019b). LDPE experienced increased brittleness along with production of cracks, cavities, and wrinkles. This study revealed that catalyst surface area was a key factor in increasing LDPE degradation (Tofa et al. 2019b). PP was treated with ZnO nanoparticles (<50 nm) activated by UV light (Razali et al. 2020). Hydroxyl radicals or superoxides were photogenerated, which rapidly oxidized polar functional groups on the polymer. Rate of PP photodegradation increased with increased reaction temperature (~50°C), which enhanced the fragmentation of microplastics. Extent of PP degradation was accelerated when treatment was conducted on small microplastics (<25 mm^2) (Razali et al. 2020).

Effectiveness of microplastic photodegradation by ZnO may be influenced by a number of variables including MP particle size, type of irradiation (UV versus visible), reaction temperature, and nanoparticle (semiconductor) shape. ZnO nanorods were irradiated by visible light in a flow-through system for degradation of PP microspheres suspended in water (Uheida et al. 2021). Hexagonal ZnO nanorods have high surface area and stability, leading to highly efficient photooxidation. Photocatalysis by ZnO reduced average PP particle volume by ~ 65% over two weeks. These results are encouraging for successful implementation of photocatalytic reactors for microplastics removal prior to water consumption or release into the environment (Uheida et al. 2021). The flow-through reactor holds potential for use in large-scale water and wastewater treatment systems; photocatalytic reactor efficiency can be scaled up by expanding the size of the photoreactor panel.

The catalytic activity of the photocatalyst can be enhanced by decorating it with selected metals, nonmetals, and biomolecules. Such modification improves optical characteristics and electron-hole pair separation, which are essential for intensifying photocatalysis (figure 10.2) (Lam et al. 2021; Varma et al. 2020; Bora and Dutta 2019).

When platinum is deposited on nano-ZnO, overall photocatalytic efficiency is augmented. The effect is associated with reduced electron-hole pair recombination and improved visible light absorption. Tofa et al. (2019a) dispersed Pt nanoparticles on ZnO nanorods for treatment of LDPE films. A 78% enhancement in absorption in the visible region was determined. Photocatalytic activity occurred primarily due to the action of superoxides and hydroxyl radicals formed during photocatalysis of LDPE. These reactive groups created wrinkles, cracks, and holes, which stimulated further photocatalytic degradation of the plastic. The ZnO-Pt catalysts exhibited ~13%

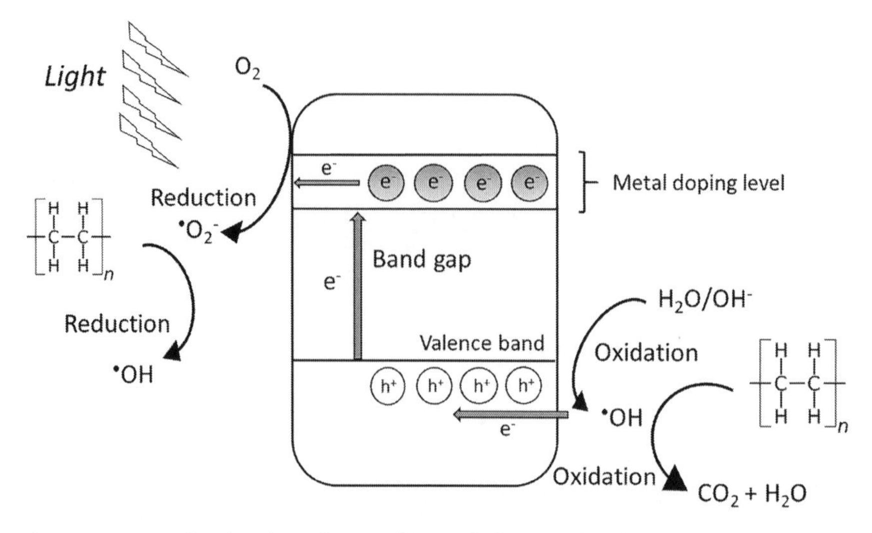

Figure 10.2 Mode of Action of a Metal-Doped Photocatalyst for Decomposition of PE Chain Showing a Reduction of Band Gap Value

Source: Adapted from Varma et al. (2020). Https://creativecommons.org/licenses/by-nc-nd/4.0.

greater potential for oxidation of LDPE compared to ZnO nanorods alone (Tofa et al. 2019a).

Other dopants improve the efficiency of ZnO, for example transition metal ions (Cu, Ni, Fe, Co). Among transition metals, iron is appealing due to its *d*-orbital electron which can overlap with the ZnO valence band. ZnO nanoparticles were decorated with Fe, and a LDPE/Fe-ZnO composite film was irradiated by sunlight (Lam et al. 2021). The modified photocatalyst possessed improved magnetic and optical properties. In addition, particle size decreased and number of catalytically active sites increased. A reduction of band gap value was also exhibited after doping with Fe. Based on weight loss data, the photocatalytic efficiency of the LDPE/Fe-ZnO film was greater than that for pure LDPE and LDPE/un-doped ZnO. This was a result of increased optical absorption and the effective separation of photogenerated charge carriers. FTIR analysis revealed the presence of carbonyl groups as LDPE degradation products.

Other Catalysts

Hydroxy-rich ultrathin BiOCl (BiOCl-X) was evaluated for photodecomposition of microspheres of HDPE (200–250 μm), Nylon-66 (2.38 mm), POM (3 mm), PP (various colors, mm range), and recycled HDPE (4 mm) (Jiang et al. 2021). BiOCl-X had an excellent capacity for photocatalytic decomposition:

the mass loss of the microplastics was 24 times greater than that by BiOCl nanosheets. The surface hydroxyl groups on BiOCl promoted the production of hydroxyl radicals, which improved degradation efficiency. Rate of photodegradation was more rapid for smaller-sized microplastics. Certain plastics experienced only minor photodegradation with this catalyst. The authors suggest that this effect may be due to the presence of antioxidants in one polymer, and another polymer originating from recycled stock. This study confirmed the importance of surface hydroxy groups in enhancing the photocatalytic decomposition of microplastics via the active generation of •OH.

$NiAl_2O_4$ spinels were evaluated for photocatalytic degradation of LDPE under visible light (Venkataramana et al. 2021). A PE sheet was partially decomposed using two different forms of spinels after 5 h. Spinels prepared by a hydrothermal method caused 12.5% PE degradation (Venkataramana et al. 2021).

Some researchers have proposed that products of polymer photodegradation be used as fuels. Photoreforming technology is a low-energy strategy where sunlight and a photocatalyst generate H_2 from an organic substrate and water at ambient temperature and pressure. The organic material (microplastic) serves as electron donor which is oxidized and decomposed to simpler organic molecules by the oxidative holes of the excited photocatalyst. In this process, the •OH radical plays only a minor role (Uekert et al. 2019). The photogenerated electrons are transferred from the photocatalyst to a co-catalyst which reduces H_2O to H_2 (Kawai and Sakata 1981; Uekert et al. 2019; Hu et al. 2021). The appropriate photocatalyst should possess a valance band sufficiently high to oxidize water or organic compounds, along with a conduction band capable of generating H_2 (Kawai and Sakata 1981). Plastics photoreforming was first studied by Kawai and Sakata (1981) who used Pt-deposited TiO_2 to transform PVC under UV irradiation. Mass analysis revealed the production of H_2 and CO_2 in the gas phase. Gaseous Cl_2 was not detected; however, the presence of the Cl^- ion was confirmed.

The use of an expensive catalyst (Pt) and reliance on UV irradiation are considered by some researchers to be drawbacks to the Kawai and Sakata (1981) method. CdS/CdO_x quantum dots were evaluated for photoreforming PLA, PET, and PUR in alkaline solution (Uekert et al. 2018). The process relied upon visible light and was operated at ambient temperature and pressure. The reactions generated pure H_2 and transformed the plastics to high-value organic products such as formate, acetate, and pyruvate. Overall conversion was $< 40\%$ for the three polymers. A key drawback to this process is that Cd is highly toxic; therefore, its utilization in a large-scale water treatment facility must be carefully managed and monitored. Uekert et al. (2019) synthesized carbon nitride/nickel phosphide (CNx/Ni_2P) to convert MPs to H_2. The nanoparticle complex was tested on PE, PET, PLA, PP, PS,

PUR, and PS-block-polybutadiene (rubber). Greatest yield of H_2 was generated from PET, PLA, and PUR. This catalyst is considered by the authors to be environmentally safe.

The carbon component of plastics (PE, PP, PVC) was transformed into C_2 fuels by photocatalysis under simulated natural conditions (Jiao et al. 2020). Single-unit-cell thick Nb_2O_5 atomic layers were synthesized and reacted with the polymers. Photodegradation of PE, PP, and PVC to CO_2 was 100% within 40, 60, and 90 h, respectively. The nascent CO_2 was further photoreduced to CH_3COOH. The overall mechanism is separated into two steps:

(1) C–C bond cleavage: O_2 and •OH radicals generated from the valence band of Nb_2O_5 trigger the oxidative C=C cleavage of polyethylene to form CO_2 (Eqs. 10.12–10.13).

$$H_2O + h^+ \rightarrow {}^{\bullet}OH + H^+ \tag{10.12}$$

$$-(CH_2CH_2)_n- \xrightarrow{{}^{\bullet}OH,\ O_2} 2nCO_2 \tag{10.13}$$

The O_2 is reduced to H_2O (Eqs. 10.14–10.16):

$$O_2 + e^- \rightarrow O_2^{\bullet -} \tag{10.14}$$

$$O_2^{\bullet -} + e^- + 2H^+ \rightarrow H_2O_2 \tag{10.15}$$

$$H_2O_2 + 2e^- + 2H^+ \rightarrow 2H_2O \tag{10.16}$$

(2) C–C coupling: the newly-generated CO_2 is catalytically reduced by the photogenerated electrons from the conduction band of Nb_2O_5 to CH_3COOH via the coupling of •COOH intermediates (Eqs. 10.17–10.19).

$$CO_2 + e^- + H^+ \rightarrow {}^{\bullet}COOH \tag{10.17}$$

$${}^{\bullet}COOH + {}^{\bullet}COOH \rightarrow HOOC\text{-}COOH \tag{10.18}$$

$$HOOC\text{-}COOH + 6e^- + 6H^+ \rightarrow CH_3COOH + 2H_2O \tag{10.19}$$

In the other half-reaction, water is oxidized to O_2 by the holes (Eq. 10.20).

$$2H_2O + 4h^+ \rightarrow O_2 + 4H^+ \tag{10.20}$$

The yield of the C_2 fuels was relatively low, so two-component photocatalysts are recommended to optimize simultaneously the C–C bond cleavage and coupling processes (Jiao et al. 2020).

PHOTOCATALYTIC REACTORS

Numerous configurations of solar photocatalytic reactors are currently in use, and demonstration and pilot studies are underway as well. These systems have been applied for the destruction of a range of organic contaminants, including endocrine-disrupting chemicals, pulp and paper effluents, textile dyeing wastes, pesticides, plasticizers, and many others (Abdel-Maksoud et al. 2016). Reactors have been effective in using both solar and artificial light. It is possible that several reactor types may be applicable for treatment of microplastics.

In the parabolic trough reactor (PTR), a transparent pipe is positioned along the surface of a parabolic light reflector and carries wastewater containing the contaminant of concern (figure 10.3). The collector is sometimes mounted on a mobile platform equipped with a solar tracking system (Abdel-Maksoud et al. 2016). A photocatalyst such as TiO_2 may be immobilized within the tubes or occur in slurry form within the pipes. The PTR system was used for treating trichloroethylene (TCE), tetrachloroethylene (PCE), chloroform, trichloroethane, carbon tetrachloride, and methylene chloride (Prairie et al. 1992). Solar detoxification was feasible only for TCE and PCE.

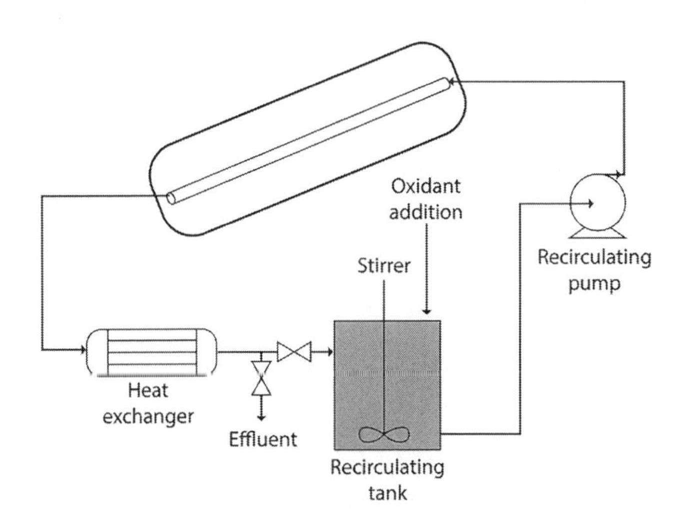

Figure 10.3 Schematic of the PTR
Source: Adapted from Abdel-Maksoud et al. (2016). Creative Commons (CC-BY) license (http://creativecommons.org/licenses/by/4.0/).

The compound parabolic concentrator (CPC) reactor consists of multiple transparent pipes mounted on a reflector (figure 10.4). The reflector is shaped with two half-parabolas intersecting beneath each tube. This shape permits any incident light to be concentrated on the transparent pipe without the need for a sun-tracking system. CPCs use direct as well as diffuse solar radiation (Bahnemann et al. 1999). The photocatalyst may be immobilized within the tubes or occur in slurry form. Martinez et al. (2014) used a CPC for decomposition of triclosan, an endocrine disruptor. A nano-TiO_2 photocatalyst was immobilized as thick films on mesoporous stones. Triclosan was removed by the solar photocatalytic reactor with efficiencies up to 74%.

The falling film photoreactor is designed with an inclined rectangular flat plate. Some designs include a covering of thin UV-transmissive glazing; in others, the surface is open to the atmosphere (Abdel-Maksoud et al. 2016). The reactor can use both direct and diffuse solar radiation. Laminar flow conditions are created to result in film thickness of a few mm (Alfano et al. 2000). When the falling film reactor operates without a cover, higher optical efficiency occurs and transmissive losses are eliminated (Lazar et al. 2012). The thin-film fixed-bed reactor, a variation of the falling film reactor, is designed to have the photocatalyst immobilized on the reactor surface (figure 10.5).

In a flat packed bed reactor, the photocatalyst is coated onto inert packing material (figure 10.6). Wastewater is pumped over the packing material while

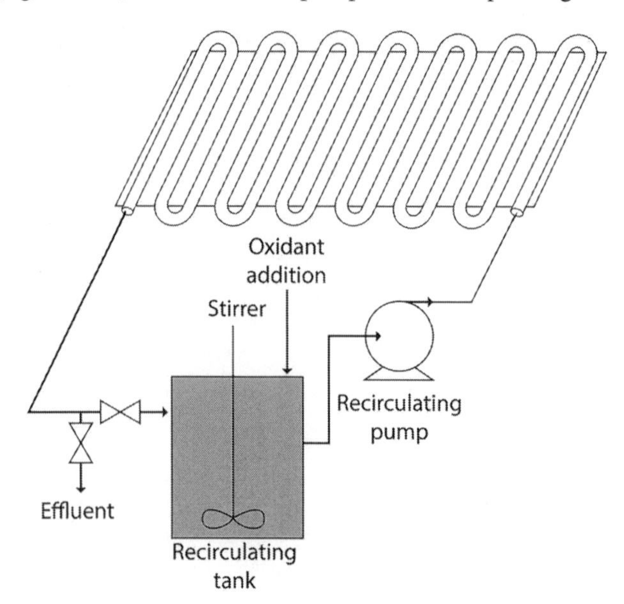

Figure 10.4 Schematic of the Compound Parabolic Reactor
Source: Adapted from Abdel-Maksoud et al. (2016). Creative Commons (CC-BY) license (http://creativecommons.org/licenses/by/4.0/).

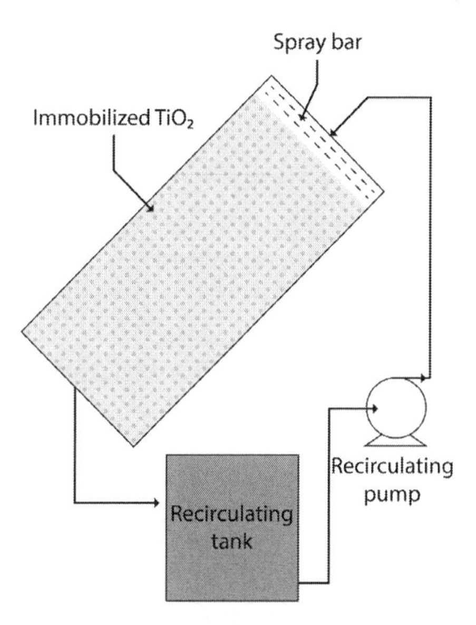

Figure 10.5 Schematic of the Thin-Film Fixed-Bed Photoreactor
Source: Adapted from Abdel-Maksoud et al. (2016). Creative Commons (CC-BY) license (http://creativecommons.org/licenses/by/4.0/).

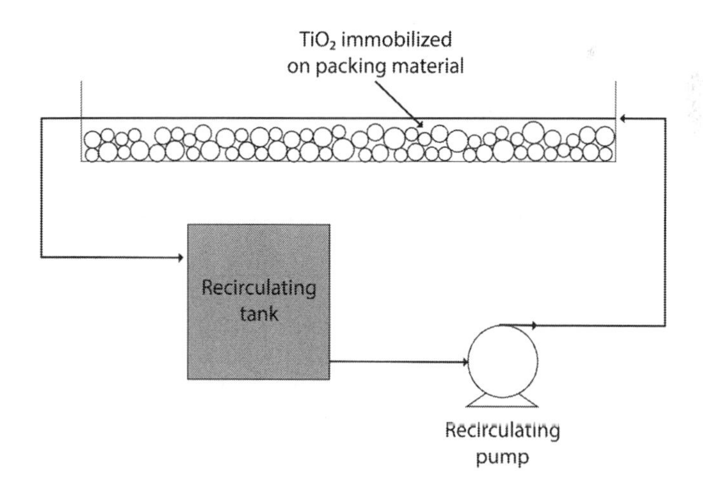

Figure 10.6 Schematic Drawing of a Packed Bed Photoreactor
Source: Adapted from Abdel-Maksoud et al. (2016). Creative Commons (CC-BY) license (http://creativecommons.org/licenses/by/4.0/).

exposed to light. Hanaor and Sorrell (2014) immobilized TiO_2 on sand as the photocatalyst in the photoreactor. The photoreactor, operated in a recirculating batch mode, was applied for destruction of *E. coli* in water.

CLOSING STATEMENTS

Photocatalysis technology provides a cost-effective option for destruction of microplastics in water and wastewater. However, this technology has its drawbacks. In many reported studies (table 10.1), excessively long periods of irradiation, whether under visible or UV light, are required to result in measurable chemical and/or physical changes in MPs (Cai et al. 2018; Ranjan et al. 2019). Additional study is needed for designing more effective photocatalysts and enhancing the efficiency of the photodegradation of MPs. Key objectives are to promote the forward reaction for generating •OH radicals, limit the reverse reaction including recombination of photogenerated electron-hole pairs, and ultimately extend the life span of photogenerated electron-hole pairs (Pham et al. 2021).

Only a limited number of photocatalysts have been evaluated for MP degradation thus far. A vast range of potential photocatalysts (e.g., carbon-based materials and metal oxides) require detailed examination (Rodríguez-Narvaez et al. 2021). The impact of irradiation type (visible vs. UV) and time on microplastic degradation also requires further study.

Heterogeneous photocatalytic technology must be tested further under varying process conditions such as pH level, temperature, use of agitation during the reaction, and use of water sampled from oceans, rivers, and secondary effluent of WWTPs. The influence of salinity and dissolved organic matter on degradation of microplastics must be addressed (Liu et al. 2021). Detailed studies on the formation of possible by-products are also needed (Lee et al. 2020).

REFERENCES

Abdel-Maksoud, Y., Imam, E., and Ramadan, A. (2016). TiO_2 solar photocatalytic reactor systems: Selection of reactor design for scale-up and commercialization—Analytical review. *Catalysts* 6:138. DOI: 10.3390/catal6090138.

Acuña-Bedoya, J.D., Luevano-Hipolito, E., Cedillo-González, E.I., Domínguez-Jaimes, L.P., Hurtado, A.M., and Hernández-López, J.M. (2021). Boosting visible-light photocatalytic degradation of polystyrene nanoplastics with immobilized CuxO obtained by anodization. *Journal of Environmental Chemical Engineering* 9(5):106208.

Alfano, O., Bahnemann, D., Cassano, A., Dillert, R., and Goslich, R. (2000). Photocatalysis in water environments using artificial and solar light. *Catalysis Today* 58:199–230.

Allé, P.H., Garcia-Muñoz, P., Adouby, K., Keller, N., and Robert, D. (2021). Efficient photocatalytic mineralization of polymethylmethacrylate and polystyrene nanoplastics by TiO_2/β-SiC alveolar foams. *Environmental Chemistry Letters* 19(2):1803–1808.

Ariza-Tarazona, M.C., Villarreal-Chiu, J.F., Barbieri, V., Siligardi, C., and Cedillo-González, E.I. (2019). New strategy for microplastic degradation: Green photocatalysis using a protein-based porous N-TiO_2 semiconductor. *Ceramics International* 45(7):9618–9624.

Ariza-Tarazona, M.C., Villarreal-Chiu, J.F., Hernández-López, J.M., De la Rosa, J.R., Barbieri, V., Siligardi, C., and Cedillo-González, E.I. (2020). Microplastic pollution reduction by a carbon and nitrogen-doped TiO_2: Effect of pH and temperature in the photocatalytic degradation process. *Journal of Hazardous Materials* 395:122632.

Bahnemann, D., Dillert, R., Dzengel, J., Goslich, R., Sagawe, G., Schumacher, H., and Benz, V. (1999). Field studies of solar water detoxification using non light concentrating reactors. *Journal of Advanced Oxidation Technologies* 4:11–19.

Beladi-Mousavi, S.M., Hermanová, S., Ying, Y., Plutnar, J., and Pumera, M. (2021). A maze in plastic wastes: Autonomous motile photocatalytic microrobots against microplastics. *ACS Applied Materials & Interfaces* 13:25102–25110.

Bora, T., and Dutta, J. (2019). Plasmonic photocatalyst design: Metal-semiconductor junction affecting photocatalytic efficiency. *Journal of Nanoscience and Nanotechnology* 19(1):383–388.

Cai, L., Wang, J., Peng, J., Wu, Z., and Tan, X. (2018). Observation of the degradation of three types of plastic pellets exposed to UV irradiation in three different environments. *Science of the Total Environment* 628:740–747.

Cao, B., Wan, S., Wang, Y., Guo, H., Ou, M., and Zhong, Q. (2022). Highly-efficient visible-light-driven photocatalytic H_2 evolution integrated with microplastic degradation over MXene/ZnxCd1-xS photocatalyst. *Journal of Colloid and Interface Science* 605:311–319.

Domínguez-Jaimes, L.P., Cedillo-González, E.I., Luévano-Hipólito, E., Acuña-Bedoya, J.D., and Hernández-López, J.M. (2021). Degradation of primary nanoplastics by photocatalysis using different anodized TiO_2 structures. *Journal of Hazardous Materials* 413:125452.

Dong, R., Hu, Y., Wu, Y., Gao, W., Ren, B., Wang, Q., and Cai, Y. (2017). Visible-light-driven BiOI-based Janus micromotor in pure water. *Journal of the American Chemical Society* 139:1722–1725.

Fadli, M.H., Ibadurrohman, M., and Slamet, S. (2021). Microplastic Pollutant Degradation in Water Using Modified TiO_2 Photocatalyst Under UV-Irradiation. *6th International Symposium on Applied Chemistry, Materials Science and Engineering*. IOP Publishing. DOI: 10.1088/1757-899X/1011/1/012055.

Hanaor, D., and Sorrell, C. (2014). Sand supported mixed-phase TiO_2 photocatalysts for water decontamination applications. *Advanced Engineering Materials* 16:248–254.

Hsiao, C.-Y., Lee, C.-L., and Ollis, D.F. (1983). Heterogeneous photocatalysis: Degradation of dilute solutions of dichloromethane (CH_2Cl_2), chloroform ($CHCl_3$), and carbon tetrachloride (CCl_4) with illuminated TiO_2 photocatalyst. *Journal of Catalysis* 82(2):418–423.

Hu, K., Tian, W., Yang, Y., Nie, G., Zhou, P., Wang, Y., Duan, X., and Wang, S. (2021). Microplastics remediation in aqueous systems: Strategies and technologies. *Water Research* 198:117144.

Ibhadon, A.O., and Fitzpatrick, P. (2013). Heterogeneous photocatalysis: Recent advances and applications. *Catalysts* 3(1):189–218.

Jiang, R., Lu, G., Yan, Z., Liu, J., Wu, D., and Wang, Y. (2021). Microplastic degradation by hydroxy-rich bismuth oxychloride. *Journal of Hazardous Materials* 405:124247.

Jiao, X., Zheng, K., Chen, Q., Li, X., Li, Y., Shao, W., Xu, J., Zhu, J., Pam, Y., Sun, Y., and Xie, Y. (2020). Photocatalytic conversion of waste plastics into C_2 fuels under simulated natural environment conditions. *Angewandte Chemie International Edition* 59(36):15497–15501.

Kawai, T., and Sakata, T. (1981). Photocatalytic hydrogen production from water by the decomposition of poly-vinylchloride, protein, algae, dead insects, and excrement. *Chemistry Letters* 10(1):81–84.

Kesselman, J.M., Weres, O., Lewis, N.S., and Hoffmann, M.R. (1997). Electrochemical production of hydroxyl radical at polycrystalline Nb-doped TiO_2 electrodes and estimation of the partitioning between hydroxyl radical and direct hole oxidation pathways. *The Journal of Physical Chemistry B* 101(14):2637–2643.

Lam, S.-M., Sin, J.-C., Zeng, H., Lin, H., Li, H., Chai, Y.-Y., Choong, M.-K., and Mohamed, A.R. (2021). Green synthesis of Fe-ZnO nanoparticles with improved sunlight photocatalytic performance for polyethylene film deterioration and bacterial inactivation. *Materials Science in Semiconductor Processing* 123:105574.

Lazar, M.A., Varghese, S., and Nair, S.S. (2012). Photocatalytic water treatment by titanium dioxide: Recent updates. *Catalysts* 2:572–601.

Lee, J.-M., Busquets, R., Choi, I.-C., Lee, S.-H., Kim, J.-K., and Campos, L.C. (2020). Photocatalytic degradation of polyamide 66; evaluating the feasibility of photocatalysis as a microfibre-targeting technology. *Water* 12(12):3551.

Liu, W., Zhang, J., Liu, H., Guo, X., Zhang, X., Yao, X., Cao, Z., and Zhang, T. (2021). A review of the removal of microplastics in global wastewater treatment plants: Characteristics and mechanisms. *Environment International* 146:106277.

Llorente-García, B.E., Hernández-López, J.M., Zaldívar-Cadena, A.A., Siligardi, C., and Cedillo-González, E.I. (2020). First insights into photocatalytic degradation of HDPE and LDPE microplastics by a mesoporous N–TiO_2 coating: Effect of size and shape of microplastics. *Coatings* 10(7):658.

Luo, H., Xiang, Y., Tian, T., and Pan, X. (2021). An AFM-IR study on surface properties of nano-TiO_2 coated polyethylene (PE) thin film as influenced by photocatalytic aging process. *Science of the Total Environment* 757:143900.

Martinez, S., Morales-Mejia, J., Hernandez, P., Santiago, L., and Almanza, R. (2014). Solar photocatalytic oxidation of Triclosan with TiO_2 immobilized on volcanic porous stones on a CPC pilot scale reactor. *Energy Procedia* 57:3014–3020.

Maulana, D.A., and Ibadurrohman, M. (2021). Synthesis of Nano-Composite Ag/TiO_2 for Polyethylene Microplastic Degradation Applications. *6th International Symposium on Applied Chemistry: Materials Science and Engineering.* DOI: 10.1088/1757-899X/1011/1/012054.

Maupin, I., Pinard, L., Mijoin, J., and Magnoux, P. (2012). Bifunctional mechanism of dichloromethane oxidation over Pt/Al_2O_3: CH_2Cl_2 disproportionation over alumina and oxidation over platinum. *Journal of Catalysis* 291:104–109.

Nabi, I., Li, K., Cheng, H., Wang, T., Liu, Y., Ajmal, S., Yang, Y., Feng, Y., and Zhang, L. (2020). Complete photocatalytic mineralization of microplastic on TiO_2 nanoparticle film. *Iscience* 23(7):101326.

Nakata, K., and Fujishima, A. (2012). TiO_2 photocatalysis: Design and applications. *Journal of Photochemistry and Photobiology C: Photochemistry Reviews* 13(3):169–189.

Nguyen, C.H., Fu, C.-C., and Juang, R.-S. (2018). Degradation of methylene blue and methyl orange by palladium-doped TiO_2 photocatalysis for water reuse: Efficiency and degradation pathways. *Journal of Cleaner Production* 202:413–427.

Olajire, A., and Mohammed, A. (2021). Bio-directed synthesis of gold nanoparticles using Ananas comosus aqueous leaf extract and their photocatalytic activity for LDPE degradation. *Advanced Powder Technology* 32(2):600–610.

Pham, T.-H., Do, H.-T., Thi, L.-A.P., Singh, P., Raizada, P., Wu, J.C.-S., and Nguyen, V.-H. (2021). Global challenges in microplastics: From fundamental understanding to advanced degradations toward sustainable strategies. *Chemosphere* 267:129275.

Pitkäaho, S., Nevanperä, T., Matejova, L., Ojala, S., and Keiski, R.L. (2013). Oxidation of dichloromethane over Pt, Pd, Rh, and V_2O_5 catalysts supported on Al_2O_3, Al_2O_3–TiO_2 and Al_2O_3–CeO_2. *Applied Catalysis B: Environmental* 138:33–42.

Prairie, M., Pacheco, J., and Evans, L. (1992). Solar Detoxification of Water Containing Chlorinated Solvents and Heavy Metals via TiO_2 Photocatalysis. *Proceedings of the 1992 ASME International Solar Energy Conference on Solar Engineering*, Maui, HI, USA, 4–8 April 1992, Vol. 1, pp. 1–8.

Ranjan, V.P., and Goel, S. (2019). Degradation of low-density polyethylene film exposed to UV radiation in four environments. *Journal of Hazardous, Toxic, and Radioactive Waste* 23(4):04019015.

Razali, N., Terengganu, U., Abdullah, W., and Zikir, N. (2020). Effect of thermo-photocatalytic process using zinc oxide on degradation of macro/micro-plastic in aqueous environment. *Journal of Sustainability Science Management* 15:1–14.

Rodríguez-Narvaez, O.M., Goonetilleke, A., Perez, L., and Bandala, E.R. (2021). Engineered technologies for the separation and degradation of microplastics in water: A review. *Chemical Engineering Journal* 414:128692.

Sanwald, K.E., Berto, T.F., Eisenreich, W., Gutiérrez, O.Y., and Lercher, J.A. (2016). Catalytic routes and oxidation mechanisms in photoreforming of polyols. *Journal of Catalysis* 344:806–816.

Shang, J., Chai, M., and Zhu, Y. (2003a). Photocatalytic degradation of polystyrene plastic under fluorescent light. *Environmental Science & Technology* 37(19):4494–4499.

Shang, J., Chai, M., and Zhu, Y. (2003b). Solid-phase photocatalytic degradation of polystyrene plastic with TiO_2 as photocatalyst. *Journal of Solid State Chemistry* 174(1):104–110.

Shi, Y., Liu, P., Wu, X., Shi, H., Huang, H., Wang, H., and Gao, S. (2021). Insight into chain scission and release profiles from photodegradation of polycarbonate microplastics. *Water Research* 195:116980.

Tian, L., Chen, Q., Jiang, W., Wang, L., Xie, H., Kalogerakis, N., Ma, Y., and Ji, R. (2019). A carbon-14 radiotracer-based study on the phototransformation of polystyrene nanoplastics in water versus in air. *Environmental Science: Nano* 6(9):2907–2917.

Tofa, T.S., Kunjali, K.L., Paul, S., and Dutta, J. (2019a). Visible light photocatalytic degradation of microplastic residues with zinc oxide nanorods. *Environmental Chemistry Letters* 17(3):1341–1346.

Tofa, T.S., Ye, F., Kunjali, K.L., and Dutta, J. (2019b). Enhanced visible light photodegradation of microplastic fragments with plasmonic platinum/zinc oxide nanorod photocatalysts. *Catalysts* 9(10):819.

Uekert, T., Kasap, H., and Reisner, E. (2019). Photoreforming of nonrecyclable plastic waste over a carbon nitride/nickel phosphide catalyst. *Journal of the American Chemical Society* 141(38):15201–15210.

Uekert, T., Kuehnel, M.F., Wakerley, D.W., and Reisner, E. (2018). Plastic waste as a feedstock for solar-driven H_2 generation. *Energy & Environmental Science* 11(10):2853–2857.

Uheida, A., Mejía, H.G., Abdel-Rehim, M., Hamd, W., and Dutta, J. (2021). Visible light photocatalytic degradation of polypropylene microplastics in a continuous water flow system. *Journal of Hazardous Materials* 406:124299.

Varma, K.S., Tayade, R.J., Shah, K.J., Joshi, P.A., Shukla, A.D., and Gandhi, V.G. (2020). Photocatalytic degradation of pharmaceutical and pesticide compounds (PPCs) using doped TiO_2 nanomaterials: A review. *Water-Energy Nexus* 3:46–61.

Venkataramana, C., Botsa, S.M., Shyamala, P., and Muralikrishna, R. (2021). Photocatalytic degradation of polyethylene plastics by $NiAl_2O_4$ spinels-synthesis and characterization. *Chemosphere* 265:129021.

Wang, H., and Pumera, M. (2020). Coordinated behaviors of artificial micro/nanomachines: From mutual interactions to interactions with the environment. *Chemical Society Reviews* 49(10):3211–3230.

Zahran, E.M., Bedford, N.M., Nguyen, M.A., Chang, Y.-J., Guiton, B.S., Naik, R.R., Bachas, L.G., and Knecht, M.R. (2014). Light-activated tandem catalysis driven by multicomponent nanomaterials. *Journal of the American Chemical Society* 136(1):32–35.

Zhang, G., Ni, C., Huang, X., Welgamage, A., Lawton, L.A., Robertson, P.K., and Irvine, J.T. (2016). Simultaneous cellulose conversion and hydrogen production assisted by cellulose decomposition under UV-light photocatalysis. *Chemical Communications* 52(8):1673–1676.

Zhang, X., Liu, Y., Deng, J., Zhang, K., Yang, J., Han, Z., and Dai, H. (2018). AuPd/3DOM TiO_2 catalysts: Good activity and stability for the oxidation of trichloroethylene. *Catalysts* 8(12):666.

Zhao, X., Li, Z., Chen, Y., Shi, L., and Zhu, Y. (2007). Solid-phase photocatalytic degradation of polyethylene plastic under UV and solar light irradiation. *Journal of Molecular Catalysis A: Chemical* 268(1–2):101–106.

Zhu, K., Jia, H., Zhao, S., Xia, T., Guo, X., Wang, T., and Zhu, L. (2019). Formation of environmentally persistent free radicals on microplastics under light irradiation. *Environmental Science & Technology* 53(14):8177–8186.

Chapter 11

Decomposition by Microorganisms

INTRODUCTION

Commercial synthetic plastics are recalcitrant in the biosphere owing to many unique characteristics including extremely long chains (i.e., high MW), complex three-dimensional structure, the presence of numerous (and sometimes toxic) additives, and general hydrophobic nature. These properties likewise limit biodegradation of microplastics (MPs) (Hadad et al. 2005; Kale et al. 2015). Still, microbial types such as bacteria and fungi are highly opportunistic and adaptive, and capable of metabolizing countless organic substrates. Microorganisms produce both endo- and exoenzymes which attack carbonaceous substrates, cleaving molecular chains and branches into smaller segments for eventual extraction of nutrients and energy by the organism (Pathak and Navneet 2017).

Compared to chemical and physical degradation technologies for MPs treatment, biodegradation has the advantages of being both environmentally benign and acceptable to the public. Microbial technologies typically tend to be of relatively low cost and have low energy requirements. Thus far, however, the relevant biological processes tend to be slow and inefficient.

Numerous efforts are underway to identify, isolate, and cultivate microorganisms capable of utilizing and decomposing synthetic MPs. In recent decades, several potentially effective bacteria and fungi have been identified (Soleimani et al. 2021; El-Sayed et al. 2021; Auta et al. 2017). MPs serve as a unique ecological niche for microorganisms: certain bacteria, fungi, and other microbial types are known to colonize MPs and utilize them as carbon and nitrogen sources (Yoshida et al. 2016; Devi et al. 2015; Shah et al. 2013). Such capabilities offer promise for large-scale application of MP removal from aquatic systems (Hu et al. 2021; Ali et al. 2021).

Microbes and their metabolic enzymes have the potential as potent tools for MPs degradation (Ghosh et al. 2019). Based on research to date, it may be feasible for microbes to degrade MPs in a dedicated reactor or a conventional wastewater treatment system.

GENERAL PROCESS

The mechanisms of MP biodegradation are complex, and the numerous degradation pathways are not yet fully identified. As MPs are polymers, mechanisms involved in degradation may be similar to those of natural polymers such as proteins, cellulose, and lignin. Biodegradation of MPs appears to occur in several distinct phases:

1) colonization of the particle surface by the microbial cell;
2) secretion of extracellular polymeric substances (EPS) and consequent formation of a biofilm;
3) microbial secretion of extracellular enzymes which cleave polymer chains;
4) metabolism of polymer segments at chain ends via progressive enzymatic action;
5) passage of water-soluble MP oligomers and monomers through the outer bacterial membrane and entrance into the cell; and
6) incorporation of small molecules into cellular biomass, and respiration and mineralization by intracellular enzymes to CO_2, H_2O (for aerobic microbes) and other products (figure 11.1) (Ebrahimbabaie et al. 2022).

Various researchers (Pasquina-Lemonche et al. 2020; Huang et al. 2008) have measured pore sizes of Gram-positive and Gram-negative bacterial cell walls to be in the low nm range. MPs are relatively large compared to microbial cell wall pores and therefore cannot pass through (Shah et al. 2008); furthermore, large particles tend to be less water-soluble than small oligomers. Biodegradation of MPs is therefore significantly more successful on fragments of relatively low MW (Andrady 2011). High-MW polymers must first be depolymerized by abiotic degradation to smaller pieces before biodegradation can occur at a practical rate (Gewert et al. 2015; Bond et al. 2018). Abiotic hydrolysis is essential for initiating the microbial degradation of many synthetic polymers like PE, PET, polycarboxylates, PLA, and their copolymers, poly-α-glutamic acids, and silicones (Balasubramanian et al. 2014; Shah et al. 2008).

In the environment, some of the key physicochemical processes that initiate plastic degradation are photodegradation, thermo-oxidative degradation,

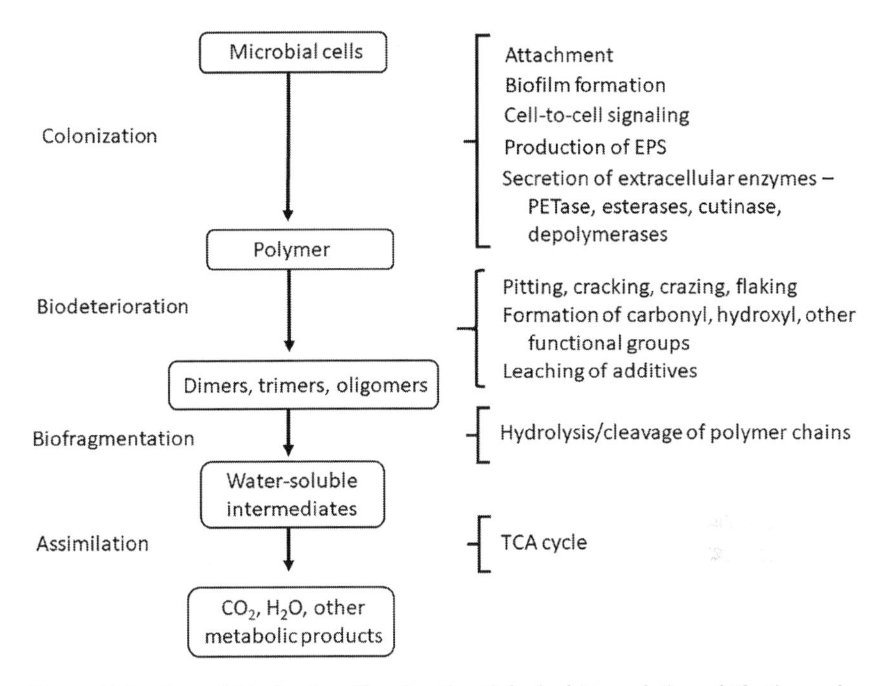

Figure 11.1 Overall Mechanism Showing the Biological Degradation of Plastics under Aerobic Conditions

and hydrolysis. These processes are, however, very slow. During abiotic degradation, carbonyl, hydroxyl, and other groups are generated that increase the hydrophilicity of the polymer, thus enhancing its availability for biodegradation (Gewert et al. 2015). The monomers, dimers, and oligomers released from the repeating units of a polymer can be mineralized by biota.

Biodegradation can proceed via two avenues, that is, aerobic and anaerobic. The aerobic pathway produces CO_2, H_2O, and heat. In the anaerobic pathway, energy is stored as CH_4; depending on MP composition and the reaction medium, other products may include CO_2, H_2S, NH_4, H_2, and H_2O (Hu et al. 2021; Jacquin et al. 2019).

In studies of microbial action on MPs, multiple types of microbes have been identified that are responsible for degradation to some degree.

BACTERIAL-DIRECTED MP DEGRADATION

Many laboratory-scale studies have evaluated the capabilities of bacteria for MP degradation (table 11.1). Researchers have evaluated mechanisms and reactions of single species or strains of bacteria; others have evaluated mixed

Table 11.1 Decomposition of MPs/Plastics by Bacterial Type, Recent Data

Microbial Type	*Genus*	*Strain*	*Source*	*MP/ Plastic Type*	*Duration (Days)*	*Results / Analytical Methods*	*References*
Bacteria	Actinobacteria (*Streptomyces, Nocardia, Rhodococcus*)	IR-SGS-T10 IR-SGS-T11 IR-SGS-Y1	Soil sample from plastic landfills	PE	60	Actinobacteria have a high ability for biodegradation of PE-based plastics. *Streptomyces* sp. IR-SGS-T10 showed the greatest reduction in weight of LDPE film (1.58 $mg \cdot g^{-1} \cdot day^{-1}$) after 60 days incubation without pretreatment. *Rhodococcus* sp. IR-SGS-T11 showed the best reduction in tensile properties of LDPE film. Results with *Streptomyces* sp. IR-SGS-Y1 indicated a significant change in structure via FTIR.	Soleimani et al. 2021
	Pseudomonas aeruginosa	–	Surface water	PE	10 20 30	Highest percentage weight loss of PE was 6.25% after 30 days at 44°C.	Mouafo Tamnou et al. 2021
	Cobetia sp., *Halomonas* sp., *Exigobacterium* sp. *and Alcanivorax* sp.	H-237, H-255, H-256, and H-265	Marine	LDPE	90	Maximum weight loss of 1.72% of LDPE film observed with isolate H-255.	Khandare et al. 2021

Bacillus	27	Mangrove sediment	PP	40	Weight loss was 4.0%.	Auta et al. 2018
Rhodococcus	36	Mangrove sediment	PP	40	Weight loss was 6.4%. Growth of isolates led to changes in pH, possibly resulting from microbial degradation.	Auta et al. 2018
Bacillus gottheilii	–	Mangrove ecosystems	PE, PET, PP, PS	40	Weight loss of 6.2, 3.0, 3.6, and 5.8% for PE, PET, PP, and PS, respectively.	Auta et al. 2017b
Pseudomonas	MYK1	Digester sludge	PLA	40	Physical disintegration and morphological changes to PLA film were observed. The specific gas production rate increased for the biodegradation of both strains under aerobic and anaerobic conditions.	Kim et al. 2017
Bacillus	MYK2	Digester sludge	PLA	40	See previous entry.	Kim et al. 2017
Ideonella sakaiensis	201-F6	Sediment, soil, wastewater, activated sludge from PET bottle recycling site	PET	60	Bacterium used PET as its major energy and C source. Produced two enzymes (glycoside hydrolases) capable of hydrolyzing PET and a reaction intermediate.	Yoshida et al. 2016

(continued)

Table 11.1 (Continued)

Microbial Type	*Genus*	*Strain*	*Source*	*MP/ Plastic Type*	*Duration (Days)*	*Results / Analytical Methods*	*References*
	Enterobacter asburiae	YT1	Plastic-eating waxworms	PE	28	Degradation of approximately 6.1%	Yang et al. 2014
	Bacillus	YP1	Plastic-eating waxworms	PE	28	Degradation of 10.7%	Yang et al. 2014
	Bacillus subtilis	MZA-75	Soil	PUR	28	The organism degraded the polyester diol portion and produced 1,4-butanediol and adipic acid. It utilizes these as a C source to mineralize the polymer. Increase in cell-associated and extracellular esterases was observed.	Shah et al. 2013
	Chelatococcus	E1	Compost	PE	80	The strain assimilated PE of very low MW and also some high-molecular weight chains.	Jeon and Kim 2013b
	Rhodococcus ruber	C208	–	PE	60	Degraded PE at 0.86% per week.	Sivan et al. 2006

	Paenibacillus amylolyticus	TB-13	Soil samples	PLA, PBS, PBSa, PCL, PES, PHBVa	14	Organism degraded poly (lactic acid), poly (butylene succinate), poly (butylene succinate-*co*-adipate), poly(caprolactone) and poly (ethylene succinate), but not poly(hydroxybutylate-*co*-valerate). However, it could not use these plastics as sole C source. Protease and esterase activities may be involved in the degradation.	Teeraphatpornchai et al. 2003
	Comamonas acidovorans	TB-35	–	PPU	8	Metabolites identified included diethylene glycol, trimethylolpropane, and adipic acid. Possibly derived from polyester segments of the PUR due to hydrolytic cleavage of ester bonds. Cell-surface-bound esterase catalyzed degradation of the polyester PUR.	Nakajima-Kambe et al. 1997
Bacterial consortia	*Enterobacter* sp nov, *Enterobacter cloacae* nov., and *Pseudomonas aeruginosa* nov. bt	DSCE01, DSCE02, and DSCE-CD03	Consortia formulated from cow dung	LDPE PP	160	Of 10 consortia formulated, potential consortium-CB3 showed greater percentage degradation (weight reduction) of 64.25 ± 2% and 63.00 ± 2% toward LDPE and PP films, respectively ($p < 0.05$) at 37°C compared to other consortia.	Skariyachan et al. 2021

(continued)

Table 11.1 (Continued)

Microbial Type	*Genus*	*Strain*	*Source*	*MP/ Plastic Type*	*Duration (Days)*	*Results / Analytical Methods*	*References*
	Bacillus sp. and *Paenibacillus* sp.		Municipal landfill sediment	PP	60	14.7% reduction in dry weight of particles. Microplastics presumably degraded by enzymatic chain scission. PE could be biologically utilized as a sole C source.	Park and Kim 2019
	Agios and *Souda*		Agios developed in laboratory; Souda from indigenous marine community	HDPE	60	*Souda* community is more efficient in decreasing HDPE weight compared to *Agios* community.	Tsiota et al. 2018
	Bacillus simplex and *Bacillus* sp.		Earthworm gut	LDPE	21	60% of the LDPE MPs initially present decayed after 21 days in presence of bacteria. Noticeable reduction in particle size of LDPE. Volatile compounds such as octadecane, eicosane, docosane, and tricosane were detected.	Huerta Lwanga et al. 2018

Indigenous marine community. Some treatments supplemented with *Lysinibacillus* sp. and *Salinibacterium* sp.		Marine	PE	180	19% reduction in weight of weathered PE samples	Syranidou et al. 2017
Exiguobacterium sp.	YT2	Mealworms (*Tenebrio molitor Linnaeus*)	PS	28	13 bacterial cultures isolated from mealworm gut. Degradation of 7.4% of PS pieces by *Exiguobacterium* sp.	Yang et al. 2015b
Bacillus cereus, B. pumilus and *Arthrobacter*		Waste disposal sites and artificial soil beds	HDPE, LDPE	14	22.41% and 21.7% weight loss in HDPE and LDPE, respectively.	Satlewal et al. 2008
Bacillus cereus and *B. sphaericus*		Marine	LDPE, HDPE	365	Non-pretreated and thermally pretreated LDPE and HDPE studied. The polymer was sole C source. Weight loss of the thermally treated LDPE and HDPE samples were 19% and 9% respectively, with *B. sphericus*; non-pretreated samples had 10% and 3.5% weight loss, respectively.	Sudhakar et al. 2007

HDPE—High-density polyethylene; LDPE—Low-density polyethylene; PE—Polyethylene; PBS—Polybutylene succinate; PBSa—Polybutylene succinate-*co*-adipate; PCL—Polycaprolactone; PES—Polyethylene succinate; PHBVa—Polyhydroxybutylate-*co*-valerate; PLA—Polylactic acid; PPU—Polyester polyurethane; PS—Polystyrene; PSS—Polystyrene sulfonate; PUR—Polyurethane; TPA—Terephthalic acid cultures derived from the gut of higher organisms, landfills, compost piles, soil, wastewater, and other media.

With few exceptions MP degradation by bacteria is a slow process, allowing for only minimal-to-moderate decomposition. The number of MP-degrading bacterial strains identified to date is small—the most commonly isolated taxa include *Bacillus*, *Pseudomonas*, *Chelatococcus*, and *Lysinibacillus* (Yuan et al. 2020; Jeon and Kim 2013; Shahnawaz et al. 2016; Pathak and Navneet 2017). Among all tested bacterial strains, *Bacillus* spp., *Pseudomonas* spp., and *Streptomyces* spp. demonstrate medium-to-high degradation efficiency toward certain MP polymers (Ali et al. 2021; Li et al. 2020; Matjašič et al. 2021).

Rhodococcus sp. strain 36 and *Bacillus* sp. strain 27 isolated from mangrove sediment caused weight loss of MPs of 6.4% and 4.0%, respectively, after 40 days (Auta et al. 2017). Weight loss of PE, PET, and PS MPs caused by *B. cereus* was 1.6, 6.6, and 7.4%, respectively, and weight reduction for PE, PET, PP, and PSMPs by *B. gottheilii* was 6.2, 3.0, 3.6, and 5.8%, respectively (Auta et al. 2018). The Antarctic soil bacteria *Pseudomonas* sp. ADL15 caused greater relative weight loss, more successful utilization, more rapid removal rate, and shorter half-life compared to *Rhodococcus* sp. ADL36 during PP decomposition (Habib et al. 2020). Actinobacteria were found to be effective in biodegradation of PE-based plastics (Soleimani et al. 2021). Streptomyces sp. IR-SGS-T10 caused the greatest weight reduction of LDPE film (1.58 $mg \cdot g^{-1} \cdot day^{-1}$) after 60 days of incubation without any pretreatment. Since successful bacterial strains have been collected from soil, it is feasible that these types may be appropriate for MP degradation in enclosed (e.g., light-limited) treatment systems such as bioreactors.

New Functional Groups and Intermediates

Bacterial attack on MPs results in the formation of new functional groups, which may render the MP susceptible to further biotic and abiotic degradation. Chemical modifications are often specific to each polymer.

Bacillus cereus and *Bacillus gottheilii* altered several functional group structures and reduced FTIR absorption peaks typical of PE, PET, and PS MPs (Auta et al. 2017). When PP MPs were treated with *Bacillus* sp. strain 27, certain C=O carbonyl bands and an O–H hydroxyl peak disappeared (Auta et al. 2018); when PP MPs were reacted with *Rhodococcus* sp. strain 36 for 40 days, an O–H band and two C=O carbonyl bands disappeared. *Pseudomonas* sp. ADL15 and *Rhodococcus* sp. ADL36 were incubated with PP MPs (Habib et al. 2020). Changes in functional groups were observed after 40 days via FTIR; it was suggested that the major alkyl group (C–H) was utilized, with possible creation of a hydroxy (–OH) group. *Acinetobacter* sp.

AnTc-1, isolated from the gut larvae of the red flour beetle *Tribolium castaneum*, was incubated in the presence of PS (Wang et al. 2020). After 60 days, ^{1}H NMR spectrograms revealed the emergence of new hydrogen-containing groups. Following a four-week exposure of PS to the gut microbiome of the land snail *Achatina fulicawere* (Song et al. 2020a), FTIR and ^{1}H NMR revealed the formation of functional groups of oxidized intermediates on MPs.

As a consequence of the above changes to functional groups, bacterial degradation results in the generation of a number of intermediate compounds (figure 11.2). *Comamonas testosterone* F4 was grown on PET as the sole C source (Gong et al. 2018); decomposition products included terephthalic

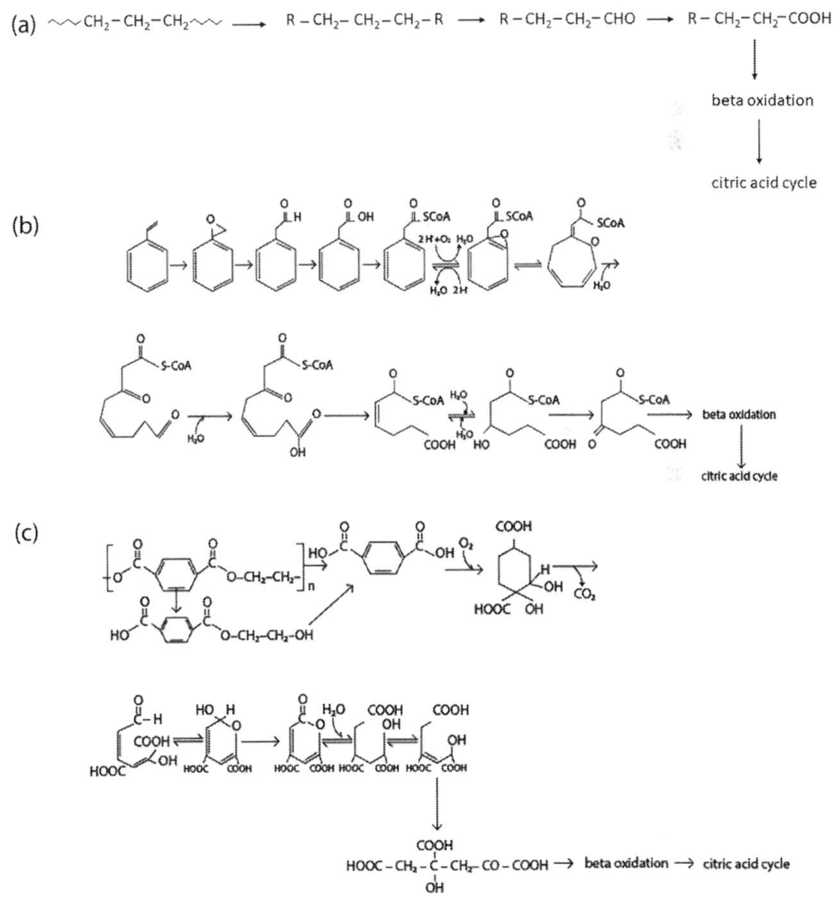

Figure 11.2 Bacterially Mediated Decomposition of Three Polymer Types: (a) PE, (b) PET, and (c) PS

Source: Jaquin et al. (2019). Creative Commons License 4.0.

acid, benzoic acid, hydroxyethyl terephthalate, and bis-(2-hydroxyethyl) terephthalate. When a mesophilic bacterial consortium was reacted with PE MPs (Park and Kim 2019), by-products after 60 days included dodecanol, nonanediol, and dencene; all have a linear carbon chain structure that presumably originated from the MP. Bacteria belonging to phyla Actinobacteria and Firmicutes partially decomposed LDPE (Huerta Lwanga et al. 2018). Volatile by-products including eicosane, docosane, and tricosane were detected, suggesting breakdown of the long C chains comprising the PE MPs. These same products plus tetracosane, pentacosane, hexacosane, and others, were produced by *Pseudomonas* spp. during PE degradation (Kyaw et al. 2012). When a consortium of microbiota was reacted with PE, a total of 12 products ranging from C_3 to C_{29} was detected (Yang et al. 2014).

Bacterial Consortia

MP degradation by individual bacterial species is limited as a consequence of limited metabolic capabilities. Different microbial types working collectively as a consortium often result in more rapid and efficient biodegradation than do individual strains. Numerous and diverse xenobiotic compounds have been degraded via cooperation of microbial assemblages (Pichtel 2017). A certain strain may be responsible for degradation of a specific molecule, or producing metabolites that encourage co-metabolic degradation (Hu et al. 2021). Others serve to remove toxic intermediate compounds (Singh and Wahid 2015; Yuan et al. 2020). Bacterial degradation of MPs may also be a result of horizontal gene transfer (Maheshwari et al. 2017) and the activities of specialized enzymes. It is also possible that ecological succession of different bacterial types occurs on the polymer surface resulting in metabolism of distinct substrates, resulting in greater overall decomposition (Pinto et al. 2019).

An indigenously developed microbial consortium degraded powdered HDPE by 22.4% and LDPE by 21.7% over two weeks (Satlewal et al. 2008). Two bacterial consortia (one via domestication, a second collected from Souda Bay, Crete) decreased the initial weight of HDPE MPs by 8% and 18%, respectively, after two months (Tsiota et al. 2018). A mesophilic bacterial consortium obtained from landfill sediment was reacted with PE MPs (Park and Kim 2019). The consortium belonged primarily to genera of class Bacilli (*Bacillus*, *Paenibacillus*, *Fontibacillus*, and *Enterococcus*) plus a genus of order Aeromonadales (*Aeromonas*) (Park and Kim 2019). Dry weight of PE MP particles was reduced by 14.7% and mean MP particle diameter by 22.8% after 60 days.

LDPE and PP were pretreated under UV and exposed to ten bacterial consortia for 160 days (Skariyachan et al. 2021). One of the consortia (so-called CB3) showed greater percentage degradation (weight reduction) toward

LDPE and PP films (64% and 63%, respectively at 37°C) compared to other consortia. Three bacterial strains that constituted CB3 were found to be novel strains, designated *Enterobacter* sp nov. bt DSCE01, *Enterobacter cloacae* nov. bt DSCE02, and *Pseudomonas aeruginosa* nov. bt DSCE-CD03. It was suggested that this consortium be scaled up for enhanced degradation of plastic polymers; a cost-effective biodigester could be designed for industrial applications using this strain as a potential inoculum (Skariyachan et al. 2021).

Consortia of gut microbiota have been examined for biodegradation of MPs. Gram-positive bacteria belonging to phyla Actinobacteria and Firmicutes were isolated from gut microbial consortia of earthworms (*Lumbricus terrestris*), and MP particle size was significantly reduced via their action (Huerta Lwanga et al. 2018). Waxworms (*Plodia interpunctella*) were capable of consuming PE films (Yang et al. 2014). Two bacterial strains isolated from the gut, *Enterobacter asburiae* YT1 and Bacillus sp. YP1, were responsible for degrading PE. Up to 10.7% ofthe PE films were degraded over 60 days in suspension culture. During ingestion of PS the gut microbiome of the land snail *Achatina fulica* experienced an increase of family Enterobacteriaceae, Sphingobacteriaceae, and Aeromonadaceae, thus suggesting their association with PS biodegradation (Song et al. 2020a). After a fourweek exposure, average PS mass decreased by 30.7%. Depolymerization of PE over 32 days was demonstrated by gut bacteria of mealworms (*Tenebrio molitor*) (Brandon et al. 2018). When mealworms were co-fed PE and PS (1:1 w/w), a significantly higher mass loss of PE occurred than for PS, indicating possible co-metabolism. These results suggest that bacteria hold promise for MP degradation, and further work is needed for identification and selection in order to optimize bacterial consortia for degradation reactions.

FUNGAL-MEDIATED MP/PLASTIC DEGRADATION

Fungi offer several advantages for MP degradation. Many are capable of adsorption to substrates, including synthetic polymers, via generation of extracellular polymers such as polysaccharides (Volke-Sepúlveda et al. 2002; Esmaeili et al. 2013). Fungal mycelia possess an extensive surface area which can proliferate on the MP surface and are capable of penetrating into the MP structure, thus promoting fungal colonization and growth (Esmaeili et al. 2013) as well as polymer decomposition. Fungi produce biosurfactants (i.e., hydrophobins), which enable them to exploit plastics and other polymers as a source of carbon and electrons (Sánchez 2020; Yuan et al. 2020). Fungal mycelia secrete extracellular enzymes such as depolymerases which are capable of decomposing polymers into low-MW fragments (Ameen et al. 2015; Ali et al. 2021).

Contaminated soil and seawater have been among the predominant sources of fungal strains for research in MP/plastic degradation (Ali et al. 2021). Fungal types documented to degrade synthetic polymers are primarily ascomycetes, followed by basidiomycetes and zygomycetes (Sánchez 2020). Early studies on plastics biodegradation employed fungal isolates such as *Phanerochaete chrysosporium* (white-rot fungi) (Orhan and Büyükgüngör 2000), *Aspergillus flavus* (El-Shafei et al. 1998), *A. niger, A. terreus, A. fumigatus*, and other *Aspergillus* strains (Manzur et al. 2004; Zahra et al. 2010) (table 11.2).

Zalerion maritimum was cultured in the presence of PE MPs measuring 250–1000 μm (Paço et al. 2017). The fungus was capable of utilizing PE, resulting in decreases in both mass and size of MPs. The biodegradation of polycaprolactone (PCL) and PVC films by *Aspergillus brasiliensis*, *Penicillium funiculosum*, and other fungal species was studied by Vivi et al. (2019). After 28 days PCL films had micropores and cracks, surface erosion, and hyphal adhesion on surfaces, with a mass loss of up to 75%. No evidence of PVC biodegradation was observed.

Compared to bacteria, the concentration of enzymes secreted by fungi is significantly higher (Gangola et al. 2019). White-rot and brown-rot fungi (*Monilinia fructicola*) contribute significantly to decomposition of polymers. Extracellular enzymes secreted by these fungi, including manganese peroxidase, lignin peroxidase, versatile peroxidase, and laccase, are known to decompose lignin and convert it to CO_2 and H_2O (Ameen et al. 2015; Ali et al. 2021; Iiyoshi et al. 1998). These enzymes also act upon synthetic polymers, as their chemical structures possess certain similarities to lignin (e.g., carbon skeleton, presence of ether bonds, aromatic rings, which are oxidized during lignin degradation) (Camarero et al. 1999; Kim et al. 2011; Ali et al. 2021). Lignin-degrading fungi IZU-154 experienced the greatest PE degradation under N- or C-limited culture conditions compared to other fungi (*Phanaerochaete chrysosporium* and *Trarnetes versicolor*) (Iiyoshi et al. 1998). Addition of Mn(II) into N- or C-limited culture medium enhanced PE degradation by the latter two species. These results suggest that mechanisms involved in PE degradation are related to ligninolytic activity of fungi.

A consortium of ascomycete strains from a mangrove wetland was capable of degrading LDPE (Ameen et al. 2015). These fungi accumulated significantly greater biomass, produced more ligninolytic enzymes, and released larger volumes of CO_2 during growth on LDPE compared to controls. Brunner et al. (2018) isolated more than one hundred fungal strains occurring on plastic debris from the shoreline of Lake Zurich, Switzerland. Four strains were capable of degrading PUR: the litter-saprotrophic fungi *Cladosporium*

Table 11.2 Decomposition of MPs/Plastics by Fungal Type, Recent Data

Genus	*Strain*	*Source*	*MP/Plastic Type*	*Duration (Days)*	*Results / Analytical Methods*	*References*
Aspergillus carbonarius and *A. fumigatus*	*Aspergillus carbonarius* MH 856457.1 and *A. fumigatus* MF 276893	Landfills in Sharqiyah Governorate, Egypt	LDPE	112	Mixed culture of two strains showed excellent weight loss % of sheets as compared to a single isolate.	El-Sayed et al. 2021
Bjerkandera adusta	TBB-03	Ohgap Mountains, South Korea	HDPE	90	Lignocellulose substrates were added to the cultures. Morphological changes after 90 days were evident via SEM: cracks formed on surfaces of HDPE samples. Various changes in Raman spectra.	Kang et al. 2019
Aspergillus flavus	PEDX3	Isolated from gut contents of wax moth *Galleria mellonella*	PE	28	HDPE degraded to lower MW.	Zhang et al. 2019
Aspergillus terreus, Aspergillus sydowii	*A. terreus* MANGF1/WL and *A. sydowii* PNPF15/TS	Rhizosphere soil of *Avicennia marina*	PE	60	Fungal strains were efficient at PE degradation based on weight loss and reduction in tensile strength (FTIR and SEM).	Sangale et al. 2019

(continued)

Table 11.2 (Continued)

Genus	*Strain*	*Source*	*MP/Plastic Type*	*Duration (Days)*	*Results / Analytical Methods*	*References*
Aspergillus brasiliensis, Penicillium funiculosum, Chaetomium globosum, Trichoderma virens, and *Paecilomyces variotii*	*A. brasiliensis* (ATCC 9642), *P. funiculosum* (ATCC 11797), *C. globosum* (ATCC 16021), *T. virens* (ATCC 9645), and *P. variotii* (ATCC 16023)	Culture collection	PCL, PVC	28	PVC showed adhesion and growth of *C. globosum* fertile structures (perithecia) after 28 d. Many micropores and cracks, pigmentation, surface erosion, and hyphal adhesion on the PCL surfaces and a mass loss of up to 75%. *Chaetomium globosum* was a pioneer in the colonization and attack of PCL. No evidence of PVC biodegradation. PVC supported the adhesion and growth of its perithecia, suggesting fungal potential for degradation.	Vivi et al. 2019
Trichoderma viride, Aspergillus nomius		Landfill soil	LDPE	45	Two of nine isolates showed best growth response in media containing LDPE. *T. viride* and *A. nomius* reduced weight of LDPE films by 5.13% and 6.63%, respectively.	Munir et al. 2018
Aspergillus oryzae	A5, 1 (MG779508).	Dump site	LDPE	112	Greatest mean weight reduction was 36.4%.	Muhonja et al. 2018
Cladosporium cladosporioides, Xepiculopsis gramínea, Penicillium griseofulvum, Leptosphaeria sp.		Lake Zurich shoreline	PE and PUR		No strain could degrade PE; however, four strains were able to degrade PUR. Growing cultures produced a visible clearance zone ("halo") as a result of enzymatic degradation (between 6–14 days of growth).	Brunner et al. 2018

Monascus ruber, *M. sanguineus*, and *M.* sp.		Soil at open dumps	PUR	14	*Monascus* sp. was the most efficient strain in PPU degradation. SEM micrographs showed complex formations between the PPU and hyphae (after 14 d). Protease, esterase, and lipase were detected. External factors such as pH, organic matter content and heavy metals may have a great effect on the enzyme biosynthesis by fungi.	El-Morsy et al. 2017
Penicillium oxalicum and *P. chrysogenum*	*P. oxalicum* NS4 (KU559906) and *P. chrysogenum* NS10 (KU559907)	Waste disposal site	HDPE and LDPE	90	Morphological changes observed on PE sheets (SEM and FTIR spectroscopy used). Better degradation of LDPE and HDPE by the isolates is attributed to their ability to form a biofilm.	Ojha et al. 2017
Zalerion maritimum	ATTC 34329	–	PE	7, 14, 21, 28	Was able to utilize PE microplastics (requiring minimum nutrients), showing decreases in both size and mass of pellets after 28 d.	Paço et al. 2017
Aspergillus tubingensis		Soil at municipal waste disposal site	PPU	60	PUR film was totally degraded into smaller pieces (SEM and FTIR).	Khan et al. 2017

(*continued*)

Table 11.2 (Continued)

Genus	*Strain*	*Source*	*MP/Plastic Type*	*Duration (Days)*	*Results / Analytical Methods*	*References*
Fusarium oxysporum, Aspergillus fumigatus, Lasiodiplodia crassispora, A. niger, Penicillium sp., *Trichoderma harzianum.*		Soil and landfill	LDPE, PUR	30, 60, 90	*F. oxysporum* and *A. niger* had greatest biodegradation efficiency of LDPE and PUR after 90 d. Morphological damage and halos around colonies on both LDPE and PUR observed.	Raghavendra et al. 2016
Aspergillus niger, A. flavus		Soil at waste site	LDPE	180	Reduction in MW of 26% (with *A. niger*) and 16% (with *A. flavus*).	Deepika and Jaya 2015
Aspergillus caespitosus, Phialophora alba, Paecilomyces variotii, A. terreus, Alternaria alternate, Eupenicillium hirayamae		Tidal water, floating debris, sediment in mangrove ecosystem	LDPE	28	Strains grew on LDPE. Biodegradation shown using SEM, CO_2 emission, and enzyme activities. Enzymatic activity (laccase, MnP, and LiP) identified after 28 d.	Ameen et al. 2015
Gloeophyllum trabeum	*G. trabeum* DSM 1398	Culture collection	PSS	20	Up to 50% reduction in molecular mass within 20 d.	Krueger et al. 2015
Aspergillus tubingensis	VRKPT1	Marine coast	HDPE	30	Efficient at HDPE degradation; virgin PE was used as C source. HDPE film became rough and cracked.	Devi et al. 2015
Aspergillus flavus	VRKPT2	Marine coast	HDPE	30	Efficient at HDPE degradation; virgin PE used as C source. HDPE film became rough and cracked.	Devi et al. 2015

Fusarium sp., *Penicillium* sp., *Aspergillus niger*, *A. Japonicus*, *A. flavus*, *Mucor* sp.		Soil	LDPE	28	In terms of weight loss LDPE was degraded by 36%, 32%, and 30% by *A. japonicas*, *Fusarium* sp., and *A. flavus*, respectively. The other strains degraded LDPE by approximately 20%.	Singh and Gupta 2014
Aspergillus sp., *Fusarium* sp.		–	LDPE	60	Efficacy of polymer degradation was analyzed by weight reduction, change in pH and CO_2 evolution. Decrease in weight of LDPE films was observed over a period of 60 d.	Das and Kumar 2014
Aspergillus terreus	MF12	Disposal site	HDPE		The strain was found to be efficient in degrading HDPE by weight loss after 30 days (SEM, FTIR spectroscopy, and GC–MS were used). Abiotic physical and chemical treatments (UV and $KMnO_4$/HCl) enhanced biodegradation of HDPE using *A. terreus* MF12.	Balasubramanian et al. 2014
Penicillium sp.		Waste disposal site	PET	28	Structural and chemical changes detected in PET powder and flakes (SEM and FTIR spectral analysis).	Sepperumal et al. 2013
Aspergillus niger (mixed culture with *Lysinibacillus xylanilyticus*)		Landfill soil	LDPE	126	Biodegradation of UV-irradiated and non-UV-irradiated LDPE films (as shown by CO_2 evolution, changes in carbonyl index and other tests) were 29% and 16%, respectively.	Esmaeili et al. 2013

(continued)

Table 11.2 (Continued)

Genus	*Strain*	*Source*	*MP/Plastic Type*	*Duration (Days)*	*Results / Analytical Methods*	*References*
Phanerochaete chrysosporium, Lentinus tigrinus, Aspergillus niger, A. sydowii		Synthesized in lab	PVC	300	A significant change in color and surface deterioration of PVC films were observed. Increase in biomass of fungal strains indicated utilization of the plastics as a C source. CO_2 production confirmed catabolic ability of fungi for PVC.	Ali et al. 2013
Mucor hiemalis, Aspergillus versicolor, A. niger, A. flavus, Penicillium sp., *Chaetomium globosum, Fusarium oxysporum, F. solani, Phoma* spp., *Chrysonilia setophila*		Soil	PE and PVC	60	*A. flavus, F. oxysporum,* and *Phoma* spp. were able to efficiently break down both plastics (FTIR spectroscopy, SEM, and changes in CO_2 production observed).	Sakhalkar and Mishra 2013
Aspergillus versicolor, Aspergillus sp.		Seawater	LDPE	7 and 17	LDPE was the sole C source. LDPE was degraded with release of CO_2. *A. versicolor* and *Aspergillus* sp. evolved approximately 4.1 and 3.9 g/L of CO_2, respectively in one week.	Sindujaa et al. 2011
Pestalotiopsis microspora (and other endophytic fungi)		Ecuadorian Amazonian plant samples	PUR	16	Isolates grew on PUR as a sole C source under both aerobic and anaerobic conditions. *P. microspora* partially degraded PUR after 16 days anaerobic incubation. A serine hydrolase was responsible for degradation of PUR.	Russell et al. 2011

Penicillium funiculosum	Disposal site	PET	84	Changes of polymeric chains detected (by FTIR spectroscopy and XPS analysis). Significant reduction in quantity of aromatic rings derived from terephthalic acid indicated. Decomposition of films by fungi occurred due to hydrolytic and oxidative enzymes.	Nowak et al. 2011
Fusarium sp.	Composted soil	Polyamide 4 (nylon 4)	60	Morphological damages observed on surface of nylon 4 films (SEM).	Tachibana et al. 2010
Aspergillus fumigatus, A. terreus, Fusarium solani	Landfill	LDPE	100	Photo-oxidation (25 days under UV-irradiation) used as a pretreatment. *A. terreus* and *A. fumigatus* utilized LDPE as a C source (SEM and FTIR).	Zahra et al. 2010
Thermomyces		Low-crystalline PET film		97% weight loss within 96 hours of reaction at 70°C. Fungal cutinase degraded PET to TPA. As degradation progressed, crystallinity of the remaining film increased due to preferential degradation of amorphous regions.	Ronkvist et al. 2009
Fusarium solani		Low-crystalline PET film		Fungal cutinase degraded PET to TPA. See previous entry.	Ronkvist et al. 2009
Penicillium citrinum	Landfill	PET	21	Polyesterase from *P. citrinum* can hydrolyze PET. Surface hydrolysis with the enzyme leads to increase in hydrophilicity.	Liebminger et al. 2007

(*continued*)

Table 11.2 (Continued)

Genus	*Strain*	*Source*	*MP/Plastic Type*	*Duration (Days)*	*Results / Analytical Methods*	*References*
Aspergillus niger and *Penicillium pinophilum*	ATCC 11797	Purchase from commercial supplier	LDPE	930	PE was thermo-oxidized (80°C) for 15 d. Significant morphological and structural changes of biologically treated MP. Mineralization of 0.57% and 0.37% by *A. niger* and *P. pinophilum*, respectively.	Volke-Sepúlveda et al. 2002
Penicillium simplicissimum YK		Soil and leaves	PE	90	PE was irradiated for 500 h with UV light. PE had lower molecular weights after 3 months; nascent functional groups aided PE biodegradation. Efficiency of PE degradation depended on growth phase of fungus.	Yamada-Onodera et al. 2001
Phanerochaete chrysosporium and *Trarnetes versicolor*	*P. Chrysosporium* IZU-154	–	PE	12	Addition of Mn(II) into N- or C-limited media enhanced PE degradation. PE degradation may therefore be related to ligninolytic activity of lignin-degrading fungi.	Iyoshi et al. 1998.

HDPE—High-density polyethylene; LDPE—Low-density polyethylene; PE—Polyethylene; PBS—Polybutylene succinate; PBSa—Polybutylene succinate-*co*-adipate; PCL—Polycaprolactone; PES—Polyethylene succinate; PHBVa—Polyhydroxybutylate-*co*-valerate; PLA—Polylactic acid; PPU—Polyester polyurethane; PS—Polystyrene; PSS—Polystyrene sulfonate; PUR—Polyurethane; TPA—Terephthalic acid *cladosporioides*, *Xepiculopsis graminea*, and *Penicillium griseofulvum*, and the plant pathogen *Leptosphaeria* sp. In a study by Russell et al. (2011), two *Pestalotiopsis microspora* isolates grew on PUR as the sole C source under both aerobic and anaerobic conditions. A serine hydrolase was proposed as the key enzyme responsible for PUR degradation.

Fungi occurring in the gut of various invertebrates have been examined for biodegradation of MPs. A PE-degrading fungus *Aspergillus flavus* PEDX3 was isolated from the gut contents of the greater wax moth *Galleria mellonella* (Zhang et al. 2020). HDPE MPs were degraded to lower MW particles by the fungus after 28 days incubation. Carbonyl and ether groups were detected on MPs using FTIR. Two laccase-like multicopper oxidase genes, AFLA 006190 and AFLA 053930, displayed up-regulated expression during the degradation process. The data suggest that these may be candidate PE-degrading enzymes.

The rate of MPs degradation by fungi appears to exceed those of bacterial strains. Regardless, bacteria are easier to grow compared to fungi, which require more stable conditions for growth. Bacteria can flourish in a wider range of environments, are more tolerant of stress (e.g., salinity, temperature, pH), and possess more diverse nutrient regimes.

OTHER MICROBIAL TYPES

Limited studies have demonstrated the capability of algal species to degrade plastic polymers. Algae can colonize the MP surface via secretion of EPS, and colonization may result in efficient degradation. The presence of polymers, including plastics, can stimulate the production of EPS (Song et al. 2020b). Filamentous blue-green algae such as *Anabaena spiroides* are able to colonize and grow on the surface of PE (Kumar et al. 2017). Certain species of diatoms and cyanobacteria have demonstrated the ability to degrade PE (Kumar et al. 2017; Moog et al. 2019; Ali et al. 2021). Two blue-green algal species, *Phormidium lucidum* and *Oscillatoria subbrevis*, were capable of degradation of low-density PE; the algae were able to use low-density PE as a carbon and energy source (Sarmah and Rout 2018).

Algae and other microbial degraders can produce H_2O_2. Extracellular superoxide and H_2O_2 were produced by five marine harmful bloom-forming algae species, including *Aureococcus anophagefferens*, *Pseudonitzschia* sp., *Heterosigma akashiwo*, *Chattonella marina*, and *Karenia brevis* (Diaz et al. 2018). Since H_2O_2 is a powerful oxidizing agent, such algal species hold promise for degradation of MPs.

FACTORS AFFECTING MICROBIAL-MEDIATED DEGRADATION OF MPs/PLASTICS

Numerous variables influence biodegradation of recalcitrant organic compounds, which must be considered for successful treatment of MPs in an engineered system such as a dedicated bioreactor.

Polymer types and environmental factors affect microbial growth and activity in bioreactors. Reactor design is likewise critical; relevant issues include size, configuration, and mode of operation. Where technically feasible, these factors must be optimized and controlled for best reactor performance (figure 11.3). Bioreactor design is discussed below.

Polymer Properties

Molecular weight is a key factor affecting rate of biodegradation of MPs—low MW of the substrate is obviously preferred. Other relevant factors include chemical composition (e.g., C–C backbone versus heteroatoms in backbone), types and amounts of functional groups, presence of additives, polymer density, tacticity, crystallinity, structural properties (branching of polymer chains; types of bonds like C–C, C–O, C–N, esters), hydrophobicity/hydrophilicity, and physical form of polymer (Gewert et al. 2015; Kale et al. 2015; Shah et al. 2008).

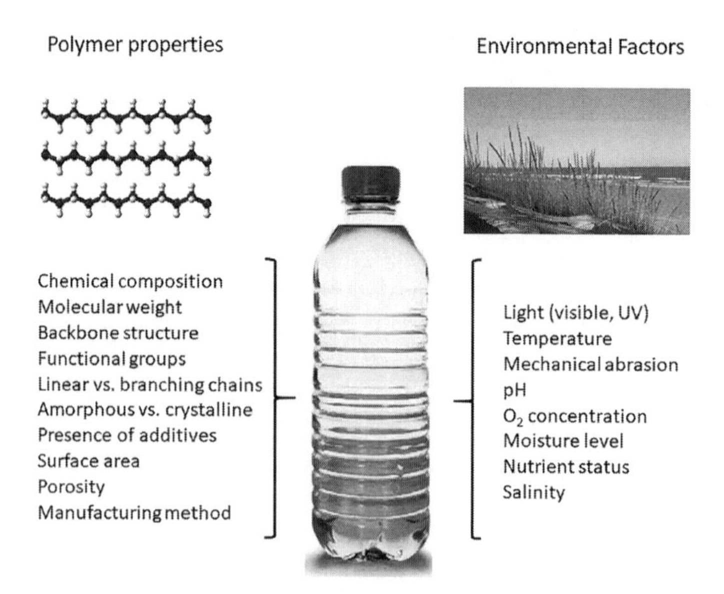

Figure 11.3 Factors Which Influence Microbial Decomposition of MPs

The polymer types in most common use can be divided into two groups (Gewert et al. 2015): (1) plastics with C–C bond (PE, PP, PS, PVC, acrylic) and (2) plastics with heteroatoms in the main chain (PA, PES, PET, PU). Plastics with the C–C backbone are more susceptible to photo-initiated and thermal oxidation, but may be more resistant to hydrolysis and biodegradation (Hu et al. 2021; Ali et al. 2021). Polymers with heteroatoms in the backbone chain tend to have better thermal stability but are prone to hydrolytic cleavage of certain linkages such as amide and ester bonds (Gewert et al. 2015; Hu et al. 2021). These polymers may be transformed to some extent through photooxidation, hydrolysis, and biological degradation (Gewert et al. 2015).

MPs that contain a greater proportion of amorphous components and less crystalline composition tend to be more susceptible to microbial action (Tokiwa et al. 2009; Castilla-Cortázar et al. 2012; Taniguchi et al. 2019). In the amorphous zone of a polymer, the arrangement of chains is less orderly compared to that of crystalline regions (see figure 2.8). The amorphous region could, therefore, provide greater area for microbial proliferation and water and nutrient retention. Crystallinity of PET particles tends to increase after microbial treatment, thus indicating the preferential decomposition of the amorphous zones (Gong et al. 2018).

The melting temperature of a polymer imparts a significant effect on enzymatic degradability. For example, the higher the melting point of polyester, the lower its biodegradability (Kale et al. 2015). Additives such as plasticizers, pigments, antioxidants, and stabilizers used in polymer manufacture can slow degradation rate and possibly be detrimental to microbial growth (Arutchelvi et al. 2008). Certain additives such as biocides are directly toxic to biota. Lastly, the physical form of the MP (e.g., pellet, foam, film, fiber, powder) may influence biodegradability of the polymer—the greater the available surface area, the more rapid the chemical and biochemical processes.

Environmental Factors

Temperature

Temperature is critical for optimizing operating efficiency of a bioreactor to achieve maximum degradation of MPs. Each microbial type possesses a temperature optimum where enzymes function most efficiently, metabolism is maximized, and a favorable rate of biodegradation is sustained.

Increased temperatures may result in increased solubility of certain polymers and increased fluidity and diffusion rates. Bioreactors are commonly designed with provision for temperature control (Tekere 2019).

Oxygen

Among the various microbial types, the aerobic heterotrophs offer the greatest potential for degradation of organics, as decomposition is most efficient—the greatest energy yield per mole of substrate is available. Microorganisms established in aqueous reactors may be supplied with oxygen by bubbling air into the base of the vessel; alternatively, oxygen-supplying compounds (e.g., hydrogen peroxide) may be injected into the reaction chamber. MP treatment processes that make use of anaerobic microorganisms (i.e., do not require free oxygen) are also feasible. Anaerobic processes are used in WWTPs for biomass conversion to useful gases, and in environmental remediation, for example, in the reductive dechlorination of contaminant molecules such as trichloroethylene.

pH

Microbial types have optimum pH regimes for survival, which are directly related to pH optima for metabolic enzyme activities. Suspension pH also influences ionic properties within the microbial cell (Tekere 2019). Most bacteria tend to grow optimally at pH 6–7.5; some bacteria, however, grow best at acidic pH regimes (*acidophiles*) or at alkaline pH (*alkaliphiles*). Fungi typically have pH optima which are lower than that of bacteria. Reactor operating pH must be established to provide the ideal pH for growth and enzyme activities.

Availability of nutrients to microorganisms is also strongly influenced by suspension pH. In the near-neutral pH range, trace metals (e.g., Cu, Co, Ni, Zn) tend to be available in micro quantities, which is generally adequate for most biota. Excess quantities of these metals, for example under acidic pH regimes, are toxic and inhibitory. Also at neutral pH, P is maximally available. A suspension pH of 5.5–7.5 is therefore generally recommended in the bioreactor.

Suspension pH is adjusted using dilute acids (mineral or organic) or bases, and stabilized with buffers. In many cases, vigorous microbial activity will alter pH. Therefore, pH must be monitored continuously and maintained within the optimum range.

Nutrients

Heterotrophic bacteria, actinomycetes, and fungi possess fairly complex nutritional requirements. Nutrients serve the following functions: providing the materials necessary for synthesis of cellular biomass; supplying the needed energy for cell growth and biosynthetic reactions; and serving as electron acceptors for energy-related reactions in the cell (table 11.3).

Table 11.3 Nutrients Required by Heterotrophic Microorganisms

Inorganics (minerals)		N, P, K, Ca, Mg, S, Fe, Mn, Cu, Zn, Co, Mo
Carbon sources		CO_2, HCO_3^-
Energy sources		Organic compounds (glucose, polysaccharides, cellulose, hemicellulose, lignin, pectic substances, inulin, organic acids)
Electron acceptors	Aerobes	O_2
	Anaerobes	NO_3^-, SO_4^{2-}, Fe^{3+}, CO_2
	Fermenters	Organic compounds
Growth factors	Amino acids	Alanine, cysteine, histidine, serine, tyrosine, etc.
	Vitamins	Nicotinic acid, riboflavin, pantothenic acid, biotin, PABA, pyridoxine, thiamine, B_{12}, folic acid, etc.
	Other	Purine bases, pyrimidine bases, peptides, etc.

Nitrogen, P, K, Mg, S, Fe, Ca, Mn, Zn, Cu, Co, and Mo are integral to the protoplasmic composition of the microbial cell; these nutrients, plus C, H, and O are required for cell synthesis. In a bioreactor, mineral nutrients may be supplied as soluble salts (for example, in commercial fertilizer). Carbon may be added in the form of biosolids or animal manures (both of which also supply mineral nutrients) or similar organic material. A balance must be established between the various mineral nutrients (N and P, for example) and carbon, or microorganisms cannot make optimum use of the carbon source.

Ideally, the MP will serve as the primary carbon source. Several bacteria and fungi have been identified that grow on plastic surfaces in a carbon-free medium (Gong et al. 2018; Russell et al. 2011). Alternatively, microbial degradation of MPs may occur during co-metabolism of soluble C sources. These may include sugars, starch, cellulose, hemicellulose, lignin, pectic substances, inulin, chitin, proteins, amino acids, and organic acids. The oxidation of these organics releases energy, and a portion is used in the synthesis of protoplasm.

Carbon dioxide, a product of both aerobic and anaerobic metabolism, is important because of its direct influence on growth. The gas is stimulatory to and often required by many heterotrophs—growth of many species will not proceed in the absence of CO_2. A portion of the CO_2 present is incorporated into the heterotrophic cell structure. The requirement for this gas in a bioreactor should not be an issue because of its continual evolution from decomposing organic compounds (figures 11.1 and 11.2).

Energy generation within a microbial cell is an oxidation process in which electrons are transported within the respiration pathway. Organic substrates (i.e., MPs) are catabolized (i.e., degraded) by microorganisms using one of three general metabolic pathways:

1) In *aerobic respiration*, the substrate is oxidized to carbon dioxide and water or other end-products using molecular oxygen as the terminal electron acceptor. Aerobic respiration occurs under highly oxygenated conditions.
2) If the oxygen supply decreases, *anaerobic respiration* may be initiated. In this mode, microorganisms metabolize hydrocarbons and use inorganic substrates as terminal electron acceptors. Common electron acceptors include nitrate, NO_3^-, sulfate, SO_4^{2-}, and Fe^{3+}. Nitrate is reduced to nitrogen (N_2) or ammonium (NH_4^+), sulfate to sulfide (S_2^-), ferrous iron (Fe^{2+}) to ferric iron (Fe^{3+}), and CO_2 to methane (CH_4).
3) Under highly reducing conditions and in the absence of inorganic electron acceptors, *fermentation* occurs. Organic compounds function as electron acceptors. Fermentation results in end-products including acetate, ethanol, propionate, and butyrate.

If anaerobic respiration or fermentation is determined to be the preferred method for MP biodegradation, the appropriate electron acceptors (e.g., NO_3^- / SO_4^{2-} / Fe^{3+}, or organic compound, respectively) must be provided to the reactor.

Organism-Related Factors

Microorganisms comprise the most diverse forms of life and have developed a range of metabolic pathways that enable them to tolerate varying and often stressful environmental conditions. Additionally, some are known to act upon and decompose xenobiotics. A range of environments including aerobic, anaerobic, acidic, alkaline, and having extremes in temperature have been utilized as sources of microbial inocula for bioremediation.

Soil and wastewater already contaminated with plastics may serve as sources of acclimatized microorganisms (Lladó et al. 2013). Various strains capable of degrading PE have been isolated from environments including mulch films, soil contaminated by crude oil, sewage sludge, landfills, and marine water. Some strains have shown the ability to utilize un-pretreated PE as a carbon source (Ru et al. 2020). Microbial species capable of metabolizing a substrate having structural similarity to MPs and/or adaptation to certain environmental conditions can be selected for MP degradation. As mentioned previously, several lignin-degrading fungi may be effective for polymer degradation. The addition of lignocellulosic wastes during biological pretreatment stage may promote the proliferation of these fungi.

In the natural environment, pollutant degradation typically occurs via complex microbial community interactions. Mixed microbial consortia are therefore used in bioreactors for degradation of target pollutants. In some cases, bioaugmentation (i.e., incorporation of exogenous microbial types recognized to carry out a desired reaction) has proven beneficial. Bioaugmentation has been successfully employed in soil remediation, sewage sludge treatment, fermentation processes, and other bio-treatment systems (Tao et al. 2018; Cycoń et al. 2017). Certain exogenous species may experience rapid growth and have high aggressiveness and fecundity, as well as great competitiveness. These properties can potentially disrupt diversity and composition of the indigenous microbial community (Tao et al. 2018). In other cases, however, concerns exist that introduced organisms may not be able to coexist with indigenous populations without being preyed upon or out-competed for resources.

Efforts are also underway to engineer recombinant bacteria with enhanced capabilities for polymer bioconversion (Mazzoli et al. 2012).

The Significance of Biofilms

A wide range of microbial types including protists, algae, bacteria, fungi, and viruses attach to MP surfaces. Such microbial colonization results in formation of biofilms that includes cell secretions and organic and inorganic particles, which comprise some rather complex ecosystems. These microenvironments harbor diverse species of both aerobic and anaerobic organisms, allowing them to survive under a range of conditions.

The ability for microbial cells to associate in structured, sessile biofilm networks offers numerous advantages over free-living planktonic cells. Aggregation of microorganisms in biofilms supports increased tolerance toward shifts in pH, salinity, nutrient levels, and temperature, as well as protection against predation and exposure to high concentrations of toxins. This array of beneficial properties offers a resilient system for biodegradation of MPs (Pichtel 2017).

Biofilms enhance the ability of cells to communicate, and may serve as sites for transfer of genetic material (Christensen and Characklis 1990). Horizontal transfer of genes on conjugative plasmids has been shown to induce planktonic bacteria to form biofilm communities, which favors additional gene transfer (Ghigo 2001). Therefore, it is possible that conjugation, and the evolution of new genetic traits, will be concentrated in biofilm communities (Stach and Burns 2002).

Biofilms alter the structure of MPs by degrading polymer chains and additives, secreting MP-degrading enzymes, and releasing metabolic by-products

(Miao et al. 2019). Biofilm-directed degradation of MPs is more complex than that observed for pure bacterial and fungal cultures and has been divided into four stages (Flemming 1998):

1) Microbes adhere to the MP surface and alter surface properties such as hydrophilicity and adhesion;
2) Enzymatic degradation accelerates leaching of additives and monomers from the MPs;
3) Enzymes or radicals of biological origin attack MPs and additives, resulting in embrittlement and loss of mechanical stability;
4) Water and microbial filaments penetrate MP surfaces with consequent decomposition and further metabolism of the MPs by microorganisms.

Interaction of microbes with MPs and formation of biofilms differ as a function of polymer type (Fotopoulou and Karapanagioti 2017). This is a consequence of the formation of different chemical functional groups during contact with UV radiation, oxygen, salinity, and other environmental influences.

Lobelle and Cunliffe (2011) investigated early biofilm formation on surfaces of PE MPs. After one week, biofilms were visible on particle surfaces and increased in size during subsequent weeks. The number of culturable heterotrophic bacteria on MPs increased from 1.4×10^4 cells·cm^{-2} in the first week to 1.2×10^5 cells·cm^{-2} at week three. During this period the hydrophilicity of particles increased significantly, and MPs began to sink from the seawater-air interface. The MPs were partially degraded due to the action of the biofilms. In a microcosm study Hossain et al. (2019) tested four fresh and two beached MPs with *Acinetobacter alcoaceticus, Burkholderia cepacia*, *E. coli*, and others. Weathered MPs had highest bacterial numbers, thus revealing the importance of surface roughness and other MP physicochemical properties in bacterial colonization.

Additional discussion of biofilms and their importance to MPs removal appears in chapter 7.

MPs DEGRADATION USING BIOREACTOR TECHNOLOGY

Application of a well-designed bioreactor is recognized as an effective means to support microbial growth and reactions in a controlled environment that provides the necessary conditions for degradation of the target contaminant (figure 11.4) (Tekere 2019). Ideally, in the reactor, a substrate is utilized by living cells or enzymes to generate a product of higher value. Products

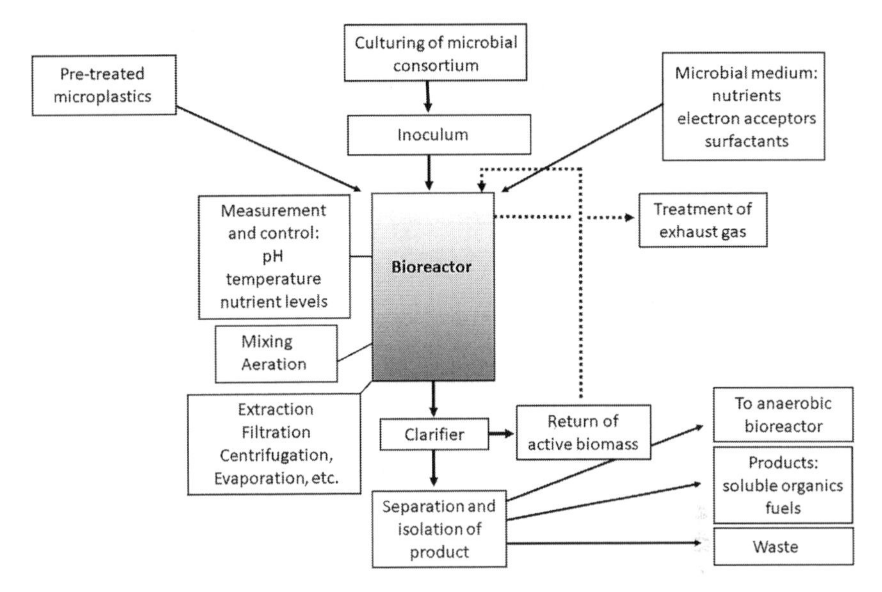

Figure 11.4 Practical Considerations for MPs Treatment in a Bioreactor
Source: Adapted from *Biotechnology and Bioengineering*, Jacob-Lopez, E., and Zepka, L.Q. (eds). Microbial Bioremediation and Different Bioreactors Designs Applied. Tekere, M. p. 1–19 (2019). Intech Open.

of MPs depolymerization could be exploited for biosynthesis of high-value chemicals via specific metabolic pathways (Ru et al. 2020).

Pretreatment of the MP Feedstock

MP decomposition by microorganisms is accelerated when MPs are first subjected to abiotic treatment such as exposure to light, heat, and/or oxidizing chemicals (Gewert et al. 2015). Sunlight (in particular UV radiation) may have a critical effect on microbial populations and enzyme activities (Gu 2003), which ultimately affect polymer biodegradation. Sunlight also initiates chemical and physical degradation of MP surfaces. According to several reports, partial biodegradation of PE can be achieved after UV irradiation, thermal treatment, and/or oxidation with mineral acid (Satlewal et al. 2008).

Abiotic processes result in scission of polymer chains and conversion to low-MW compounds, which can subsequently be metabolized by microorganisms (Andrady 2011). However, the process is slow (Ali et al. 2021).

In an early study (Albertsson et al. 1987), PE was incubated in water and soil. Treatment with UV light and/or oxidizing agents resulted in creation of carbonyl groups. A strain of *Penicillium simplicissimum* YK experienced better growth on PE which was irradiated with UV light for 500 hours; little growth occurred when the plastic was not irradiated (Yamada-Onodera et al. 2001). Pretreatment with UV light and incubation with HNO_3 at 80°C for six days caused functional groups (carbonyl, carboxyl, and ester) to form on PE surfaces, consequently increasing the hydrophilicity of the polymer. After three months incubation with the fungus, lower MW of PE was measured (Yamada-Onodera et al. 2001).

Figure 11.5 presents a suggested pretreatment train for MP wastes via physicochemical degradation.

The Bioreactor

The bioreactor system consists of the reactor vessel, sensors, control system, and software to monitor and control the conditions taking place within the reactor (Kaur and Dev Sharma 2021). The vessel system includes:

1) agitation for mixing of cells with medium including nutrients and electron acceptors;
2) aeration (to provide O_2 for aerobic populations);
3) baffles, to prevent vortex formation and improve aeration efficiency;
4) regulation of variables including temperature, pH, pressure, aeration, nutrient feeding, liquid level, etc.;
5) sterilization and maintenance of sterility; and
6) withdrawal of cells/medium (for continuous processes) (Kaur and Dev Sharma 2021).

Continuous Stirred Tank Bioreactor

Many bioreactor designs are available, including the bubble column, airlift, fluidized bed, continuous stirred tank reactor (CSTR), and others. The CSTR is

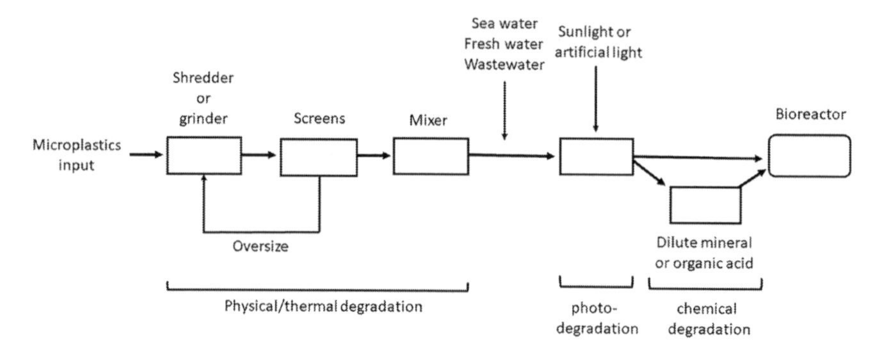

Figure 11.5 Pretreatment Train for a Bioreactor

the most common type of aerobic bioreactor in use today and may be applied for MPs decomposition. The CSTR system consists of a cylindrical vessel with motor-driven central shaft that supports one or more agitators (impellers) (figure 11.6). Gas is typically introduced below the mixing impeller. The MPs-enriched wastewater to be treated is added at the top of the reactor. The agitators provide mixing of microbial biomass and substrate, and also serve as gas-dispersing tools and provide high values of mass transfer rates (Tekere 2019).

Effective mixing in the bioreactor is essential; poor mixing may negatively affect microbial process efficiency (Tekere 2019). Mixing prevents thermal stratification, helps maintain uniform conditions in the reactor, and ensures good contact between microbial populations and growth media.

Hydraulic retention times (HRT) required to achieve the necessary biodegradation goals in the bioreactor must be calculated and optimized. Long HRTs result in inefficient substrate loading and processing. Shorter HRTs may not allow microorganisms to degrade the pollutant effectively, and can result in microbial washout from the system (Chelliapan et al. 2011).

A CSTR was used to treat hydrocarbon-rich industrial wastewaters using an acclimatized microbial consortium (Gargouri et al. 2011). After an experimental period of 225 days, the process was shown to be effective in decontaminating the wastewater—COD was reduced by up to 95%.

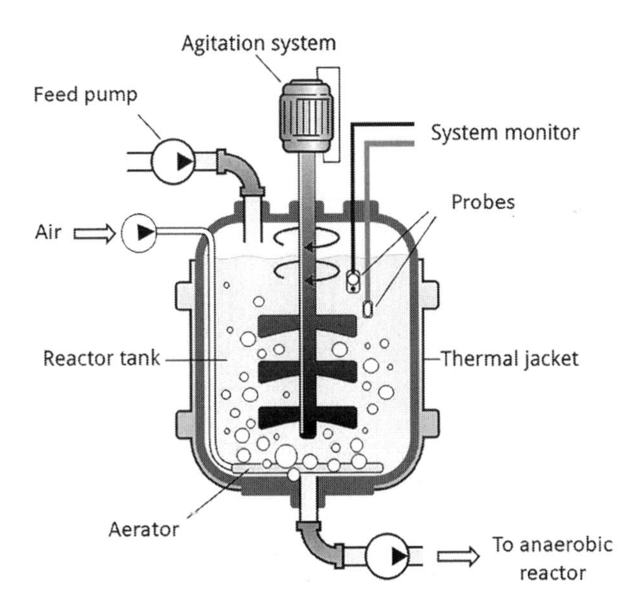

Figure 11.6 Generalized Schematic of a Proposed CSTR for MPs Decomposition
Source: Adapted from G. Yassine Mrabet (own work), CC BY-SA 3.0, https://commons.wikimedia.org/w/index.php?curid=8301774, Creative Commons Attribution-Share Alike 3.0 Unported license.

Advantages of the CSTR include efficient gas transfer to the active biomass, good mixing of contents, and flexible operating conditions. These bioreactors are also commercially available. A disadvantage of the stirred tank bioreactor is that mechanical agitation requires energy input; furthermore, stirring can cause shear strain on microbial cells (Tekere 2019).

Anaerobic Phase

The residue from the aerobic reactor, including unreacted MPs and intermediate compounds, can be transferred to an anaerobic vessel for additional biotreatment of the MPs and their decomposition products. A new microbial consortium, possibly containing methanogenic bacteria, can be added. A portion of the MPs may be converted to biogas including methane (CH_4), carbon dioxide (CO_2), and other gases.

OTHER INNOVATIVE BIOLOGICAL TECHNOLOGIES

Composting is primarily an aerobic biological process where a diverse consortium of microorganisms, acting concurrently, decompose organic substrates to a stable, humus-like material, CO_2, and H_2O. The most active players in composting are bacteria, actinomycetes, fungi, and protozoa. These organisms are naturally present in soil, plant debris, and other organic materials. Many composting systems are available for waste decomposition, both in closed and open systems. Composting can be adapted for decomposition of MPs.

Hyperthermophilic composting (*h*TC) has recently been developed to exploit the activity of hyperthermophilic bacteria for MP degradation. Common pile temperature in hyperthermic composting is approximately 90°C, which is about 20–30°C higher than that of conventional thermophilic composting (Yu et al. 2018). *h*TC was applied for in situ biodegradation of 200 tons of sludge-based MPs (Chen et al. 2020). MPs identified in the pile were primarily PE, PET, PE, and PP, at an estimated concentration of 7.4×10^4 items·kg^{-1}. After 45 d of *h*TC treatment, 43.7% of the MPs was eliminated from the pile. The extreme temperatures in *h*TC resulted in generation of functional groups such as C=O and C–O thus improving hydrophilic characteristics of the MPs. Several water-soluble degradation products were detected in the residue of the compost. Bacteria from the genus *Thermus* (54.2%), *Bacillus* (24.8%), and *Geobacillus* (19.6%) presumably played an important role in biodegradation of MPs under *h*TC conditions. Abundance, diversity, and activation of many bacteria or enzymes decreased at 85°C (Chen et al. 2020).

Periphytic biofilms are an environmentally benign method for MP biodegradation by virtue of their ubiquity in nature. Shabbir et al. (2020) assessed biodegradation of PE, PP, and PET MPs using a periphytic biofilm. After 60 d incubation with glucose as co-substrate, biodegradation increased by 9.5–18.0%, 5.9–14.0%, and 13.2–19.7% for PP, PE, and PET, respectively, compared to biofilm alone. Results from MiSeq sequencing revealed that in natural biofilms the most prevalent phyla were *Deinococcus-thermus* > *Proteobacteria* > *Cyanobacteria*. It was concluded that changes in microbial community structure after addition of the carbon sources altered the extent and pathways of biodegradation. Under controlled conditions, this technology holds potential for aquatic environments polluted with MPs, including wastewater treatment systems (Shabbir et al. 2020).

CLOSING STATEMENTS

Certain bacterial and fungal types have been identified that degrade MPs; however, the weight-loss rate tends to be low. Furthermore, microbial-mediated degradation is a relatively long process, often requiring months to years (Yoshida et al. 2016; Jeon and Kim 2013; Muhonja et al. 2018). Such rates are clearly impractical in a municipal facility dedicated to water or wastewater treatment. As MPs are only moderately biodegradable, degradation cannot be efficiently carried out by the microbial types currently identified. Successful systems for biodegradation of MPs must address identification and support of bacterial and fungal strains that use MPs efficiently as a C source.

Degradation of MPs by bacterial consortia holds promise by virtue of the potential interactions among diverse microbial types and the participation of varied enzymes. Additional consortia must be identified, along with relevant enzymes and mechanisms involved in MP decomposition. The ecological interactions and requirements of the key participating organisms must be better understood in order to optimize growth and degradation activity.

Several simple and low-cost technologies such as *h*TC technology may be highly successful in MP destruction and deserve further study. Likewise, both aerobic and anaerobic bioreactors may be effective. Such degradation would be optimized by incorporating chemical and/or physical pretreatment of MPs.

Given that no microbial method degrades MPs completely within a reasonable interval, research must address combining biodegradation with other methods such as advanced oxidation processes.

REFERENCES

Albertsson, A.-C., Andersson, S.O., and Karlsson, S. (1987). The mechanism of biodegradation of polyethylene. *Polymer Degradation and Stability* 18(1):73–87.

Ali, S.S., Elsamahy, T., Al-Tohamy, R., Zhu, D., Mahmoud, A.-G.Y., Koutra, E., Metwally, M.A., Kornaros, M., and Sun, J. (2021). Plastic wastes biodegradation: Mechanisms, challenges and future prospects. *Science of the Total Environment* 780:146590.

Ameen, F., Moslem, M., Hadi, S., and Al-Sabri, A.E. (2015). Biodegradation of low density polyethylene (LDPE) by mangrove fungi from the red sea coast. *Progress in Rubber, Plastics and Recycling Technology* 31(2):125–143.

Andrady, A.L. (2011). Microplastics in the marine environment. *Marine Pollution Bulletin* 62(8):1596–1605.

Arutchelvi, J., Sudhakar, M., Arkatkar, A., Doble, M., Bhaduri, S., and Uppara, P.V. (2008). Biodegradation of polyethylene and polypropylene. *Indian Journal of Biotechnology* 7:9–22.

Auta, H., Emenike, C., and Fauziah, S. (2017). Screening of Bacillus strains isolated from mangrove ecosystems in peninsular Malaysia for microplastic degradation. *Environmental Pollution* 231:1552–1559.

Auta, H.S., Emenike, C.U., Jayanthi, B., and Fauziah, S.H. (2018). Growth kinetics and biodeterioration of polypropylene microplastics by *Bacillus* sp. and *Rhodococcus* sp. isolated from mangrove sediment. *Marine Pollution Bulletin* 127:15–21.

Balasubramanian, V., Natarajan, K., Rajeshkannan, V., and Perumal, P. (2014). Enhancement of in vitro high-density polyethylene (HDPE) degradation by physical, chemical, and biological treatments. *Environmental Science and Pollution Research* 21:12549–12562.

Bátori, V., Åkesson, D., Zamani, A., Taherzadeh, M.J., and Horváth, I.S. (2018). Anaerobic degradation of bioplastics: A review. *Waste Management* 80:406–413.

Bond, T., Ferrandiz-Mas, V., Felipe-Sotelo, M., and Van Sebille, E. (2018). The occurrence and degradation of aquatic plastic litter based on polymer physicochemical properties: A review. *Critical Review in Environmental Science and Technology* 48(7–9):685–722.

Brandon, A.M., Gao, S.-H., Tian, R., Ning, D., Yang, S.-S., Zhou, J., and Criddle, C.S. (2018). Biodegradation of polyethylene and plastic mixtures in mealworms (larvae of *Tenebrio molitor*) and effects on the gut microbiome. *Environmental Science & Technology* 52(11):6526–6533.

Briceño, G., Levio, M., González, M.E., Saez, J.M., Palma, G., Schalchli, H., and Diez, M.C. (2020). Performance of a continuous stirred tank bioreactor employing an immobilized actinobacteria mixed culture for the removal of organophosphorus pesticides. *3 Biotech* 10:252. DOI: 10.1007/s13205-020-02239-9.

Brunner, I., Fischer,M., Rüthi, J., Stierli, B., and Frey, B. (2018). Ability of fungi isolated from plastic debris floating in the shoreline of a lake to degrade plastics. *PLoS ONE* 13(8):e0202047.

Camarero, S., Sarkar, S., Ruiz-Dueñas, F.J., Martínez, M.J., and Martínez, A.T. (1999). Description of a versatile peroxidase involved in the natural degradation of lignin that has both manganese peroxidase and lignin peroxidase substrate interaction sites. *Journal of Biological Chemistry* 274(15):10324–10330.

Castilla-Cortázar, I., Más-Estellés, J., Meseguer-Dueñas, J.M., Ivirico, J.E., Marí, B., and Vidaurre, A. (2012). Hydrolytic and enzymatic degradation of a poly (ε-caprolactone) network. *Polymer Degradation and Stability* 97(8):1241–1248.

Chelliapan, S., Wilby, T., Yuzir, A., and Sallis, P.J. (2011). Influence of organic loading on the performance and microbial community structure of an anaerobic stage reactor treating pharmaceutical wastewater. *Desalination* 271(1–3):257–264.

Chen, Z., Zhao, W., Xing, R., Xie, S., Yang, X., Cui, P., and Wang, S. (2020). Enhanced in situ biodegradation of microplastics in sewage sludge using hyperthermophilic composting technology. *Journal of Hazardous Materials* 384:121271.

Christensen, B.E., and Characklis, W.G. (1990). Physical and chemical properties of biofilms. In: Characklis, W.G., and Marshall, K.C., editors. *Biofilms*. New York: John Wiley & Sons, pp. 83–138.

Cycoń, M., Mrozik, A., and Piotrowska-Seget, Z. (2017). Bioaugmentation as a strategy for the remediation of pesticide-polluted soil: A review. *Chemosphere* 172:52–71.

Devi, R.S., Kannan, V.R., Nivas, D., Kannan, K., Chandru, S., and Antony, A.R. (2015). Biodegradation of HDPE by *Aspergillus* spp. from marine ecosystem of Gulf of Mannar, India. *Marine Pollution Bulletin* 96(1–2):32–40.

Diaz, J.M., Plummer, S., Tomas, C., and Alves-de-Souza, C. (2018). Production of extracellular superoxide and hydrogen peroxide by five marine species of harmful bloom-forming algae. *Journal of Plankton Research* 40(6):667–677. DOI: 10.1093/plankt/fby043.

Ebrahimbabaie, P., Yousefi, K., and Pichtel, J. (2022). Photocatalytic and biological technologies for elimination of microplastics in water: Current status. *Science of the Total Environment* 806:150603.

El-Sayed, M.T., Rabie, G.H., and Hamed, E.A. (2021). Biodegradation of low-density polyethylene (LDPE) using the mixed culture of *Aspergillus carbonarius* and *A. fumigates*. *Environment, Development and Sustainability* 23:14556–14584.

El-Shafei, H.A., Abd El-Nasser, N.H., Kansoh, A.L., and Ali, A.M. (1998). Biodegradation of disposable polyethylene by fungi and *Streptomyces* species. *Polymer Degradation and Stability* 62(2):361–365.

Esmaeili, A., Pourbabaee, A.A., Alikhani, H.A., Shabani, F., and Esmaeili, E. (2013). Biodegradation of low-density polyethylene (LDPE) by mixed culture of *Lysinibacillus xylanilyticus* and *Aspergillus niger* in soil. *PLoS ONE* 8(9):e71720.

Flemming, H.-C. (1998). Relevance of biofilms for the biodeterioration of surfaces of polymeric materials. *Polymer Degradation and Stability* 59(1–3):309–315.

Fotopoulou, K.N., and Karapanagioti, H.K. (2017). Degradation of various plastics in the environment. In: Takada, H., and Karapanagioti, H.K., editors.

Hazardous Chemicals Associated With Plastics in the Marine Environment, the Handbook of Environmental Chemistry. Cham (Switzerland): Springer International.

Gangola, S., Joshi, S., Kumar, S., and Pandey, S.C. (2019). Comparative analysis of fungal and bacterial enzymes in biodegradation of xenobiotic compounds. In: *Smart Bioremediation Technologies*. Academic Press, pp. 169–189. DOI: 10.1016/b978-0-12-818307-6.00010-x.

Gewert, B., Plassmann, M.M., and MacLeod, M. (2015). Pathways for degradation of plastic polymers floating in the marine environment. *Environmental Science: Processes & Impacts* 17(9):1513–1521.

Ghigo, J.M. (2001). Natural conjugative plasmids induce bacterial biofilm development. *Nature* 412:442–445.

Ghosh, S., Qureshi, A., and Purohit, H.J. (2019). Microbial degradation of plastics: Biofilms and degradation pathways. In: Kumar, V., Kumar, R., Singh, J., and Kumar, P, editors. *Contaminants in Agriculture and Environment: Health Risks and Remediation*. Haridwar (India): Agro Environ Media.

Glass, J.E., and Swift, G. (1990). Agricultural and synthetic polymers: Biodegradability and utilization. U.S. Department of Energy, Office of Scientific and Technical Information. https://www.osti.gov/biblio/7175073.

Gong, J., Kong, T., Li, Y., Li, Q., Li, Z., and Zhang, J. (2018). Biodegradation of microplastic derived from poly (ethylene terephthalate) with bacterial whole-cell biocatalysts. *Polymers* 10(12):1326.

Gu, J.-D. (2003). Microbiological deterioration and degradation of synthetic polymeric materials: Recent research advances. *International Biodeterioration & Biodegradation* 52(2):69–91.

Habib, S., Iruthayam, A., Yunus Abd Shukor, M., Alias, S.A., Smykla, J., and Yasid, N.A. (2020b). Biodeterioration of untreated polypropylene microplastic particles by Antarctic bacteria. *Polymers* 12:2616. DOI: 10.3390/polym12112616.

Hadad, D., Geresh, S., and Sivan, A. (2005). Biodegradation of polyethylene by the thermophilic bacterium *Brevibacillus borstelensis*. *Journal of Applied Microbiology* 98(5):1093–1100.

Hossain, M.R., Jiang, M., Wei, Q., and Leff, L.G. (2019). Microplastic surface properties affect bacterial colonization in freshwater. *Journal of Basic Microbiology* 59:54–61.

Hu, K., Tian, W., Yang, Y., Nie, G., Zhou, P., Wang, Y., and Wang, S. (2021). Microplastics remediation in aqueous systems: Strategies and technologies. *Water Resources* 198:117144.

Huang, K.C., Mukhopadhyay, R., Wen, B., Gitai, Z., and Wingreen, N.S. (2008). Cell shape and cell-wall organization in Gram-negative bacteria. *Proceedings of the National Academy of Sciences* 105(49):19282–19287.

Huerta Lwanga, E., Thapa, B., Yang, X., Gertsen, H., Salánki, T., Geissen, V., and Garbeva, P. Decay of low-density polyethylene by bacteria extracted from earthworm's guts: A potential for soil restoration. *Science of the Total Environment* 624(2018):753–757.

Iiyoshi, Y., Tsutsumi, Y., and Nishida, T. (1998). Polyethylene degradation by lignin-degrading fungi and manganese peroxidase. *Journal of Wood Science* 44(3):222–229.

Jacquin, J., Cheng, J.G., Odobel, C., Pandin, C., Conan, P., Pujo-Pay, M., Barbe, V., Meistertzheim, A.L., and Ghiglione, J.F. (2019). Microbial ecotoxicology of marine plastic debris: A review on colonization and biodegradation by the "plastisphere." *Frontiers in Microbiology* 10:865.

Jeon, H.J., and Kim, M.N. (2013). Isolation of a thermophilic bacterium capable of low MW polyethylene degradation. *Biodegradation* 24(1):89–98.

Kale, S.K., Deshmukh, A.G., Dudhare, M.S., and Patil, V.B. (2015). Microbial degradation of plastic: A review. *Journal of Biochemical Technology* 6(2):952–961.

Kaur, I., and Dev Sharma, A. (2011). Bioreactor: Design, functions and fermentation innovations. *Research & Reviews in Biotechnology & Biosciences* 8(2):116–125.

Kim, S., Chmely, S.C., Nimlos, M.R., Bomble, Y.J., Foust, T.D., Paton, R.S., and Beckham, G.T., (2011). Computational study of bond dissociation enthalpies for a large range of native and modified lignins. *Journal of Physical Chemistry Letters* 2:2846–2852.

Kumar, R.V., Kanna, G.R., and Elumalai, S. (2017). Biodegradation of polyethylene by green photosynthetic microalgae. *Journal of Bioremediation and Biodegradation* 8(1):1000381.

Kyaw, B.M., Champakalakshmi, R., Sakharkar, M.K., Lim, C.S., and Sakharkar, K.R. (2012). Biodegradation of low-density polythene (LDPE) by *Pseudomonas* species. *Indian Journal of Microbiology* 52(3):411–419.

Li, J., Song, Y., and Cai, Y. (2020). Focus topics on microplastics in soil: Analytical methods, occurrence, transport, and ecological risks. *Environmental Pollution* 257:113570.

Lladó, S., Gràcia, E., Solanas, A.M., and Viñas, M. (2013). Fungal and bacterial microbial community assessment during bioremediation assays in an aged creosote-polluted soil. *Soil Biology and Biochemistry* 67:114–123.

Lobelle, D., and Cunliffe, M. (2011). Early microbial biofilm formation on marine plastic debris. *Marine Pollution Bulletin* 62(1):197–200.

Maheshwari, M., Abulreesh, H.H., Khan, M.S., Ahmad, I., and Pichtel, J. (2017). Horizontal gene transfer in soil and the rhizosphere: Impact on ecological fitness of bacteria. In: *Agriculturally Important Microbes for Sustainable Agriculture*. Berlin: Springer, pp. 111–130.

Manzur, A., Limón-González, M., and Favela-Torres, E. (2004). Biodegradation of physicochemically treated LDPE by a consortium of filamentous fungi. *Journal of Applied Polymer Science* 92(1):265–271.

Matjašič, T., Simčič, T., Medveščék, N., Bajt, O., Dreo, T., and Mori, N. (2021). Critical evaluation of biodegradation studies on synthetic plastics through a systematic literature review. *Science of the Total Environment* 752:141959.

Mazzoli, R., Lamberti, C., and Pessione, E. (2012). Engineering new metabolic capabilities in bacteria: Lessons from recombinant cellulolytic strategies. *Trends in Biotechnology* 30(2):111–119.

Miao, L., Wang, P., Hou, J., Yao, Y., Liu, Z., Liu, S., and Li, T. (2019). Distinct community structure and microbial functions of biofilms colonizing microplastics. *Science of the Total Environment* 650:2395–2402.

Miloloža, M., Cvetnić, M., Kučić Grgić, D., Ocelić Bulatović, V., Ukić, S., Rogošić, M., Dionysiou, D.D., Kušić, H., and Bolanča, T. (2022). Biotreatment strategies for the removal of microplastics from freshwater systems. A review. *Environmental Chemistry Letters.* DOI: 10.1007/s10311-021-01370-0.

Moog, D., Schmitt, J., Senger, J., Zarzycki, J., Rexer, K.H., Linne, U., Erb, T., and Maier, U.G. (2019). Using a marine microalga as a chassis for polyethylene terephthalate (PET) degradation. *Microbial Cell Factories* 18(1):171.

Muhonja, C.N., Makonde, H., Magoma, G., and Imbuga, M. (2018). Biodegradability of polyethylene by bacteria and fungi from Dandora dumpsite Nairobi-Kenya. *PLoS ONE* 13(7):e0198446.

Orhan, Y., and Büyükgüngör, H. (2000). Enhancement of biodegradability of disposable polyethylene in controlled biological soil. *International Biodeterioration and Biodegradation* 45(1–2):49–55.

Paço, A., Duarte, K., da Costa, J.P., Santos, P.S., Pereira, R., Pereira, M., and Rocha-Santos, T.A. (2017). Biodegradation of polyethylene microplastics by the marine fungus *Zalerion maritimum. Science of the Total Environment* 586:10–15.

Park, S.Y., and Kim, C.G. (2019). Biodegradation of micro-polyethylene particles by bacterial colonization of a mixed microbial consortium isolated from a landfill site. *Chemosphere* 222:527–533.

Pasquina-Lemonche, L., Burns, J., Turner, R.D., Kumar, S., Tank, R., Mullin, N., Wilson, J.S., Chakrabarti, B., Bullough, P.A., Foster, S.J., and Hobbs, J.K. (2020). The architecture of the Gram-positive bacterial cell wall. *Nature* 582:294–297.

Pathak, V.M., and Navneet. (2017). Review on the current status of polymer degradation: A microbial approach. *Bioresources and Bioprocessing* 4(1):1–31.

Pichtel, J. (2017). Biofilms for remediation of xenobiotic hydrocarbons – A technical review. In: Ahmad, I., and Husain, F.M., editors. *Biofilms in Plant and Soil Health.* Berlin: Springer.

Pichtel, J. (2019). *Fundamentals of Site Remediation for Metal- and Hydrocarbon-Contaminated Soils*, 3rd ed. Lanham (MD): Bernan Press.

Pinto, M., Langer, T.M., Hüffer, T., Hofmann, T., and Herndl, G.J. (2019). The composition of bacterial communities associated with plastic biofilms differs between different polymers and stages of biofilm succession. *PLoS ONE* 14(6):e0217165.

Quintero, J.C., Lu-Chau, T.A., Moreira, M.T., Feijoo, G., and Lema, J.M. (2007). Bioremediation of HCH present in soil by the white rot fungus *Bjerkandera adusta* in a slurry batch bioreactor. *International Biodeterioration & Biodegradation* 60(4):319–326.

Ru, J., Huo, Y., and Yang, Y. (2020). Microbial degradation and valorization of plastic wastes. *Frontiers in Microbiology* 11(442):1–20.

Russell, J.R., Huang, J., Anand, P., Kucera, K., Sandoval, A.G., Dantzler, K.W., and Koppstein, D. (2011). Biodegradation of polyester polyurethane by endophytic fungi. *Applied and Environmental Microbiology* 77(17):6076–6084.

Sánchez, C. (2020). Fungal potential for the degradation of petroleum-based polymers: An overview of macro-and microplastics biodegradation. *Biotechnology Advances* 40:107501.

Satlewal, A., Soni, R., Zaidi, M., Shouche, Y., and Goel, R. (2008). Comparative biodegradation of HDPE and LDPE using an indigenously developed microbial consortium. *Journal of Microbiology and Biotechnology* 18(3):477–482.

Shabbir, S., Faheem, M., Ali, N., Kerr, P.G., Wang, L.-F., Kuppusamy, S., and Li, Y. (2020). Periphytic biofilm: An innovative approach for biodegradation of microplastics. *Science of the Total Environment* 717:137064.

Shah, A.A., Hasan, F., Hameed, A., and Ahmed, S. (2008). Biological degradation of plastics: A comprehensive review. *Biotechnology Advances* 26(3):246–265.

Shah, Z., Krumholz, L., Aktas, D.F., Hasan, F., Khattak, M., and Shah, A.A. (2013). Degradation of polyester polyurethane by a newly isolated soil bacterium, *Bacillus subtilis* strain MZA-75. *Biodegradation* 24(6):865–877.

Shahnawaz, M., Sangale, M.K., and Ade, A.B. (2016). Rhizosphere of *Avicennia marina* (Forsk.) Vierh. as a landmark for polythene degrading bacteria. *Environmental Science and Pollution Research* 23(14):14621–14635.

Singh, L., and Wahid, Z.A. (2015). Methods for enhancing bio-hydrogen production from biological process: A review. *Journal of Industrial and Engineering Chemistry* 21:70–80.

Skariyachan, S., Taskeen, N., Kishore, A.P., Krishna, B.V., and Naidu, G. (2021). Novel consortia of Enterobacter and Pseudomonas formulated from cow dung exhibited enhanced biodegradation of polyethylene and polypropylene. *Journal of Environmental Management* 284:112030.

Soleimani, Z., Gharavi, S., Soudi, M., and Moosavi-Nejad, Z. (2021). A survey of intact low-density polyethylene film biodegradation by terrestrial Actinobacterial species. *International Microbiology* 24(1):65–73.

Song, C., Liu, Z., Wang, C., Li, S., and Kitamura, Y. (2020b). Different interaction performance between microplastics and microalgae: The bio-elimination potential of *Chlorella* sp. L38 and *Phaeodactylum tricornutum* MASCC-0025. *Science of the Total Environment* 723:138146.

Song, Y., Qiu, R., Hu, J., Li, X., Zhang, X., Chen, Y., and He, D. (2020a). Biodegradation and disintegration of expanded polystyrene by land snails *Achatina fulica*. *Science of the Total Environment* 746:141289.

Stach, J.E.M., and Burns, R.G. (2002). Enrichment versus biofilm culture: A functional and phylogenetic comparison of polycyclic aromatic hydrocarbon-degrading microbial communities. *Environmental Microbiology* 4(3):169–182.

Syranidou, E., Karkanorachaki, K., Amorotti, F., Franchini, M., Repouskou, E., Kaliva, M., and Corvini, P.F.-X. (2017). Biodegradation of weathered polystyrene films in sea water microcosms. *Scientific Reports* 7(1):1–12.

Taniguchi, I., Yoshida, S., Hiraga, K., Miyamoto, K., Kimura, Y., and Oda, K. (2019). Biodegradation of PET: Current status and application aspects. *ACS Catalysis* 9(5):4089–4105.

Tekere, M. (2019). Microbial bioremediation and different bioreactors designs applied. In: Jacob-Lopez, E., and Zepka, L.Q., editors. *Biotechnology and Bioengineering*. Intech Open, pp. 1–19.

Tekere, M., Mswaka, A.Y., Zvauya, R., and Read, J.S. (2001). Growth, dye degradation and ligninolytic activity studies on Zimbabwean white rot fungi. *Enzyme and Microbial Technology* 28(4–5):420–426.

Tokiwa, Y., Calabia, B.P., Ugwu, C.U., and Aiba, S. (2009). Biodegradability of plastics. *International Journal of Molecular Sciences* 10(9):3722–3742.

Tsiota, P., Karkanorachaki, K., Syranidou, E., Franchini, M., and Kalogerakis, N. (2018). Microbial degradation of HDPE secondary microplastics: Preliminary results. Paper Presented at the Proceedings of the International Conference on Microplastic Pollution in the Mediterranean Sea. Springer Water. Cham (Switzerland): Springer. DOI: 10.1007/978-3-319- 71279-6_24.

Volke-Sepúlveda, T., Saucedo-Castañeda, G., Gutiérrez-Rojas, M., Manzur, A., and Favela-Torres, E. (2002). Thermally treated low density polyethylene biodegradation by *Penicillium pinophilum* and *Aspergillus niger. Journal of Applied Polymer Science* 83(2):305–314.

Wang, Z., Xin, X., Shi, X., and Zhang, Y. (2020). A polystyrene-degrading acinetobacter bacterium isolated from the larvae of *Tribolium castaneum. Science of the Total Environment* 726:138564.

Yamada-Onodera, K., Mukumoto, H., Katsuyaya, Y., Saiganji, A., and Tani, Y. (2001). Degradation of polyethylene by a fungus, *Penicillium simplicissimum* YK. *Polymer Degradation and Stability* 72(2):323–327.

Yang, J., Yang, Y., Wu, W.-M., Zhao, J., and Jiang, L. (2014). Evidence of polyethylene biodegradation by bacterial strains from the guts of plastic-eating waxworms. *Environmental Science & Technology* 48(23):13776–13784.

Yoshida, S., Hiraga, K., Takehana, T., Taniguchi, I., Yamaji, H., Maeda, Y., and Oda, K. (2016). A bacterium that degrades and assimilates poly(ethylene terephthalate). *Science* 351(6278):1196–1199.

Yu, Z., Tang, J., Liao, H., Liu, X., Zhou, P., Chen, Z., Rensing, C., and Zhou, S. (2018). The distinctive microbial community improves composting efficiency in a full-scale hyperthermophilic composting plant. *Bioresource Technology* 265:146–154.

Yuan, J., Ma, J., Sun, Y., Zhou, T., Zhao, Y., and Yu, F. (2020). Microbial degradation and other environmental aspects of microplastics/plastics. *Science of the Total Environment* 715:136968.

Zahra, S., Abbas, S.S., Mahsa, M.-T., and Mohsen, N. (2010). Biodegradation of low-density polyethylene (LDPE) by isolated fungi in solid waste medium. *Waste Management* 30(3):396–401.

Zhang, J., Gao, D., Li, Q., Zhao, Y., Li, L., Lin, H., and Zhao, Y. (2020). Biodegradation of polyethylene microplastic particles by the fungus *Aspergillus flavus* from the guts of wax moth *Galleria mellonella. Science of the Total Environment* 704:135931.

Chapter 12

Final Remarks

Plastics comprise a family of remarkable materials that meet countless functional and aesthetic needs of modern society. Since the last century, global production of plastics has soared, from approximately 1.5 million tons per year in the 1950s to over 368 million tons today (PlasticsEurope 2020).

Consumption of plastic products worldwide continues its upward trajectory as populations grow and many nations embrace consumer-based lifestyles. Unfortunately, many developed and developing countries do not practice comprehensive management of municipal solid waste. As a consequence, plastic debris is entering waterways at an accelerated rate and poses an obvious hazard to human health and the environment. Based on current practices, it is inevitable that the volume of microplastics (MPs) in the biosphere will increase due to release of primary MPs and fragmentation of macro waste.

There is an urgent need for governments to enact policies on plastics management and develop effective strategies to control plastic pollution. In 2015 the US Congress passed the Microbead-Free Waters Act which banned the manufacture of PE microbeads. As of this writing, certain US states are formulating stronger microbeads bills without ambiguous language for biodegradable beads or exceptions for over-the-counter medications. Other nations have banned microbeads from cosmetics including Canada, the United Kingdom, certain EU Member States, India, New Zealand, the Republic of Korea, and Taiwan. Such regulations focus only on primary MPs, particularly microbeads from personal care products, however. It is essential for governments to formulate policies that address secondary MPs—specifically to trace their sources and enact corresponding regulations to prohibit their release and dispersal (Hu et al. 2021).

Certain consumer groups, researchers, and industries advocate the substitution of conventional synthetic plastics with biodegradable plastics (e.g., polyhydroxybutyrates, polylactic acid, polybutylene adipate terephthalate, and starch- and chitosan-based blends) (Fojt et al. 2020) as a means of reducing long-term pollution from macro- and microplastics. Applications for biopolymers include agriculture, horticulture, certain forms of packaging, and consumer goods, among others. These so-called bioplastics can be degraded by microorganisms in the local environment within a relatively short time. Unfortunately, however, bioplastics suffer from their share of practical limitations and cannot be expected to replace their synthetic counterparts in the foreseeable future.

Other strategies have been proposed to manage MPs, such as encouraging greater recycling rates of solid wastes, monitoring and preventing marine litter and MPs from marine sources (e.g., fisheries, off-shore shipping) (UNEP 2017), restricting use of certain plastics that rapidly fragment into MPs (European Commission 2018), and limiting consumption of single-use plastic products (ECHA 2019; UNEP 2018).

Several approaches to management of MPs pollution are feasible. This book has addressed current progress in MP recovery and remediation/transformation strategies. The features and drawbacks of each technology have been presented.

In the case of aquatic environments already contaminated by MPs, a limited number of techniques for their recovery are already in use, and others are being researched at laboratory and pilot-scale. For example, many municipal drinking water and wastewater treatment plants experience high removal efficiency of MPs in various unit operations. However, massive volumes of MPs continue to be released into receiving water systems from these facilities. Research is underway for the design of effective methods for specifically filtering out MPs in drinking water treatment plants (Hu et al. 2021).

Coastal and benthic sediments have been identified as long-term sinks for MPs. Methods originally developed for mining and mineral processing industries are being evaluated at laboratory-scale for removal of MPs from sediment; unfortunately, large-scale remediation of historic MP contamination is not practical due to long-term dispersal of particles, potential ecological damage, and cost.

MPs have been identified as an environmental threat for several decades; only recently, however, have technologies been investigated that specifically degrade MPs. A vast number of studies and strategies to remediate macro- and microplastics are based on microbial degradation. Biodegradation can decompose MPs in an environmentally acceptable manner, but relatively long times are required to attain complete remediation. Different polymer types will furthermore respond differently to microbial action and require further study. Likewise, the potential toxicity of polymer additives to biotreatment must be investigated.

Certain AOP systems such as Fenton-like processes and catalytic wet oxidation are highly effective and can rapidly convert MPs to smaller organic molecules. These products can even serve as substrates for microbial utilization. However, such advanced technologies usually necessitate high energy inputs, and requirements for reagents may be considerable. Photocatalytic treatments can convert MPs into value-added products such as fuels. Research on these promising and innovative technologies is still in its early stages and needs further development to identify designs that can be effectively utilized on the industrial scale.

MPs are not the headline-grabbing pollutants that cause great angst among regulatory agencies, scientists, and citizens as was the case for mercury in the 1960s, PCBs and chlorinated dioxins during the 1970s and 1980s, and PFAS more recently. Regardless, there is agreement that MPs are a pervasive environmental contaminant with unknown health effects. Decades after the introduction of the term "microplastic" and following extensive research, myriad questions remain about MP contamination of the global environment. Engineered solutions for the capture of MPs contamination in freshwater and marine ecosystems, as well as destruction techniques continue to be researched and evaluated. However, the MPs problem will not simply disappear due to the appearance of a technological fix. No solution to MPs pollution is safer, more comprehensive, low-energy, and cost-effective than conscientious waste management at the source, that is, by the consumer.

REFERENCES

ECHA. (2019). ECHA ANNEX XV Restriction Report—Proposal for a Restriction: Intentionally Added Microplastics [accessed 2022 November 16]. https://echa.europa.eu/documents/10162/05bd96e3-b969-0a7c-c6d0-441182893720.

European Commission. (2018). A European Strategy for Plastics in a Circular Economy [accessed 2022 November 16]. https://ec.europa.eu/environment/circular-economy/pdf/plastics-strategy-brochure.pdf.

European Commission. (2019). Microplastics: New Methods Needed to Filter Tiny Particles From Drinking Water, 529 [accessed 2022 November 16].https://ec.europa.eu/environment/integration/research/newsalert/pdf/microplastics_drinking_water_529na1_en.pdf.

Hu, K., Tian, W., Yang, Y., Nie, G., Zhou, P., Wang, Y., Duan, X., and Wang, S. (2021). Microplastics remediation in aqueous systems: Strategies and technologies. *Water Research* 198:117144.

UNEP. (2017). Draft Resolution on Marine Litter and Microplastics [accessed 2022 November 16]. https://leap.informea.org/sites/default/files/unea-resolutions/UNEPEA.3.20-EN.pdf.

UNEP (United Nations Environment Program). (2018). Legal Limits on Single-Use Plastics and Microplastics: A Global Review of National Laws and Regulations.

Acronyms Used in This Book

A^2O	Anaerobic-anoxic-aerobic (activated sludge processing method)
ABS	Acrylonitrile butadiene styrene
ADWTP	Advanced drinking water treatment plant
Alum	Aluminum sulfate
AOP	Advanced oxidation process
ATH	Aluminum trihydrate
BOD	Biochemical oxygen demand
CAS	Conventional activated sludge
COD	Chemical oxygen demand
CWAO	Catalytic wet air oxidation
DAF	Dissolved air flotation
DDL	Diffuse double layer
dw	Dry weight
DWTP	Drinking water treatment plant
EC	Electrocoagulation
EPS	Extracellular polymeric substances
EPS	Expanded polyethylene
FTIR	Fourier-transform infrared spectroscopy
HALS	Hindered amine light stabilizers
HCH	Hexachlorocyclohexane
HDPE	High-density polyethylene
HGT	Horizontal gene transfer
K_{OW}	Octanol-water partition coefficient
$K_{P/W}$	Polymer-water distribution coefficient
LOD	Limit of detection
LDPE	Low-density polyethylene
MBR	Membrane bioreactor

MDR	Multi-drug resistant (bacteria)
MF	Microfiber
MF	Microfiltration
MP	Microplastic
MPSS	Munich Plastic Sediment Separator
NOM	Natural organic matter
NP	Nanoplastic
PA	Polyamide
PAC	Polyaluminium chloride
PAH	Polycyclic aromatic hydrocarbon
PAM	Polyacrylamide
PBDE	Polybrominated diphenyl ether
PCB	Polychlorinated biphenyl
PCCPs	Personal care and cosmetic products
PE	Polyethylene
PES	Polyester
PET	Polyethylene terephthalate
PFASs	Perfluoroalkyl substances
PFOA	Perfluorooctanoic acid
PMMA	Poly(methyl methacrylate)
POM	Polyoxymethylene
POP	Persistent organic pollutant
PP	Polypropylene
PS	Polystyrene
PVC	Polyvinyl chloride
PUR	Polyurethane
Pyr-GC/MS	Pyrolysis–gas chromatography–mass spectrometry
RBC	Rotating biological contactor
RO	Reverse osmosis
ROS	Reactive oxygen species
RSF	Rapid gravity sand filter
SEM	Scanning electron microscopy
SEM-EDS	Scanning electron microscopy–energy dispersive x-ray spectroscopy
SS	Suspended solids
TEM	Transmission electron microscopy
TrOC	Trace organic contaminants
TRWP	Tire and road wear particles
UF	ultrafiltration
WAO	Wet air oxidation
WAS	Waste activated sludge
ww	Wet weight
WWTP	Wastewater treatment plant

Glossary of Terms

Abiotic Non-living, chemical, or physical component of the environment.

Acidophile Microorganism which grows best under acidic pH regime.

Activated sludge Biological method for treatment of urban wastewater.

Additive Chemical that is incorporated with a polymer mixture or applied externally once the material is formed to modify certain physical and/or chemical properties; includes enhancing the desirable characteristics of the product or reducing unwanted properties.

Aerobic Requiring free O_2 for survival.

Aerobic respiration Mechanism of microbial respiration where substrate is oxidized to CO_2 and H_2O or other end-products using molecular O_2 as terminal electron acceptor. Aerobic respiration occurs under highly oxygenated conditions.

Aging *See* weathering.

Alkaliphile Microorganism which grows best under alkaline pH regime.

Amorphous Polymer which contains chains that are unordered; non-repeating or non-crystalline.

Anaerobic Requiring an absence of free O_2.

Anaerobic respiration Mechanism of microbial respiration where substrate is oxidized to CO_2 and H_2O or other end-products using NO_3^-, SO_4^{2-}, and Fe^{3+} as electron acceptors. Occurs in the absence or near-absence of molecular O_2.

Antioxidants Group of chemicals added to plastic to restrict oxidative degradation.

Antistatic agents Group of chemicals added to plastic to dispel static electrical charges. Typically hygroscopic, they attract moisture and dissipate static charge. Also known as antistats.

Baekelite First fully synthetic polymer, attributed to Belgian-American chemist Leo Hendrick Baekeland. Produced by condensation of phenol and formaldehyde.

Bioaccumulation Gradual accumulation of a substance such as a hazardous chemical in an organism. Absorption occurs at a rate faster than elimination from the organism.

Biocide Chemical agent added to a plastic to inhibit bacterial establishment and growth.

Biofilm Consortium of microorganisms enclosed within self-produced extracellular polymeric substances comprising a complex structure of carbohydrates, proteins, and DNA.

Bioreactor Engineered vessel that supports microbial growth and reactions in a controlled environment that provides the necessary conditions for degradation of a target contaminant.

Biotic Relating to living organisms.

Blocking The adherence of two adjacent plastic films to each other.

Bongo net Set of two plankton nets mounted adjacent to each other, having a long funnel shape. Both nets are enclosed by a cod-end for collecting particulate matter. Each net typically has a different mesh width to collect different-sized materials. The bongo net is pulled through water by a research vessel.

Box corer Marine geological sampling tool for soft sediments in lakes or oceans. Deployed from a research vessel and designed to minimize disturbance of sediment surface during sampling.

Cellulose Naturally occurring carbohydrate polymer.

Chain-growth polymerization *See* addition polymerization.

Chromophore Chemical element or compound that absorbs light at a specific wavelength, thus imparting color to a molecule.

Collectors Organic surfactants which form a hydrophobic coating on a particle surface in the froth flotation pulp, thus creating optimal conditions for attachment of hydrophobic particles to air bubbles.

Colorants Chemicals are added to plastics in order to improve appearance. Some provide shielding against UV light.

Concentrate The lightweight product of the froth flotation process that is transported to the overflow.

Condensation polymerization Plastic manufacturing process where two different monomers react to form a new chemical group.

Copolymer Polymer having two or more different monomer units in their chains.

Corona Gas molecules in proximity to a discharge electrode become ionized and conductive.

Crystallinity State where polymer chains have a definite geometric form.

Diffuse double layer Sphere of electrical charge surrounding a charged colloidal particle which occurs at the interface between the solid phase and the aqueous phase. Consists of an inner fixed layer of charge (Stern layer) surrounded by a thin mobile layer (diffuse layer).

Dioxin *See* chlorinated dibenzodioxin.

Direct oxidation Ozone reacts with organic compounds via electrophilic addition to double bond.

Elastomer Polymer that can be stretched and deformed under tension; quickly rebounds to its original shape and size when tension is removed.

Electrostatic separation Separation of particles based on their attraction or repulsion under the influence of an externally applied electrical field.

Elutriation Technology involving the application of a fluid stream counter-current to direction of sedimentation. Elutriation separates microplastics from sediment based primarily on differences in density.

Endoenzyme Enzyme that functions within the cell in which it was produced.

Exoenzyme Enzyme secreted by a cell and functions outside of that cell.

Extracellular polymeric substances (EPS) Natural polymers of high MW secreted by microorganisms into their environment. Composed primarily of polysaccharides and proteins, but include other macromolecules such as DNA, lipids, and humic substances. EPS constitutes 50% to 90% of the total organic matter of a biofilm.

Fallout Diverse mix of radioactive particles that fall to earth following a nuclear detonation. Consists of fission products, irradiated soil, and water. May take from seconds to decades to return to earth.

Fenton's reagent Solution of hydrogen peroxide (H_2O_2) with ferrous iron as a catalyst. This reagent is used to oxidize a wide range of organic substances.

Fermentation Mechanism of microbial respiration where substrate is oxidized to CO_2 and other end-products including acetate, ethanol, lactic acid, propionate, and others. Occurs under highly reducing conditions and in the absence of inorganic electron acceptors. Organic compounds function as electron acceptors.

Fixed film process An activated sludge process in wastewater treatment where the active microbial culture is attached to a solid support during decomposition of BOD.

Froth flotation Technology involving physical separation of fine particles from a solids/water suspension based on the ability of air bubbles to adhere to particle surfaces.

Frothers Short chain surfactant compounds which are active at the air-water interface. Consist of a polar head and hydrocarbon tail.

FTIR (Fourier-transform infrared spectroscopy) Method of IR spectroscopy where samples interact with IR radiation. Infrared radiation is absorbed, reflected, and transmitted through the sample and into a detector. Fourier Transformation converts the output produced by the detector into a sample spectrum that can be interpreted and used to identify the sample.

Fulvic acid Low-MW component of soil organic matter. Has high content of carboxylic acid, and hydroxyl and phenolic groups.

Furan *See* polychlorinated dibenzofuran.

Gangue Unwanted material recovered during particle separation process; impurities in the product.

Glass transition Transition in amorphous polymers from a hard and brittle ("glassy") state to a viscous or rubbery state as temperature increases.

Gravity concentration Method of creating conditions where particles of different densities, sizes,and shapes move relative to each other while under the influence of gravity.

Grab Sample A small representative sample (air, water, soil) collected at a single point in time.

Gyre A large, rotating ocean current. In the context of plastics pollution, a gyre is a massive vortex where solid waste accumulates; several such gyres exist worldwide.

Heat stabilizer Chemical added to plastic to provide protection against heat-induced decomposition.

Heterochain polymer Polymer containing a number of other elements may be linked in the backbone chain including oxygen, nitrogen, sulfur, and silicon

Heterogeneous catalyst Catalyst that occurs in a different phase from the reactants.

Hindered settling	Sedimentation of particles at a slower speed, resulting from interactions with nearby particles compared to the speed of settling of a single particle.
Homochain polymer	Polymer that has solely carbon atoms along the backbone chain.
Homogeneous catalyst	Catalyst that is soluble in the same solution as the reactants.
Horizontal gene transfer	Movement of genetic material between organisms other than by transmission of DNA from parent to offspring. Bacteria in close contact exchange genetic information via plasmid transfer from donor to recipient cell.
Humic acid	Highly complex and amorphous collection of phenolic and linear hydrocarbons originally derived from plant and animal matter and converted by microorganisms. Soluble in dilute NaOH but coagulated by acidification of alkaline extract.
Humic material	Complex, high-MW polymers of organic matter synthesized by microbial reactions. Has high content of carboxylic and phenolic groups, with small amounts of aliphatic- and enolic-OH groups.
Humin	Insoluble component of soil organic matter that remains after extraction of other organic components that are soluble in dilute NaOH.
Hydrolysis	Splitting of molecule via reaction with water.
Hydrophilic	Water-loving; tendency to dissolve in or be wetted by water.
Hydrophobic	Water-fearing; tending to repel water.
Incineration	Process carried out under highly controlled conditions for the destruction and/or volume reduction of various waste types.
Ion diffuse layer	Outer layer of the diffuse double layer of a charged particle where ions move freely due to diffusion.
Jigging	Gravity concentration method that is applied for separation of solids based primarily on differences in density.

Lamellae Molecular chains that can fold and be bundled into ribbon-like configurations about 10 μm thick, 1 μm long, and 0.1 μm wide.

Landfill Engineered excavated facility designed for long-term disposal of waste to land.

Launder *See* concentrate.

Macromolecule Large-chained molecular structure; also known as a polymer.

Marine snow Organic-rich aggregates measuring approximately > 200 μm, composed of fecal pellets, phytoplankton, microbial biomass, particulate organic matter, and various inorganics.

Melting transition Transition between a crystalline solid and a liquid.

Mesoplastics Plastic particles which are intermediate in size between microplastics and macroplastics. Approximate size 5–25 mm.

Microbeads Microplastics that are spherical or irregular in shape and formulated as ingredients in personal care and cosmetic products (PCCPs), paints, and industrial abrasives.

Microcapsules *See* microbeads.

Microfibers Fine synthetic filaments composed of polymers including nylon, polyester, rayon, PET, PP, acrylic, or spandex.

Microplastic Small plastic fragments occurring in oceans and other ecosystems and in organisms; particles having a size < 5 mm.

Monomer Simple molecule which can form covalent linkages with other monomers to form a larger polymer chain.

Mulch film Plastic thin plastic sheeting spread over high-value crops to prevent growth of weeds. A source of microplastics in terrestrial ecosystems.

Nanocapsules *See* microbeads.

Nanoplastics Plastic particles measuring less than 1 μm in size.

Nanospheres *See* microbeads

Natural rubber Polymer obtained from rubber plant *Hevea brasiliensis* and other plants. Consists almost exclusively of the *cis*-1,4 polymer of isoprene (C_5H_8) which is produced in the latex of the plant.

Neuston net Net used to collect particulate matter occurring in the water surface microlayer. Consists of a towing line and bridles, nylon mesh net, and a cod end.

Nile red Dye used for rapid identification of microplastics in environmental samples.

Non-humic material Recognizable organic molecules released from cells of plant and animal residues such as proteins, amino acids, sugars, and starches.

Nurdles Primary microplastics that are the raw materials for manufacture of plastic products.

Nylon-6,6 A polyamide produced from the condensation of the monomers adipic acid and 1,6- hexanediamine.

Oligomer Low MW polymer generally containing 2–10 repeat units.

Oxidation potential Tendency for a chemical species to be oxidized at standard conditions. Measured in volts.

Pathogenic Disease-causing.

Photocatalysis Technology where a nanoscale metal oxide and metal oxide complex semiconductors are employed for enhanced degradation of an organic contaminant.

Photolysis Process where absorption of light leads directly to splitting of chemical bonds.

Photooxidation Process where light is absorbed by a polymer and results in formation of radicals that induce oxidation at various locations within the polymer.

Photoreforming Technology where sunlight and a photocatalyst generate H_2 from an organic substrate and H_2O at ambient temperature and pressure.

Pinning Process where an electrically charged particle adheres to a conducting surface against the force of gravity.

Plankton net Net used for collecting samples of particulate matter in standing bodies of water. Consists of a towing line and bridles, nylon mesh net, and a cod end. Can be used for both vertical and horizontal sampling.

Plastic A material that is easily molded or shaped. A synthetic product made from various organic polymers that can be molded into shape while soft and subsequently set to a rigid or elastic form.

Plasticizer Chemical added to plastic that reduces intermolecular forces in polymers, making it more flexible; increases the workability and flexibility of a polymer.

Plastisphere The unique microbial communities which colonize marine plastics; these communities differ markedly from surrounding seawater communities.

Polychlorinated dibenzodioxins Group of long-lived toxic and persistent halogenated organic compounds. Produced inadvertently by incineration of certain wastes and other processes. Commonly (but inaccurately) referred to as dioxins.

Polychlorinated dibenzofurans Group of long-lived toxic and persistent halogenated organic compounds. Produced inadvertently by incineration of certain wastes and other processes. Commonly (but inaccurately) referred to as furans.

Polymer Macromolecule composed of multiple repeating units (monomers) where the backbone is connected by covalent bonds.

Polymerization Chemical process of joining monomers in a chain to form a polymer.

Polymer-water distribution coefficient Concentration of a solute sorbed per unit mass of solid polymer divided by the concentration of solute in the aqueous phase.

Primary microplastic Microplastic manufactured for specific commercial uses.

Primary sludge Sludge solids which settle during primary treatment of wastewater.

Primary treatment Physical process that removes settleable, suspended, and floatable materials from influent wastewater within settling tanks.

Protein Natural polyamide composed of long chains of amino acids.

Pyrolysis-GC/MS Analytical method where an unknown sample is thermally decomposed under well-established conditions following which the evolved components are separated and identified.

Raman spectroscopy A light scattering technique where a sample interacts with a laser that produces a measurable shift in scattered radiation. The Raman shift created allows for chemical identification of the sample.

Ragging In jigging technology, a bed of non-cohesive heavy particles having a density intermediate between the densities of particles being separated.

Reactive oxygen species Highly reactive molecules formed from O_2. Examples include hydroxyl radicals (•OH), superoxide radical anions (O2•⁻), and perhydroxyl radicals (HO_2•) to degrade and mineralize organic pollutants including microplastics.

Recalcitrant Resistant to decomposition.

Recycling Process of converting wastes into new materials and objects. May include the recovery of energy from wastes.

Resin Organic compounds that convert into rigid polymers by the process of curing. Resins typically contain reactive end groups such as acrylates or epoxides.

Scission Breakage of a chemical bond.

Sea surface microlayer Top 1000 micrometers of the ocean surface. The critical boundary layer where the exchange of matter occurs between atmosphere and ocean. Chemical, physical, and biological properties of this microlayer differ substantially from water just a few centimeters below.

Sea swash zone Part of a beach where water washes up after an incoming wave has broken.

SEM A scanning electron microscope employs a beam of high-energy electrons to create images of the surface morphology of a sample. This image is generally produced as result of the collection of secondary electrons produced from interaction of the high-electrons with the surface of the sample.

Secondary microplastics Microplastics created from the fragmentation and degradation of macroplastics and primary microplastics.

Secondary treatment Biological decomposition of dissolved and suspended organic matter in wastewater. Carried out by a consortium of heterotrophic aerobic microorganisms under controlled conditions.

Semiconductor Crystalline solid which is intermediate in electrical conductivity between a conductor and an insulator.

Slip agents Chemical added to a plastic to reduce the coefficient of friction of the surface of a polymer. Provides lubrication to the film surface.

Slip Measure of interaction between adjacent polymer surfaces, or between polymer surface and processing equipment. Also known as surface friction.

Slipping plane *See* ion diffuse layer.

Step growth polymerization *See* condensation polymerization.

Stern layer Component of the diffuse double layer of a charged particle where oppositely charged ions are bound tightly to the particle surface.

Striatum corneum Outermost layer of the epidermis.

Suspended growth An activated sludge process in wastewater treatment where the active microbial culture is suspended in a vessel during decomposition of BOD.

Sweep coagulation Process where newly formed metal precipitates entrap colloidal particles and produce a sludge blanket.

Tacticity Manner in which substituent groups are arranged along the polymer chain. Tacticity patterns are three-dimensional and influence the physical and chemical properties of the plastic.

Tailings The sediment minerals that do not float during froth flotation and which are removed via pumping.

Tertiary treatment Final wastewater treatment stage for effluent prior to release to the aquatic environment. Typically, a chemical treatment unit operation.

Thermoplastic Polymer containing linear or lightly branched chains that are capable of sliding past one another when exposed to increased temperature and pressure.

Thermoset Polymer that contains extensive crosslinked networks that do not soften when heated.

Trawl Method of fishing (or collection of particulate matter) that involves pulling a fishing net through water behind a boat.

True flotation Attachment of target particles to air bubbles during froth flotation.

Unzipping Mechanism of depolymerization where end monomer repeat units are released one by one.

UV stabilizer Chemical added to plastic which absorbs and dissipates the energy from UV rays as heat.

Vulcanization Process where elastomers such as natural rubber are cross-linked by heating with sulfur.

Weathering Deterioration of plastic articles during exposure to sunlight, oxygen, natural organic matter or other substances and stimuli. Also known as aging.

Index

About the Authors

John Pichtel is a professor of environment, geology and natural resources at Ball State University in Muncie, Indiana, where he has been on the faculty since 1987. He received PhD in Environmental Science at Ohio State University, MS in Soil Chemistry/Agronomy at Ohio State, and BS in Natural Resources Management at Rutgers University. His primary research and professional activities have embraced remediation of contaminated environments, management of hazardous and municipal wastes, and environmental chemistry. He was twice awarded the Fulbright Fellowship, and is a Certified Hazardous Materials Manager. He has served as a consultant in hazardous waste management projects and conducted environmental assessments and remediation research in the United States, the United Kingdom, Ireland, Finland, Poland, and Italy.

Parisa Ebrahimbabaie

Mathew Simpson is a PhD candidate in Environmental Science at Ball State University, and is researching the environmental behavior and fate of microplastics in aquatic systems. He has collected, identified, and studied microplastics from high alpine aquatic systems (Nepal) and beach sediment (continental US). He has extensive experience with the theory and techniques of FTIR. He is a co-principal investigator at the Ball State University Hazardous Waste Research Laboratory.